Thomas Belden Butler

A concise analytical and logical development of the atmospheric

system

Thomas Belden Butler

A concise analytical and logical development of the atmospheric system

ISBN/EAN: 9783337276003

Printed in Europe, USA, Canada, Australia, Japan

Cover: Foto ©berggeist007 / pixelio.de

More available books at **www.hansebooks.com**

A CONCISE

ANALYTICAL AND LOGICAL DEVELOPMENT

OF THE

ATMOSPHERIC SYSTEM,

AND OF THE

ELEMENTS OF PROGNOSTICATION,

BY WHICH THE WEATHER MAY BE FORECASTED,

ADAPTED TO THE PRACTICAL MIND OF THE COUNTRY.

"Truth is mighty and will prevail."

By THOMAS B. BUTLER.

AUTHOR OF THE "PHILOSOPHY OF THE WEATHER."

Revised Edition.

PUBLISHED AT NORWALK, CONNECTICUT,
BY ANDREW SELLECK.
1870

PRINTED BY
CASE, LOCKWOOD & BRAINARD,
HARTFORD, CONN.

INTRODUCTION.

THIS BOOK is the result of *habitual personal observation* and continued *investigation* and *reflection* for half a century. It will be found, on critical and thorough examination, to be substantially what its title page imports.

It was no light matter to undertake to brush away the mystery and superstition which have enveloped the subject, and make it comprehensible by practical men; and a much more serious thing to attempt to meet, expose, and refute the prevalent false theories by which the subject is persistently obscured; and the reader should understand and appreciate my motives. He will find them stated in several places in the volume; and particularly on pages 367, 8, and 9, but a brief explanation is proper here.

My early days were spent at Wethersfield, near the banks of the Connecticut. At that point on the river there are two large coves, and extensive meadows containing many creeks, and it is a favorite stopping place for water-fowl in their spring and fall migrations up and down the river. They were then much more numerous than now. At the age of 12 I had access to a small fowling piece, and became passionately fond of hunting them. I was permitted to indulge my passion within certain limits, and at prescribed hours of certain days, and I soon began to feel interested to know what the weather would then be, and to watch it, and enquire respecting it of some retired and weatherwise sea captains who lived in the neighborhood. I soon discovered also that the fowl seemed to forecast the weather, and change their haunts in advance of the atmospheric changes, and I directed my attention to that also, in order to know where to find them without wasting my limited time in fruitless search. And thus I went on observing and studying the matter, and reading every book I could find relating to it, until I became imbued with a love of it, and continued the habit after I left home to acquire and practice the profession of medicine to which it was germain, and after I changed my profession in consequence of feeble health. In 1856, I was pursuaded to publish a small volume, entitled the " Philosophy of the Weather." It was prepared hastily while engaged in the full practice of the law, and was not as logical and full as I might have made it. But it contained the germs of this development, and had a considerable sale, doing much good. In 1859 I listened to a lecture by Prof. Henry, at Springfield, which rehashed the Halley fallacies with the Espyan improvement, and which was endorsed by Prof. Loomis and others. That lecture, and the endorsements and the debate which followed, in connection with other facts which will appear, convinced me, irrepressibly, that material progress in the development of the science by the present generation of meteorologists, while holding to those *fundamental fallacies*, was IMPOSSIBLE, and that as they were in a position to control in a great measure the scientist mind, it was my duty to develop it to the comprehension of the *practical mind* of the country before I should pass away.

The preparation of the work was commenced the following year, and has been deliberately and carefully pursued, during periods which could be honestly devoted to it without interfering with the performance of my judicial duties. And, being thus the result of long-continued personal observation and investigation, and deliberately prepared, and I may truly add, without any view to pecuniary advantage or scientific reputation, I think I may assure the journalists and professional and practical men of the country, and that *they may accept the assurance,* that it is, so far as it professes to go, a *truthful development*, and opens out the *only true way* to a *perfect* development of the science, and is worthy of a *careful study* and a *candid appreciation.* Much labor and considerable money have been bestowed and expended in the preparation of the work. I do not expect compensation for one, or a full return of the other. But it would be a satisfaction to know that it is appreciated by those for whose benefit it has been prepared.

The book is *necessarily* aggressive. Error must be exposed that truth may prevail. The new and fundamental fact, that all the atmospheric phenomena result from the operation of a SYSTEM of *organizations* created and controlled by *ascertainable* and *intelligible laws* could not be demonstrated without conflict with fundamental errors in Meteorology, and the men who make it a specialty. That conflict involves the accuracy of their public teachings, and the worthiness of their reputations as meteorologists, and will excite their pride of opinion and opposition. But with such opposition disinterested scientists, who love and seek truth, and the community should not sympathize. The subject is of great general and practical interest to all, and *all should sympathize with, and sustain every honest and disinterested effort to develop it*. New facts and deductions should be promptly and candidly examined and tested, by whomsoever discovered, and however boldly and forcibly presented; the truth should be received, and error, when fully exposed, abandoned. By such a course, honestly and perseveringly pursued, this subject can be speedily and fully developed, and take a place among the sciences.

Meteorology, as now taught in lecture, essay, and treatise, is not a science. It is not "a knowledge of laws, principles, and relations, deductively or inductively" attained. It consists of one fundamental assumption, made by Halley in 1686, before there were facts known from which it could be deduced, and independent descriptions of the varied atmospherical phenomena, with some special theoretic assumption to explain each of them. It recognizes no organization of the atmosphere, or system, and no laws or principles but those which belong to temperature, evaporation, condensation, and the ascent of heated air, and the mechanical effects which they are assumed to produce. It is substantially all theory and assumption—uncertain and conflicting—with little or nothing of deduction or logic; and the inquiry is never made whether there is an atmospheric system, and what are its laws, but *what is your theory?* and meteorological progress is limited to the discovery and description of new facts, and the *invention* of some plausible explanatory *theory* to reconcile each fact with the fundamental assumption of Halley.

That fundamental theory of Halley, on which the almost innumerable and conflicting theories since invented hinge, assumes that the atmosphere is an aerial ocean resting upon the earth, and that all the phenomena which occur in it are produced, here and there, *casually*, by the rays of the sun heating the earth beneath it, and thereby a part of the air, causing the heated portion to rise, whereby "*commotions*' and their accompanying phenomena are mechanically produced,—as boys raise bubbles of carburetted hydrogen in ponds and streams by thrusting in sticks and stirring the mud at the bottom. This theory suggests the idea of an aerial *Puddleology,* but it suggests, and admits of no *organization,* general or special, of the atmosphere, nor any system, principles, or laws; it suggests only a difference of result from differences in the nature of the surfaces to be heated, or their relative exposure to the heat of the sun. In aquatic Puddleology, the bubbles and the "commotion" they produce, are in proportion to the quantity of gas that is disengaged. In aerial Puddleology according to the theory, the "commotion" and its effects are proportionate to the quantity of air heated, and the extent of the heating.

Reader, that fundamental theory and all its associated and special theories are baseless and untrue. The atmosphere is *systematically* ORGANIZED, with *general* and *special organizations,* and from their operation and laws all the phenomena result. Meteorology *is a science*, but not *the* science you are taught.

To show that this is so, and also the *truth* and *justice* of all the preceding statements of this introduction, I will anticipate, and allude to a few facts taken from the mass of like tenor to be found in the text. I ask for them your candid attention.

I. The ascensive force of confined heated air, when raised to a temperature of 100° above the surrounding atmosphere, is but about $\frac{1}{3}$ of an ounce to the cubic foot. This, as shown at page 150, was the ascertained result of the experiments of the Mont-golfiers in their ascensions with balloons containing heated air. It is the rule given by Mr. Wise in his "History and Practice of Aeronautics," and substantially the result of my own experiments with an upright tin box, seven feet long and one foot square, open at the bottom and raised on legs, with an angular thermometer in-serted into it through an aperture, and the box covered alternately by thin pieces of pasteboard of different weights,—the contained air being heated by kerosene lamps placed beneath and in it. If there be not that contrast of temperature between the confined and surrounding air, there will not be that extent of force—for the force is proportionate to the contrast. It is obvious at a glance, that *this force is utterly inadequate to produce any of the claimed effects*, and this is the ascensive force of *confined* air. Un-confined air is subject to the laws of diffusion and expansion, and its force is very much less and comparatively infinitessimal. It will not lift a revolving paper toy, which is suspended above the hot stove, from the pivot on which it revolves. In view of these facts, there cannot be a grosser absurdity than to attribute the gale, the storm, or the hurricane, or any of the atmospheric phe-nomena to such an *infinitessimal force*.

II. Over and upon an area of about one thousand miles in ex-tent, either way, in the tropics to the eastward of the Windward Islands of the West Indies, the temperature of the air ranges dur-ing the summer and early fall at about 82°, and the water from 80° to 82°; and over and upon that area, and generally near the center of it, east of the Windward Islands, nearly all the violent hurricanes which devastate the north Atlantic, and our own coast, originate. (See the chart of Mr. Redfield on page 66.) When they commence they are small, never probably exceeding fifty miles in diameter, and they widen and enlarge as they progress to the Northwest and curve to the Northeast. (See figure 19 on page 67.) Between the temperature of the air where they origi-nate, and that which adjoins them when they commence, there cannot be *any* appreciable contrast, or *any* ascensive force in

operation. The same fact is observable in our own country. In
May, 1860, a destructive hurricane, forty miles wide, formed in
and sped up the valley of the Ohio at the rate of seventy miles
an hour, destroying a great amount of property, and where it
commenced, and along by the side of its path, there was no contrast of temperature—absolutely none, and the temperature of the
air nowhere exceeded 76°, or that of the earth 60°. (The record
of that hurricane will be found on pages 69–77.) It is incontrovertibly true, therefore, that the most violent storms and hurricanes commence and progress where there is no contrast of temperature to create an ascensive force, *and no such force can possibly be in operation.*

III. Winter storms, precisely like our winter snow storms in
every element and characteristic, are found high up within the
Arctic circle, in mid-winter, where the earth and sea are locked
up by frost and covered with snow, where the sun has not shone
at all for weeks and the thermometer ranges in fair weather at
from 30° to 40° below zero, and is not above zero, (except when
it moderates to storm,) within 1,000 miles of the locality. (The
record of two such storms will be found on page 195.) No possible ascensive force of heated air could *create* such a storm or
continue it till it arrived there.

IV. The trade winds of the tropics on the summer side of the
central belt of rains, all around the earth, do not always blow, as
assumed by Halley and taught by Meteorologists, from colder to
warmer surfaces, but on the contrary, on the summer side of the
belt of rains, from warmer, and in many places *very hot ones,*
toward and over those which are much cooler. In our summer
the Northeast trades blow from the range of deserts which extend
from Northwestern India to the Atlantic Ocean, and on which
the temperature of the earth at mid-day ranges from 130° to 160°,
and the air, in the shade, from 112° to 115°, toward and over surfaces where neither the earth, the air, or the water, rise above
84°. This fact is also incontrovertibly shown in the text, and is
utterly inconsistent with the persistently taught Halley theory.

V. The tornado is the most violent of the "commotions," as

Meteorologists term them. The force exerted by one described by Professor Stoddard, was estimated by him to be equal to the force of a body of air moving with a velocity of 340 miles per hour; and that of another, described by Professor Loomis, was estimated by him to be equal to a body of air moving at the rate of 682 miles per hour. Professor Henry, in his Patent Report Essays, and his Springfield lecture, attributed this "commotion," as he termed it, to an unstable or tottering equilibrium of the surface atmosphere, caused by an abnormal contrast of temperature between the upper and lower strata of the atmosphere, the former being colder, and the latter warmer than usual—which unstable equilibrium causes a narrow portion, coming in contact "with a heated spot or slight elevation," to rise, the ascent causing condensation, and the liberation of heat and the further heating of the air, whereby the ascent of the mass is continued. He made no attempt to support the assumption by evidence, and its absurdity is apparent. The extraordinary and awful violence at the surface of the earth, he attributed to the momentum caused by the "continued action of the impelling force," which he attempted to illustrate by the effect of a continued blast upon the arrow in an air-gun The assumed impelling force is compounded of "*tottering equilibrium*" and the *heat* of a *heated spot*, or the *trip* of a *slight elevation*, and as they must act from instant to instant upon *different* and *successive portions* of air, the parallel between their action and that of the *continuing* blast upon the *same* confined arrow utterly fails. The same want of parallelism would exist if the acquisition of momentum was attributed to the ascent of the air heated by condensation, for the action of the force must be constantly changed to new portions of air. Moreover, the tornado of the land and the water-spout of the sea are confessedly identical in their nature and cause, *and there is no heated spot or slight elevation at sea.* I understood Professor Loomis to endorse this theory at Springfield, but in his recent treatise he attributes the tornado to the formation of an eddy in the inblowing wind of a storm. How the wind, blowing in at the rate of 30 or 40 miles an hour, can create an eddy in which the air will move at the rate of 682 miles an hour, he does not and cannot say. These theories agree in assuming that the tornado commences at the surface of the earth, but are alike absurd on their face and demonstrably untrue.

The tornado and water-spout do not commence at the surface of the earth, but at the base of an overlying cloud, and extend thence downward to the earth, and withdraw upward again when spent —ending at the cloud, which is never perforated. This also is incontrovertibly true, and disposes of "tottering equilibrium" and "eddy." The reader will find the evidence of its truth in cases cited on pages 309 to 337 of the text. Every reliable statement relative to the inception of a tornado which I have seen, has been of a similar tenor to those there copied. This fundamental fact is not only fatal to their theories respecting the tornado, but to the calorific theory.

VI. The only other fact to which I will allude in this introduction, is of the most conclusive nature possible. It is that *observation shows*, and *every one may see for himself, that no such ascent ever takes place.* Nothing could be more readily seen if it in fact occurred.

The atmosphere is divided into strata, or stories, by the upper or counter trade current, and each of the strata has its particular kind of clouds, which are peculiar in *appearance* and *function.* This the reader will find fully developed in the second Chapter. The scud or peculiar clouds of the surface story *never rise under any circumstances into the other stories.* Of this truth any person who will may satisfy himself by a half dozen glances upward every day for a few months. I have watched their course for fifty years, often getting up several times in a night to watch a storm, and I KNOW the fact to be so. In the "Philosophy of the Weather" it was stated that no such ascent had been discovered during an observation of thirty years, and I could fill pages with affidavits of intelligent men who, since that publication, have watched with like result, but it cannot be necessary. No man, not even Mr. Espy, pretends to have ever seen such an ascent during a storm, and every man may see, if he will, that it never exists.

Mr. Espy says that in some instances he has pointed out to others an inblowing wind beneath an approaching storm, and an outspreading of the cirrus in a different direction above, but he does not pretend to have seen such inblowing wind ascend, and the blowing of the wind, and the spreading out of the cirrus, are perfectly distinct and independent operations.

Now reader, making every reasonable allowance for pride of opinion and prejudice, I must believe that no man, who is honest with himself, can resist the logic of these facts; whether considered separately, or collectively, or in connection with other facts, of like tenor and equal force, which may be found in the text, or with the mass of facts which show the falsity of the theories, by showing that the atmosphere is organized, and that all the phenomena result from the operation of its general and special organizations and their laws.

The table of contents discloses the plan of the work, but a concise statement of it, and the purpose and connection of the different chapters, in this place, will assist the reader to comprehend them.

Assuming the received view of meteorologists, that all the atmospheric phenomena are the result of mechanical " commotions " in an aerial ocean, produced at the bottom by the action of the sun's rays in heating the earth and atmosphere and causing the latter to rise, to be demonstrably incorrect ; and that the atmosphere is systematically organized, with general and special organizations, which produce the various states and phenomena, I proceed to state, in anticipation, three propositions, as the result of rigid induction from the facts which will be presented.

First, That the normal state of the atmosphere in the temperate zones, is calm, fair weather.

Second, That the changes from that state, and all the *states* and changes constituting the weather, are produced by the passage of successive *organizations* commonly called storms, but which, for reasons stated in the text, I term " CONDITIONS."

Third, That the CONDITIONS result from the operation of a great, central, permanent organization, and originate in it, or in an atmospheric current called the counter, or upper trade, which is produced by and sent forth from that central organization, out over the temperate zones.

I then proceed to show that there are seven alternating and changing *states* of the atmosphere, and to describe them, and illustrate the manner in which they are produced, by a history and description of two passing CONDITIONS,—one, a summer belt of showers,—the other, an autumn southeast storm.

The plan of the work then, as developed in the first chapter, is: *first*, to examine and fully understand the *states* of the atmosphere and their changes, constituting the weather ; *second*, to trace those states and changes to the passing CONDITIONS, and show the manner in which those *conditions* produce them ; *third*, to analyze and examine the passing *conditions*, and ascertain their elements, mode of operation, and effects ; and *fourth*, to ascertain the place and manner in which those *conditions* are formed and organized in, or by the influence of the permanent, central, organized *condition*, and the paths they pursue ; finally, to analyze and examine that central condition, its organization, mode of operation and effects, and thus, step by step, commencing with the *simple states*, and ascending to the *great, central, fundamental condition*, develop the atmospheric system, and the modes of operation by which the states and phenomena constituting the weather, are produced.

The second chapter is digressive, in order to develop the fact that the atmosphere consists of *strata* or *stories;* that they are constituted by the interposition of the equatorial current, or upper trade, proceeding from the central condition ; that the clouds of the different stories differ in *function* as well as in *form;* and to give a clear and explicit description of the clouds of the different stories, and their distinguishing characteristics and functions, with a view to a correct analysis and understanding of the various conditions.

In the third chapter I return from the digression to an analytical examination of the conditions and a classification of them, so far as such classification is possible, and a description of their distinctive characteristics, and the localities where one or the other are generally to be found.

The fourth chapter, in continuation, is mainly devoted to an examination of the paths which the conditions pursue in passing over this continent, and to show that there are three distinct and diverse systems; that each system has its focal path; that those focal paths change their location with the seasons, and that the climatology of the country, in all its varying characteristics, results from these diverse systems and the changes of their paths. In that chapter I have further examined, analytically and logically, the movements and mode of operation of the passing conditions, and the manner in which they produce the states of the atmosphere, (including special and prevalent winds, and rainy and dry seasons,) in different localities, at different seasons of the year. *That chapter contains a key for the explanation of every climatological fact peculiar to any section of the continent.*

Having thus fully analyzed and examined the conditions peculiar to the temperate zones, their elements, organization, and mode of operation in producing the states, I next proceed, in chapter fifth, to a critical examination of the permanent, fundamental, central condition, and its elements, organization, mode of operation, and effects. That examination is too extended to admit of even a brief synopsis here. The chapter ends with a resumè of all the facts and inductions contained in it.

In the sixth chapter I return to the temperate zone, and to a more critical examination of its climatological peculiarities, especially its irregularities, and the *cycles* of drouth and drench, and of cold and hot seasons, and endeavor to trace them to their cause.

After having thus developed, as matter of fact, the atmospheric system, and the manner in which the various phenomena are produced, I proceed, in the seventh chapter, to deduce and consider the elements of prognostication, by which an isolated observer can forecast the weather. This chapter is also extended and full, and a synopsis of it cannot be given. It is practical and instructive, as well as logical beyond anything to be found in our language, or any other. The reader should read carefully the synopsis of the chapter in the table of contents, and at its head, and possess himself of the salient points. If he has fully comprehended the developments of the previous chapters, he will have no difficulty in comprehending the means at our command for prognosticating the weather, as fully described in that important chapter.

In the eighth and last chapter, I have inquired, analytically, into the laws of the conditions, and the motive force of the system. This chapter, also, is too full and extended to admit of synopsis here.

From this brief outline, and a careful reading of the table of contents, the reader will learn the plan of the work, and the analytical and logical character of the development, and be enabled to understand and appreciate it as he proceeds.

I have not deemed it important to preface the work with any of the minor details of the science. All that are of material consequence will come in view, and be explained appropriately as we proceed.

There is one matter, however, that should be fully understood, and that is, that the winds are uniformly named according to the points of the compass *from* or *through which* they blow. The following figure represents the mariner's compass. If the reader will imagine that he stands in the center of the card, that the wind is blowing toward him, and observe *from* or *through what point it passes to reach him*, he will know why the winds are described as they are, and have no difficulty thereafter.

CONTENTS.

CHAPTER I.

CHAPTER II.

CHAPTER III.

CHAPTER IV

OF THE SYSTEMS OF CONDITIONS ON THE CONTINENT OF NORTH AMERICA AND IN THE UNITED STATES.

CHAPTER V.

THE GENERAL ORGANIZATION OF THE ATMOSPHERIC SYSTEM.

Origin of a class of conditions in the central zone—Central belt of precipitating cloud—encircles the earth—Mean width about 500 miles—chart of it and its connections—Situation in August, north of the equator—Polar zones of rain connected with it—Situation of polar zones in August—On each side of the central belt an area covered by dry trade-wind called N.E. and S.E. trades—these areas also surround the earth. The central belt of cloud and rain, the two areas of dry trade-wind and the two polar zones of rain make five permanent and connected parts of a general system.—All as a connected whole have a northern and southern annual transit—All commence their movement together to the south about the first of August, and reach their southern position about the first of February—Chart showing the position in February of the five parts, to wit the central belt, the two trade-wind zones, and the two polar zones—The connected whole commences its northern transit early in February, and reaches its northern position in August. This organized whole to be particularly examined. Some preliminiary facts to be considered. Transit more extended north and south some years than others—Some sections which are not covered by either zone of rains during the transit—such sections constitute a class of deserts—Such as New Holland and Kalahari in the Southern Hemisphere—Arabia, Egypt, Sahara, Colorado, &c., in the Northern Hemisphere—There are also arid areas in the polar zones—Principal rivers of Africa and South America rise under the central belt of rains. Central belt passes over some places twice a year—Monsoons. Examination of the great central condition—Constituted by or composed of a central belt of condensation and two wings of winds, the trades—Analogous in this respect to the conditions of the polar zones—Analogous also in all its essential elements. Critical examination of its elements. Examination of the trade-winds and their character. Examination of the central belt of condensation and its character—Contains no vortex—Theory that it does, a mere assumption—originally made by Halley in 1686—never proved by any direct evidence, nor capable of such proof—Every fact in nature bearing upon it, adverse to it—earth, air, and water under the belt of rain colder than in the trades everywhere—fundamental base of the theory therefore untrue. Review of evidence on the point. Fact undeniable, undenied, yet not regarded.—Coast wind of California, not an exception—that wind does not reach the valleys which are assumed to cause it. The air as heated in the atmosphere has not the ascensive force attributed to it—review of the evidence on this point—impossible therefore that

CHAPTER VI.

CHAPTER VII.

PROGNOSTICATION.

val not to be expected, although conditions of more or less intensity pass frequently—they sometimes pass on a given day of the week, for several weeks with or without an intervening condition—no certainty as to the character and intensity of the next condition—nor any reliance on planetary influence,—only reliance upon changes of state produced by passing conditions. Signs, proverbs and maxims not founded on or connected with those changes of state, empirical and worthless. The inquiry is for prognostic inferences derivable from changes of state. Certain other elements and contingencies to be considered, viz : location of observer, season of the year, and situation of the year in the decade. Recapitulation of the changes of state induced by the conditions from which prognostic inferences are to be drawn. *Weight* of the atmosphere, how measured—various instruments examined. Barometer the principal one to be relied on—its indications—its mean elevation generally upon the earth—difference of elevation in different localities—mean elevation least where conditions are most frequent and intense. Examination of the fact in relation to this country. Places and circumstances under which it ranges highest and lowest. Allowance to be made for elevations above the level of the sea—rules for determining that allowance. No fair weather standard for the barometer—each observer must fix one for himself—standard must vary with the transits of the focal path and deduction must be made for altitude. Rules for fixing such standard—collection of rules used in England, for forecasting weather by barometer—critical examination of these rules. Prognostic inferences to be drawn from *temperature*—before northeasters— before belts of showers—in relation to the probability of snow or rain. Exceptional warm periods in winter without passing conditions—how produced—thermometer to be consulted in reference to the continuance of storms—Sudden changes of our climate—their magnitude and how produced—importance of forecasting and regarding them. Prognostic inferences to be drawn from the *winds* and their changes. Winds not felt at the surface discoverable by sounds and scud. All fresh, earnest winds created by conditions—Importance of this class of prognostic inferences. Prognostic inferences drawn from *clearness* or *cloudiness*—their character and importance. Examination of the maxims founded on them. Resumé of the appearances of the sky from which prognostic inferences may be drawn. Prognostic inferences from the existence of *humidity*—Evaporation and hygrometry—Devices by which humidity is measured. Contrast between the English climate and ours, in relation to humidity Examination of weather signs founded on humidity. Prognostic inferences from *rain*, *hail, and snow*—examination of weather proverbs in relation to them. Rules for determining whether a storm will precipitate snow or rain. Prognostic inferences from the seventh element or *electric* state of the at-

CHAPTER VIII.

The organization and motive force of the system invisible—its existence recognized in its effects—a knowledge of it to be acquired by inference from the nature of the organizations and their actions. Elements of the tornado and their mode of operation—place where it originates—manner of formation and form—its appearance and substance—its apparent manner of action in the air—its action and mode of operation in contact with the earth—estimates of the force employed—various descriptions of the power of that force—operates in two lateral lines or currents. The right hand current crosses the center in advance of the left hand, curves backwards and ascends over it. The left hand current curves behind the other and rises over it,—both together constituting the whirl in the air. These currents thus existing and operating, constitute the law of the tornado. This law of the small is also the law of the great, and is traceable through all the varied atmospheric organizations—Traced first in the belts of showers and in elliptical storms of the Northern Hemisphere—in those of the Southern Hemisphere—nearly all storms of that Hemisphere of this character. Critical examination of Col. Reid's chapter on the "Gales of High Southern Latitudes"—in every instance he purports to describe the northern half of the gale—southern half of a revolving gale not experienced—all the gales described by him, were, with a single exception, elliptical—that exception a straight line southeaster, corresponding to our straight line northeasters. "Ojo" of the Spaniards, not an indication of the center of the storm—constituted by a clear interval after the storm clouds have all passed by, and before the scud of the fair weather wind obscure the sky. The same kind of storm's eye visible with us, at least ten times a year,—never seen in the center of the storm.—All the descriptions cited by Col. Reid, belonged to elliptical storms with lateral winds—described as such by Mr. Mildrum. Extended examination of the theories of Redfield and Espy. Certain amount of truth in both

CHAPTER I.

Definition of the weather—Its normal state in the temperate zones, still, clear
and dry—All the states which are a departure from it, result from organized conditions—The conditions result from the laws of a general organization—The conditions are organized—Constitute a series—are successive—
are varied—within five miles of the earth, and move or pass over it—the different states produced by the different parts as they pass—Description of
the particular states—weight—temperature—wind—cloudiness—humidity
—precipitation—electrical state—History and description of a belt of showers
in August—manner in which it produced the changes of state—Description of an autumn S. E. condition—manner in which it produced the states
—various other facts bearing upon the subject.

THE weather is defined to be "the *state* of the atmosphere with
respect to heat or cold, wetness or dryness, calm or storm, clearness or cloudiness, rain, snow, hail, fog, etc." Thus defined it
means the existing *state* of the atmosphere in the several particulars named in any particular locality, and the *changes* which are
there occurring, or which may occur, from one state to another.
A separate and particular description of these states is not important in this stage of our inquiry, for we are all more or less familiar
with them, and they will be often under consideration.

The *normal* (regular) *state* of the weather in the temperate
zones, and of course over most of our country, when undisturbed
by direct or indirect influence from the centre of the system in
the tropical zone, is *still, clear, dry weather.* There are no independent atmospheric arrangements or influences *originating* in it,
and none *exerted upon it,* which can disturb that normal state,
except those which *emanate* from the *basis* of the atmospheric
system in the *tropics.* Meteorologists tell you that the sun, *heating
the land* of the temperate zone produces the various changes of

2

state and all the other phenomena, but they are mistaken and mislead you. The hottest sun heated lands are changeless and arid. In the San Joachim Valley, of California, at Millerton, the average summer mean temperature is about one hundred degrees and that mean sometimes rises to 104° with unbroken fair weather during the whole period. The influences emanating from the tropical base of the system, then operate to the north of it. In their absence the *sun heats up* the land, but *in vain*. Those influences return in winter and bring changes and rain upon it again. And this is true of nearly one-fourth of the continent. Every fact in nature properly understood and comprehended, tends to prove the proposition I am stating. I cannot allude to others now without anticipating, but I can assure you that this *truth* will *appear* as we proceed *too clearly* to be *questioned* by any mind which is *honest with itself*. Accept as true then my first fundamental proposition, that the normal *state* in the temperate zones is still, clear, fair weather.

My second proposition is that every other one of the enumerated or known states and changes of the weather, and all the phenomena connected with them, result from, or are *incident* to, and a *part* of some one of several *distinct, peculiar, organized atmospheric conditions*, limited and circumscribed in character, although frequently covering a considerable area.—which are formed in the atmosphere of the tropical zone, and pass from thence on to and over our zone, or, are formed in our zone of materials and by influences emanating from the base of the system in the tropical zone.

My third proposition is, that all the conditions referred to, result from the operation of fixed and intelligible laws, pertaining to a GENERAL AND INTELLIGIBLE ORGANIZATION of the atmosphere, or ATMOSPHERIC SYSTEM, which has its base in the tropical zone.

These three propositions cover the whole ground Reversing the order of description, we have an ORGANIZED ATMOSPHERIC SYSTEM, created and controlled by laws which in their operation produce limited, distinct and ORGANIZED CONDITIONS, and those conditions in their operations and movements produce the simple STATES and PHENOMENA which constitute "THE WEATHER."

And now reader, if you sincerely wish to understand the system from the operation of which you cannot escape, and which influences in a hundred ways, for good or evil, your daily life and business and happiness, discard, or hold in abeyance, all opinions and impressions derived from the teachings of professed meteorologists, or received theories, and give me a candid attention and reasonable confidence, while I trace for you as *matter of fact* the manner in which the STATES are produced by the CONDITIONS,—the *elements* and *mode of operation* of the conditions ;—and then investigate and unfold to you the elements and laws of the GENERAL SYSTEM by which the *conditions* are produced—leaving theoretic discussion for the close of our inquiry.

The various states and phenomena constituting the weather then, for any considerable period, as a month, a season, or a year, which are a departure from fair weather, result from or are produced by, a series of organized—*varied*—*successive*—*passing*—*atmospheric* CONDITIONS. This is our second, fundamental proposition. It is complex, and I wish all its elements to be fully understood and perfectly impressed upon the memory.

In the first place that the conditions are *organized* portions or "*conditions*" of the atmosphere. I use the term *conditions* for the want of a better. *Condition* and *state* are sometimes used synonymously, but you *must not confound them.* I use the term *state* to denote a part or element of a CONDITION and I use the term *condition* to denote the *entire*, limited *organization* of which the *states* are incidents or parts. And I say they are organized because each distinct variety is a distinct, circumscribed body, and has its distinct parts and elements, with their relative movements and operations, producing specific actions and results, and *that* constitutes an *organization.*

These atmospheric conditions occur *frequently* though irregularly, and therefore constitute during any considerable period, a SERIES.

They follow after one another at *intervals*, more or less regular and are therefore SUCCESSIVE.

There are several kinds of them, differing more or less from each other in their elements, and in the states or phenomena which they produce ; and sometimes one succeeds, and sometimes another, and therefore they are VARIED.

They embrace or involve a certain distinct part of the atmosphere, near the earth—generally within five miles of it—and that part, so involved and organized into a distinct condition, *moves or passes over it.* In the United States, north of lat. 35°, these conditions pass over the country from the westward to the eastward, and *produce* the different *states* or phenomena and their *changes*, as the *different parts* or elements of the *organized condition* pass *over* or *near* us. Though many of them are *storms*, in the common acceptation of the word, that term is not sufficiently comprehensive to describe them, for they are not always or necessarily stormy, or attended by the precipitation of snow or rain ; nor, when they do precipitate, is there always rain or snow at all the places where the changes of state occur. Thus they are PASSING and *produce the states* as they pass ; and they frequently pass by, at the south or north, but near enough nevertheless to affect the states of the weather to some extent, according to circumstances to be considered.

These are the *general* elements of our second proposition. Let us now look *specifically* at the various particulars in which these conditions severally affect or produce changes in the state of the atmosphere at any particular place in the country, as they pass over or near it.

FIRST, they affect the Atmosphere in respect to *weight* as measured by the *Barometer ;*

SECOND, in respect to *temperature*, as felt and as measured by the *Thermometer ;*

THIRD, in respect to the *movements of that part* which is in contact with the surface of the earth, from calm to windy, or wind from one direction to wind from another ;

FOURTH, as to *clearness* or *cloudiness ;*

FIFTH, in respect to *humidity*, as felt and as measured by the various contrivances termed *Hygrometers ;*

SIXTH, in respect to *precipitation*, producing rain, hail or snow.

SEVENTH, they affect the electrical state of the atmosphere, thereby affecting the feelings of men and animals, and producing lightning and thunder and other electrical phenomena.

But in order to a perfect and clear understanding of the character and scope of this second proposition, I must anticipate somewhat, and give you a *history* and *description* of one or two *passing conditions*, (in relation to which, my data are perfect and will not be disputed) and the changes of *state* which they produced.

On the 3d day of August, 1859, I left home at evening, for the purpose of attending the meeting of the American Association for the advancement of Science, at Springfield, Mass., and going from thence to Saratoga. The weather for several days had been pleasant, and the enveloping state had been a normal fair weather one ; but at evening there were indications of a change, and that a disturbing condition was coming on. My barometer, which hangs about 40 feet above the level of the sea, and which had stood during the continuance of the fair weather condition at about 30.1 inches, began to fall. There had been during the afternoon, a *fresh breeze* from the *southward*. The thermometer did not drop as rapidly at nightfall, as upon the preceding fine days. The air seemed more humid and sultry although the sky was entirely cloudless. These were indications, though not then decisive, of the approach of a belt of showers or a showery CONDITION from the westward, which as it approached had begun to exert its influence upon the atmosphere with which we were then in contact.

I staid that night with a friend, Col. Porter of Hartford. Waking about three o'clock in the morning, I heard the rattling of the shutters and the sound of the wind which had continued through the night and freshened to a stiff breeze. *As quick as thought the entire programme of the weather for the next three days, was as clearly understood by me as if it had all happened and become history.* At the breakfast table Col. Porter was felicitating himself upon the quantity of hay he had cut and cured, during the preceding fine days, and the prospect of cutting the last of it

during that day.   I inquired of him whether he had housed what
he had made; He replied that he had not.   Learning that he had
a considerable number of tons thus cured and in heaps, I advised
him to let his machines rest and set all hands to housing it, for it
was sure to rain within 24 hours.   He was loth to believe me
for the sky was still cloudless, and the wind, he said, had blown
the same way the afternoon before.   But I satisfied him that *that*
was a part of the indication and he did follow my advice, and
saved at least a hundred dollars in the quality of his hay, and the
expense he would have incurred in re-drying it.   His letter ac-
knowledging these facts is before me.

Passing up the valley in the afternoon, I saw many fields of
hay, cured, unhoused and imperfectly cocked—the owners still
engaged thoughtlessly, in cutting and curing, although the cloud-
bank of the approaching rain was becoming visible low down in
the northwest.   The rain reached Springfield in the night, with
lightning and thunder and continued into the forenoon of the next
day.   The following diagram exhibits approximately the character of the condition which then passed over New England, and as
it was then approaching side foremost from the westward.

FIG. 1.

In that condition, as generally happens in those of its class in the summer months, the lightning and thunder and heavy wind and rain were in, and under the *eastern part* of the belt, or, to be more accurate the southeastern part or side, and were first experienced. Probably the southeastern third of the belt, covering thirty or forty miles of its width, was composed of clouds of that character. After they had passed over to the east, another third or more of the width, continued to precipitate rain as it passed, without thunder or lightning or much wind—like a continuous and gradually diminishing rain. The remaining portion of the belt of clouds was more or less broken, gradually thinning and melting toward the western edge.

The following view taken by daguerreotype looking northwest at a different time, but when a similar condition was approaching and had become visible in the northwest, will show approximately the appearance of this at sundown on the 4th at Springfield.

FIG. 2.

During the continuous rain in the morning of the 5th, and while the Association was gathering for a general meeting, the late venerable Prof. Silliman came to me in the hall, stating that the Association was invited to go to Amherst the next day, and asked me if I could tell him what the weather would be. I replied: " Certainly ! It will be fair, with a stiff, cool, northwest wind." He seemed interested and asked me how I knew there would be *such a wind.* I told him I knew from the character of the *southerly wind* of the preceding day, and I explained to him the *nature* of the condition which was *then passing* over us, and the manner in which it had produced the various phenomena and changes of the preceding forty-eight hours, and that a *corresponding northwest wind* on the western side was an *essential part* of the condition, and *must* necessarily pass over us *in turn.* At the opening of the general meeting, the question whether the Amherst invitation should be accepted came up, and with it the inquiry what the weather was likely to be. Prof. S. stated that he " had inquired of a distinguished Meteorologist and was assured that it would be pleasant, with a stiff, cool, northwest breeze." As such winds are uncommon at that hottest season of the year, there were some sneers and a few smiles of incredulity, but the prediction was generally well received. That prediction was fulfilled to the letter, and I have before me a letter from the venerable Professor certifying to these facts.

During the day, the lecture of Prof. Henry, to which I have heretofore alluded and shall again allude, was delivered. In the afternoon I took the cars for Albany. A gentleman from New York,—Hon. Frederick Conklin, introduced himself to me and also inquired for the grounds upon which I based the prediction, and I explained them to him in like manner. I told him moreover that we should run out from under the western portion of the belt which then covered us, before we reached Albany, and meet the northwest wind. We did so run out near the state line, and we met that northwest wind though light, under a clear sky, as we left the cars and stepped upon the ferry-boat at East Albany.

The following view taken also at a different time, by daguerreotype, but when a southwest condition of this class was clearing off from the northwest, moving to the eastward, will give you a sufficiently accurate idea of the appearance of the western edge of the belt we are considering, as seen before the train ran out from under it, at the state line between Massachusetts and New York. The view was taken looking to the northwest. But you will have opportunities to *see* the *whole operation* several times a year if you will observe as you should do.

FIG. 3.

The occurrence and general character of that atmospheric condition, being thus conclusively proved, let us go back and note the changes and phenomena, which were produced by it in its passage.

First, there was a precedent fall of the barometer for more than thirty-six hours; Second, there was a rise of the thermometer, and increased sultriness, especially in the evening and night, and through the day preceding the rain; Third, there was a fresh, snappish, earnest, southerly wind, gradually increasing for more than twenty-four hours before the belt of rain reached the place. Fourth, the atmosphere grew humid and misty and close, and there were patches of cloud called scud floating in the southerly wind toward the approaching belt of showers; and the moisture stood upon the tumblers and gathered upon the stones, and upon every cold object. Fifth, there was a change in the electrical state of the atmosphere and doubtless (altho' I do not know the fact) heads ached that were accustomed to ache when thunder showers were approaching, and rheumatic joints, and corns, and bones that had once been broken, were also cognizant of its approach. And so when the belt of showers reached the place, thunder and lightning and heavy rain were experienced, and then as the belt passed, the quiet, continuous and diminishing rain, and after that, the curtain of spent, melting and gradually thinning clouds of the western portion of the belt continued to pass over, and after all these had passed, there followed the cool, dry, equally fresh, snappish and *earnest* northwest wind of its western side. And when that had passed in turn, the place was again enveloped in a still, clear, normal, fair-weather state and so remained until the 13th, when another similar condition passed.

The condition described, moved slowly. It appeared and began to precipitate at Buffalo on the 3d, at 7.30 P. M. At Auburn, according to the Register of Dr. Taylor, the rain began to fall between midnight and 4 A. M., of the 4th, distance about 115 miles.

At Amherst at 4 P. M. of the 4th. At Springfield on the evening of the same day. At Cambridge, Mass., on the 5th, the time of day not ascertained, and at New Bedford, at 4.30 of the same day. It was therefore about 48 hours moving from Buffalo to Cambridge and New Bedford, a rate of progress eastward (varying in different localities and at different times of the day) from 12 to 15 miles an hour.

There was a previous similar condition but of little intensity, which was at Buffalo on the 1st, and Amherst on the 2d. It passed Harvard on the 3d, as is shown by an overcast sky and the other phenomena, although it was spent and did not precipitate either at Harvard or New Bedford, and the northerly wind was not produced. And there was still another of considerable intensity during the first half of August, which reached Buffalo at 6.30 P.M. of the 11th, and reached Amherst on the night of the 12th, and Harvard College and New Bedford on the 13th. All this the records which follow this page will show. I have often passed through such belts. The following is an instance:

On the 14th day of July, 1870, Rev. Dr. Thomas S. Childs, Dr. Samuel Lynes, my wife and myself left Norwalk about 10 A.M. for Saratoga, by the Housatonic route. The weather had been unusually hot and dry, and some of the party being invalids, we were anxious respecting the heat of the day. The early morning was without sign of change. But as we ran along the shore to Bridgeport we observed a south-southwest breeze, of more than usual strength and humidity for that hour of the day, and saw that the northern sky, low down, was paled by a misty cirrus. From these indicia I inferred that we should meet and pass through a belt of showers on our way, and have a comfortable day, and so told my companions. As we ran north on the Housatonic road the breeze freshened, and the sun began to be obscured by the formless misty cirrus of the advancing belt. When we reached New Milford, the southeastern edge of the expected belt, with its layer of distinct cirrus overhead, and its layer of forming stratus beneath, became visible low down in the northwest. As we passed on the sun became more and more obscured, the breeze stronger, and the belt of clouds more visible, until we passed beneath the southeastern edge of it at Barrington in Massachusetts. That belt was about fifty miles wide, and of moderate intensity at that point, its bands ranging from S.W. to N.E., but its northern part was very intense and destructive at Montreal on the 13th, and in Maine on the 14th, as the newspapers stated. The showers were not continuous, but scattered beneath it. We passed between two of them in crossing the belt, receiving a slight rain from the corner of one, and emerged from under the belt between the State line and Albany, meeting the cool, refreshing northwest wind, with its scud, on the other side, which we enjoyed till we arrived at Saratoga. That belt of showers, with its wings of wind and its extended curtain of cloud, gave us a comfortable day and an agreeable ride, and my companions watched our progress toward and through it with great interest. We subsequently found that it was over Saratoga in the forenoon, and at Albany about noon. We passed through it between two and four o'clock in the afternoon.

The foregoing statement is correct.

T. S. CHILDS,
SAMUEL LYNES, M. D.

PLACE OF OBSERVATION, BUFFALO, COUNTY OF ERIE, STATE OF NEW YORK, LATITUDE 42° 50' LONGITUDE 1° 55', HEIGHT ABOVE THE SEA, 600 Ft.—Aug. 1859.

Day of Month	Thermometer in the open air				Rain and Snow				Clouds — 8 A.M.				Clouds — 2 P.M.				Clouds — 9 P.M.				Winds — 8 A.M.		Winds — 2 P.M.		Winds — 9 A.M.		Day of Month
	8 A.M.	2 P.M.	9 P.M.	Mean	Time of beginning of rain or snow	Time of ending of rain or snow	Amount of rain or melted snow, in inches	Depth of snow, in inches	Amount of cloudiness	Kind of clouds	Velocity	Direction	Amount of cloudiness	Kind of clouds	Velocity	Direction	Amount of cloudiness	Kind of clouds	Velocity	Direction	Direction	Force	Direction	Force	Direction	Force	
1	69	72	70	70.3	P.M.	A.M. 8.30	.113		10	Nim.			4	cu.	2		3	cir. st.	2		S.W.	2	S.W.	2	S.W.	2	1
2	66	78	71	71.6					5	cir.cu.	2		4	cu.	3		3	cu.			N.E.	2	E.	2	E.	1	2
3	72	78	73	73.3	7.30	9.	1.538		4	cir.cu.	2		8	cu. st.	2	S.	10	Nim.			S.W.	2	S.W.	2	S.W.	2	3
4	68	72	70	70.	12.30	P.M. 2.10	.088		10	Nim.	4		8	cir.cu.	2		0				N.W.	2	N.W.	2	S.W.	3	4
5	69	74	66	69.6					7	cir.cu.	2		8	Nim.	3		0				S.W.	2	S.W.	2	W.	2	5
6	66	77	69	70.6					3	cir.	2		0				3				S.W.	1	S.W.	3	S.W.	2	6
7	71	76		73.3					0				4	cir.cu.	2		0				S.W.	1	S.W.	3	S.W.	1	7
8	66	78	69	71.					0				3	cir.	1		0				S.W.	1	S.W.	2			8
9	69	81	75	75.					3	cu.			0				0				S.	2	S.	2	E.	1	9
10	74	87	78	79.6	P.M. 6.30	A.M. 9.	.277		10	Nim.			5	cu.	2		0				S.E.	2	S.E.	2			10
11	75	83	73	77.					10	St.			5	cir.cu.	2		8	cir.cu.			S.E.	1	S.E.	2	E.	2	11
12	70	75	73	72.3					0				3	cu. st.	3		0				S.E.	1	E.	2	S.E.	2	12
13	69	76	73	72.6					0				0				0				N.E.	1	N.E.		E.	2	13
14	67	76	69	70.6					0				4	cu.			0						N.				14
15	70	79	72	73.6					5	cir. st.							0				N.	1	N.	2	N.	1	15

PLACE OF OBSERVATION, AMHERST, COUNTY OF HAMPSHIRE, STATE OF MASSACHUSETTS, LATITUDE 42° 22' 17", LONGITUDE 72° 34' 30", HEIGHT ABOVE THE SEA, 267 Ft.—Aug. 1859.

Day of Month	Thermometer in the Open Air 7 A.M.	2 P.M.	9 P.M.	Mean	Rain and Snow — Time of beginning of rain or snow	Time of ending of rain or snow	Amount of rain or melted snow, in inches	Depth of snow, in inches	Clouds 7 A.M. Amount of cloudiness	Kind of clouds	Veloc.	Direc.	2 P.M. Amount of cloudiness	Kind of clouds	Veloc.	Direc.	9 P.M. Amount of cloudiness	Kind of clouds	Veloc.	Direc.	Winds 7 A.M. Direction	Force	2 P.M. Direction	Force	9 P.M. Direction	Force	Day of Month
1	60.5	77.2	66.8	68.2	P.M.	P.M. 7.30	.3		10	st.	7	S.W.	3	st.	6	S.W.	7	st.			S.S.E.	2	S.S.E.	2	S.S.E.	3	1
2	69.	81.8	66.	72.3	6.		.402		7	st.	7	S.E.	3	cu.	6	S.W.	5	st.			S.E.	2	S.W.	2	S.	1	2
3	68.	81.7	70.5	73.4	Th S	4 P.M.			10	st. fog		S.E.	5	cu.	3	S.W.	5	st.			S.E.	3	S.W.	3			3
4	71.5	84.8	72.	76.1	(4 Th S 4 to		.402		10	st.			7	st.			7	st.			N.W.	1	N.W.	4	W.	1	4
5	68.8	74.5	68.5	70.5	7 & in nig't)	night	2.822		10	nim.			9	st.			1	st.			S.W.	1	N.W.	3	N.W.	1	5
6	64.	74.	61.4	66.5					1	st.			1	cu.st.			0				N.W.	3	W.	2	W.	1	6
7	59.7	79.4	64.5	68.					0				2	cu.			1				N.E.	1	S.E.	1	S.	2	7
8	59.7	76.3	64.0	66.7					3	fog.			1	cir.			0				S.E.	2	S.E.	2	S.E.	2	8
9	57.8	78.	66.7	67.5					10	fog.			1	cu.			0				S.E.	1	S.W.	1	S.E.	1	9
10	63.2	79.4	67.	69.9					10	fog. st.			9	st.			5	st.	3	S.W.	S.W.	1	S.W.	1	S.S.	2	10
11	61.1	65.	65.		night	A.M. 10.	1.121		8	st.	5	S.W.	8	cir.st.	4	W.	0				W.	1	N.W.	1	W.	1	11
12	67.1	74.8	67.	69.6					10	nim.			2	cir.cu.			5	st.			W.	1	W.	1	W.	1	12
13	64.2	72.8	66.	67.3					10	fog.			3	cu.	3	N.E.	1	st.			N.W.	1	E.	1	E.	1	13
14	67.5	82.4	74.	74.6					8	cu & st			3					cir.			N.W.	3					14
15	69.8	76.6	64.	70.1																							15

PLACE OF OBSERVATION, HARVARD COLLEGE, CAMBRIDGE, COUNTY OF MIDDLESEX, STATE OF MASSACHUSETTS

LATITUDE 42° 22' 48", LONGITUDE 71° 7' 40", HEIGHT ABOVE THE SEA, —— Ft.—Aug. 1859.

Day of Month	Thermometer 7 A.M.	Thermometer 2 P.M.	Thermometer 9 P.M.	Mean	Rain: Amount of rain or melted snow in gauge, in inches	Clouds 7 A.M. Amount	Clouds 7 A.M. Kind	Clouds 2 P.M. Amount	Clouds 2 P.M. Kind	Clouds 9 P.M. Amount	Clouds 9 P.M. Kind	Wind 7 A.M. Direction	Wind 7 A.M. Force	Wind 2 P.M. Direction	Wind 2 P.M. Force	Wind 9 P.M. Direction	Wind 9 P.M. Force	Day of Month
1	64.	82.	65.			0		0		8	cir. st.	S.W.	0	S.W.	3	S.W.	4	1
2	67.	83.	70.			10	nim.	9	cu.	3	nim.	S.W.	1	E.	0		0	2
3	71.	87.	74.			10	nim.	0	cu.	0	nim.	E.	1	S.	2		0	3
4	74.	86.	76.	78.67	.247	3	cir. st.	4	cu. st.	10	nim.		2		6	S.W.	8	4
5	72.	76.	70.		.912	10	nim.			10	nim.	W.	0	N.W.	5	W.	1	5
6	65.	76.	66.			5	cir. st.	0	haze.	3	cir. st.	W.	3	N.W.	2		0	6
7	63.		71.			0		2	cir.	0	cir.	W.	2	E SE	1		0	7
8	63.	76.	68.			0		1	cir.cu.	0	cir.	N.	1	S.	1	S.	1	8
9	62.	78.	68.			0		6	cu.	1	cir. cu.	N.	1	S.E.	1	S.W.	1	9
10	68.	83.	66.			0		4	cu.	5	cir.cu.	S.W.	1			S.W.	1	10
11	64.	76.	67.			6	cir.cu.	10	cu.	0		S.W.	0	N.			0	11
12	65.	73.	66.		.171	8	fog.	3	cir.	10	mist.		0	N.	1	N.	1	12
13	68.	73.	71.			10	rain.	1	cir.	3	cir.cu. st.	W.	1	N.W.	1		0	13
14	68.	84.	60.			0				0		W.	1	E.	1	N.E.	1	14
15	67.	73.	60.			3	cir.cu.					E.	1	N.E.	2	N.E.	1	15

PLACE OF OBSERVATION, NEW BEDFORD, COUNTY OF BRISTOL, STATE OF MASSACHUSETTS, LATITUDE 41° 39' LONGITUDE 70° 56', HEIGHT ABOVE THE SEA, about 90 Ft.—Aug. 1859.

Day of Month	Thermometer in the open air				Rain and Snow				Clouds 7 A.M.				Clouds 2 P.M.				Clouds 9 P.M.				Winds 7 A.M.	Winds 2 P.M.		Winds 9 P.M.	
	7 A.M.	2 P.M.	9 P.M.	Mean	Time of beginning	Time of ending	Amount of rain or melted snow, in inches	Depth of snow, in inches	Amount of cloudiness	Kind of clouds	Velocity	Direction	Amount of cloudiness	Kind of clouds	Velocity	Direction	Amount of cloudiness	Kind of clouds	Velocity	Direction	Direction	Direction	Force	Direction	Force
1	64.	74.5	64.5	67.7					1	cir. st.	1	N.W.	8	cir. st.	1	W.		haze.		N.	wsw	ssw	2	wsw	1
2	68.5	73.	68.	69.8					10	fog.			10	cira↑g			10	fog.			S.W.	ssE	1	ssw	1
3	68.	81.	70.	73.0	night	night	.02		10				4	cir. st.	1	W.	4	haze.		S.W.	S.W.	S.W.	2	S.W.	1½
4	71.5	78.	71.5	73.7	4½ PM	night	1.40		10	cir. st.	13	wsw	9½	cir.	3	S.W.	9	cir.	8		S.W.	S.W.	8½	S.W.	8
5	71.5	80.	68.	73.2					9½				9½	cir. st.	2	S.W.	10	st.			S.W.	S.W.	1½	w by S	2
6	85.5	77.	63.	68.5					9	st			1.5				5	cir. st.		S.	wsw	w by N		wsw	2
7	63.5	76.	68.5	69.3					1			S.	0				9				wsw	S.W.	1	wsw	2
8	64.	77.	64.	68.3					0				1	cir. cu.	1	N.	0				wsw	E.	4	S.	4
9	65.	79.	64.5	69.5					½	haze.			8	cir. cu.	1	NNE.	0	haze.	1	N.W.	wsw	S by E	1	w by S	8
10	61.	76.	64.	68.3					10			N.	2	cum.		N.	8	haze.			N.	ssE	1	S.W.	1
11	61.	75.	66.	67.3	night	P.M.			8	fog.			10	cir. st.	5	W.	9	cir. st.			S.W.	ssE	2	S.E.	1
12	69.	72.	69.	71.		9.30	1.20		10	rain.			9.5				10	rain.			N.W.	S.W.	1	S.W.	1
13	71.	72.	70.5	71.2					3				4				1				N.W.	S.W.	1	N.W.	1
14	68.	80.	73.	73.					5				5								N.E.	N.E.	3	N.E.	2
15	68.	72.	62.	67.3																					

Having thus taken a general view of that distinctive, passing condition, and the states and changes of the weather, and the other phenomena produced by it, it may be well to look at another condition of the *same class* which occurred late in the Autumn in a different part of the country, and passed over an extensive surface, constituting an Autumn southeaster, in respect to which my data are perfect, but too voluminous for insertion.

FIG. 4.

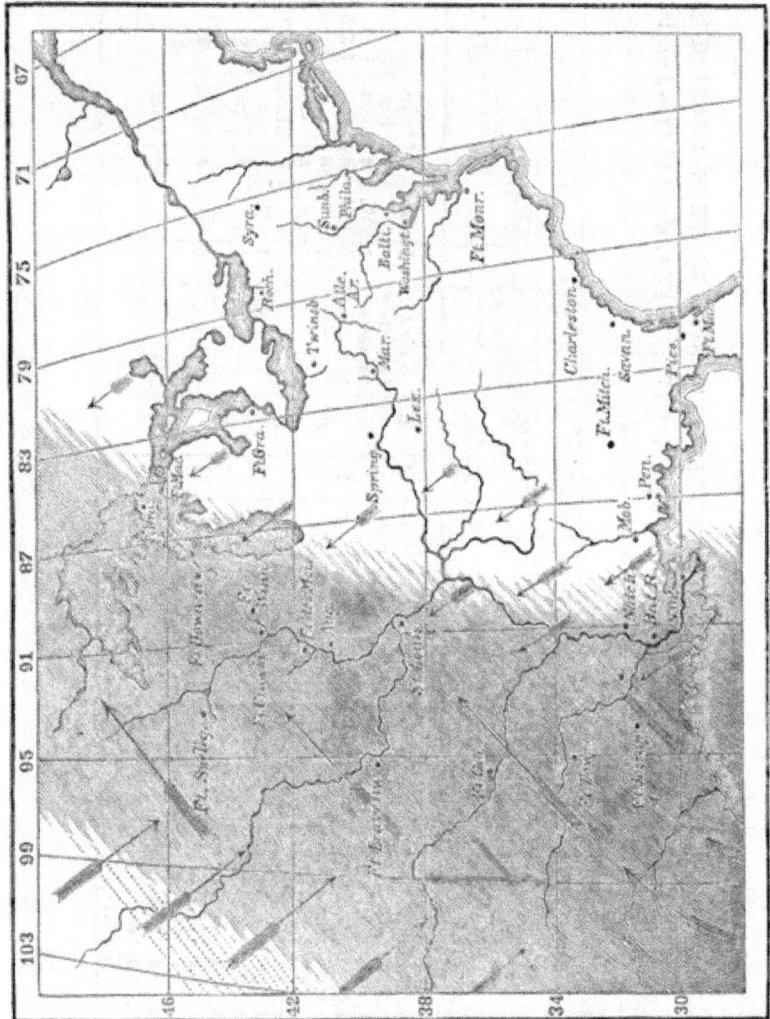

The foregoing diagram exhibits the position of such a southeaster, which entered upon the continent over Texas, and curving to the northeastward had crossed the upper part and arrived at the lower part of the Mississippi river, and was drifting to the eastward to cover the entire territory of the United States and Canada, east of that river, and pass off on to the Atlantic. It will be seen that it had the same lateral winds as the one I have described, but as the southeast wind was much stronger, (indeed, it blew a gale,) its direction was more at right angles with the axis of the storm. There will be noticed on this diagram, arrows in the body of condensation or central portion of the storm, indicating that that portion was moving to the northeast and this was true of it, and is true of all the various conditions, as a rule. There are some exceptions which will be hereafter noted. It will be further observed that the dotted lines on the northwest portion of the storm, indicate that portions of it had moved up to the northeast and left the surface uncovered, showing the manner in which such storms generally " clear off," from the northwest. It should also be noticed that the shading, down at the Gulf, indicates the accession of additional portions upon its eastern side, to which we shall hereafter more particularly refer.

Let us now consider ourselves as having been at Springfield, Ohio, over which this storm passed in its movement to the eastward and northeastward. There we should have noticed the same order of events except that there was no thunder or lighting—the lateral winds were much stronger—the body of condensation broader—the rain longer continued—and snow fell wherever, and to the extent, that the northwest wind blew in under the western edge of the storm. In the showery condition which we have described as passing over Springfield, Mass., the southerly wind blew in under the belt of clouds, and obliquely across it, nearly to its western edge, but the northwest wind did not blow under the belt of cloud, or at least, not with any force. In the condition we are now describing, the southeast wind also blew at right angles under and *across* the belt of condensation, but the northwest wind blew in *under that southeast wind*, a small part of the way, and turned the rain to snow.

The *states* and *changes* which then occurred at Springfield were as follows: Anterior to the approach of the storm, within influencing distance there was a fair, pleasant, still and normal day, sometimes called a weather breeder, with a high barometer. As soon as the influence of the storm reached the place and was felt, the barometer began to fall, the air to move towards the storm, and the wind to freshen from the southeast and fill with scud—the thermometer rose—the air grew damp—approaching cloud condensation (cirrus) was seen in the west and northwest, and men and animals felt sensibly the approach of the storm. In the belt of showers which we have described, the commencement of the fall of rain was nearly coincident with the arrival of the eastern abrupt edge of the cloudy portion of the condition over the place. In the condition we are now describing, the cloudiness extended from one to two hundred miles to the eastward of the rain, gradually thickening from the eastern edge to the part where rain was falling, and the western edge terminated abruptly with the fall of snow. In all other respects and indeed in all their essential features, the two conditions were alike, and belong to the same general class. The August condition would have been wider and like the autumn one if it had occurred later in the season; and so the autumn one would have been narrower, and a similar belt of showers, if it had occurred in summer.

The number of this class of conditions (taking a period of ten years) which occur in each year, at any given point east of the 95th meridian, will not vary, on an average, much from thirty, and if anything exceed it. All, or nearly all of our thunder showers, so called, are *contained in* and are *a part* of them. In the course of nearly fifty years habitual and close observation in different parts of the country, I have not seen a dozen, *single*, *isolated* thunder showers. Such isolated showers may be frequent in Europe, but that most of them occur there also, as parts of just such passing conditions, is demonstrable by their records. Slight showers sometimes occur on the eastern or western edge of the belt before the main body has arrived or after it has passed, which seems to be isolated, but they are a part of the condition. There was such an one at Amherst, August 2d, 1859, as appears by the tables.

Of the thirty or more, at least one-fifth do not precipitate at all over the eastern part of the continent. Whether because their energy is spent before they reach us, or because they never were sufficiently intense in their character, it is not always easy to determine. Probably sometimes from one cause and sometimes the other. That of August 1st, 1859, precipitated lightly at Buffalo and Amherst, but not at Cambridge or New Bedford. These feeble conditions are most common during summer droughts. They are perfectly distinct—have *all* the *elements* and go through *all* the *motions* of the most intense conditions of their class, but feebly and deceptively. They excite and disappoint hopes and are the cause and foundation of the proverb that "*all signs fail in a dry time.*" But I shall refer to this class of conditions again and note other occasional peculiarities.

Before leaving this branch of our enquiry I wish to put on record two or three facts, mainly for reference hereafter. On my way to Saratoga from Springfield, I saw frequently fields where cured hay was lying in various conditions, imperfectly cocked, in the windrow, and spread around, and in several instances partly loaded and left when the owner was overtaken by the rain. Tens of thousands of dollars were lost by farmers in New York and New England by that single rain, in the value of their hay and the cost of redrying it, which would have been saved if they had understood the thirty-six hours *perfectly intelligible warning* which preceded it. It was not my fault that they did not understand it. In 1856 I published in the "Philosophy of the Weather," a full account of the condition with a diagram and directions how to understand and prepare for it. The following is a part and a small part of that description, but it contains the gist of the matter:

"This class of storms, or belts of showers, present the following succession of phenomena in summer:

1. Still warm weather, one or more days.

2. Fresh southerly wind, one or more days; if more than one, dying away at the S. W., at nightfall, but continuing into the evening of the day before the belt of condensation arrives.

3. Belt of condensation, with or without rain or showers, with the easterly wind blowing axially, if the condensation is heavy and the belt wide ; westerly, if the condensation is feeble or the belt narrow—the clouds moving about E. N. E.

4. Cooler air, light N. W. in summer, heavy N. W. in autumn, winter, and spring.

And, the next period—

5. Still warm weather or light airs.

6. Southerly wind, fresh.

7. Belt of condensation.

8. Cool northerly wind.

And so on, successively, unless broken in upon by some other class.

Sometimes these periods are exceedingly regular, at other times the other classes prevail. I have much reason to believe that this is the *normal, periodic* provision for condensation of our portion of the northern hemisphere, and probably of every other where rain falls regularly in the summer season, and that the other classes are exceptions, as the hurricanes are exceptions to the normal condition of the weather every where. Perhaps in some seasons, during the northern transit, the exceptions may equal the rule, but I do not now remember such a season. In other years nearly all the storms are of this character. Thus, Dr. Hildreth (in Silliman's Journal for 1827,) speaking of the year 1826, in a note to his register of that year, says: 'There have been, this year, an unusual number of winds from N. or N. W. Nearly every rain the past summer has been followed with winds from the northward, when in many previous summers, the wind shifted to the southward after rain.' "

When preparing that book I took great pains to enlist some of our college Professors in the observation of this class of conditions. I succeeded to some extent by telegraphing and calling personally, and pointing out the succession of phenomena as they passed to the late respected Prof. D. Olmstead of Yale College. But when apparently satisfied of the truth of the matter, he advised a publication through the Smithsonian, and said if not so published, no matter what the facts might be, they would be *ignored.* I replied that if such was the state of things, I should publish at my own expense; for if there was to be an *Aristocracy* of science, or an *Autocracy in any branch of it*, in this country, it would have no aid from me. To test the matter I sent one of the first copies to Prof. Henry, and although it contained the full description alluded to, and fifty other new facts of great practical interest to farmers, and during the next three years he published several elaborate essays which were scattered broadcast, at government expense, in the agricultural part of the Patent office report, in which he professed to give to the country *all the facts* then known bearing upon the subject, there was no allusion to the book or the facts.

I also sent in August, 1856, one hundred and fifty copies of the book, by express, to the Association for the advancement of Science then in session at Albany, a sufficient number to supply every member then present with a copy. They were duly delivered, three days before the adjournment, but were never acknowledged, distributed or returned. The box was received and opened, but closed again, and taken to the premises of the President of the Association, and there remains, safely kept, so that the books can do no mischief to theories, or reputations. On my enquiring afterwards, a lame apology was given by the president for their non-distribution *then*, but *none has been given for their non-distribution or non return since.* Prof. Olmstead was a true prophet.

If those books had been distributed, the body of the Association, who are doubtless candid and honest enquirers for truth, would have examined them, and anticipated and understood the phenomena which occurred at Springfield, as I did.

With this general view of the *states* of the atmosphere and *the manner in which they are produced* by one class of the passing conditions, and the general view also of the *character* of *that class of conditions* in their intensity and feebleness, and in summer and autumn, I will close this chapter. We have yet to analyze and understand the organization of the long, central, cloudy and precipitating *body* of the condition and its "*wings of wind*," but before we can do so intelligently and understandingly we must examine and understand the *stories of the atmosphere* and their *cloud furniture*.

CHAPTER II.

The stories of the Atmosphere—clouds of the different stories—different character and function—Three of those stories—constituted by the interposition of the upper counter trade—course of that trade—varies in quantity and altitude—constitutes the middle story—rate of progress—open to observation—three kinds of cloud in the first or surface story, mist, scud and fog—mist common to all the stories—Description of fog—low fog—high fog—Scud practically of great importance. Nomenclature of Howard—does not include the scud—Description of the clouds by Howard—Founded on form and structure alone—Location, function, and use not regarded—Howard's descriptions for that reason imperfect—Nimbus not a distinct form and should be discarded—Return to consideration of scud—they form and float in the surface winds—characteristics of the N. W. scud—characteristics of the N. E. scud—of the S. E. scud—of the S. W. scud Clouds of the second story—all of the stratus forms—Three forms of stratus, viz: Cirro-stratus, cumulo-stratus and stratus—Description of them—constitute the rain-bearing clouds—A single form of cloud known to the upper story, the cirrus—description of the cirrus.

"The *stories of the atmosphere!*" And has the atmosphere stories?

It has; and as distinct and peculiar as the stories of your dwelling.

And are the clouds of the different stories peculiar and unlike?

Yes, in character and function, and as much so as the furniture of your parlor and bed-chamber and attic.

And are the stories constant and permanent?

Substantially so over the United States east of the meridian of 100°; and in some other portions of both hemispheres; but that is for our consideration when we come to examine the organization of the General System.

There are three of these stories. They are constituted by the interposition of a current of warm, rain-bearing air from the tropics. It enters over the United States from the southward in large volume, at different points, at different seasons of the year. It moves to the northwest in the tropics and curves between 25°

and 35° lat., at different seasons of the year, moving afterwards to the northeastward. It varies in quantity. Sometimes not more than a thousand or two feet in depth, at others from 6 to 10,000 feet—"a river in the air." Where it enters upon the continent its inferior surface may generally be half a mile to a mile above the earth. Gradually descending it will be nearer at 40° N. lat., and in the western states. It seems to be elevated again somewhat in passing the Alleghanies, and the easterly wind blows in under it in greater volume after it has passed than elsewhere.

That current constitutes the *middle* story. All below it is the surface story,—all above it the upper story. Various names have been given that middle story. In the "Philosophy of the weather" I called it the "counter trade." Sir John Tyndal in a recent work calls it the "upper trade." It is more generally known as the "Equatorial current." For convenience I shall call it the "*trade* current" or "*trade* story." It may be seen in a large proportion of the days of the year, and known by its elevation, direction, and the character of the clouds which form and float in it. At the latitude of 40°, it moves when unexcited at the rate of about twelve miles an hour, but its rate of motion is sometimes less and very often much more; during intense storms, 24 miles or more an hour. Meteorologists attribute its change of direction and easterly progression to the rotation of the earth on its axis. If the reason assigned were not intrinsically absurd, the fact that it moves slowly sometimes and at others very rapidly, (as the motion of the earth is uniform,) would seem to be a conclusive refutation of the theory. But I am to waive all theoretic discussion till *all the facts* have been passed in review. The difference in its rate of progress and that of the conditions also under different circumstances, is open to the observation of all, and I commend it to your constant attention.

There are three and but three kinds of condensation or cloud furniture, found in the surface story; *mist, scud* and *fog*. Of these the scud and fog are peculiar to that story. The mists occur in all the stories by day and by night. They have been met

with by aeronauts when invisible from the earth, and they are
often sufficient to dim the light of the sun on a cloudless day. As
they are nearly transparent, that is their only effect. Fog and
cloud are of similar constitution. They have form, are not transparent—and reflect the light ; and since the time of De Saussure and
Kratzenstein who experimented and investigated the subject with
great care in the lowlands and upon the mountains, the opinion
generally received has been, that both clouds and fogs are alike
composed of vesicular vapor, and that the vesicles are hollow.
Those meteorologists and others measured them, and tabulated
their measurements (see Kaemtz's Meteorology, page 111.)
This has been recently questioned in this country, but not on reliable evidence or authority. I do not propose to discuss the
question. The fact that fog and cloud differ from mist is unques-
tionable, and it is not material in this connection, whether owing
to the *size* of *globules* of water, or to hollow vesicles. Clouds
in one story and mist in another may be observed and contrasted
at least twenty times a year. Moreover, the fact that fogs are
sometimes vesicular is patent to the naked eye. I have myself
observed and you may observe it. After being out in a morning
fog which forms in the valleys in still, clear weather, I have observed and you may observe the small vesicles adhering to the
nap of your hat or your coat, and you may see them rising from
the surface of rivers in white lines and wisps, like the fibro cirrus
of which I shall hereafter speak.

There are two kinds of fog peculiar to the surface story. Both
form in still, clear weather, and disappear in the early part of the
day. The first or low fogs rest on the surface. They form in
the valleys and over streams and rivers, rising to the height of
from one to two hundred feet, being very dense and obscuring the
vision of those whom they envelope. Their upper surface is often
irregular, presenting rounded elevations like those which are generally seen upon the upper surface of clouds. They are a splen-
did addition to the landscape when over looked from an elevation
like the Catskill Mountain House, just after sunrise, like so many
white ribbons stretched north and south over the valleys to the

eastward. The high fog is practically of more consequence. It also forms in the night,—generally towards morning, in the upper part of the surface story, from 1500 to 2000 feet above the earth. It obscures the sun as effectually as cloud, and induces those who do not understand it to suppose that it is going to rain, but intelligent, practical men who do understand it, will tell you that it is nothing but high fog, and will clear away by 10 o'clock. It is generally dissipated by that time or soon after, and I think if you will reflect, most of you will remember the peculiar intensity of the sun's rays, as, upon some such occasion it first shone upon you after the fog was dissipated. It is of practical importance to distinguish this species of fog from cloud. It forms when the surface story is still and there is no wind. It is nearer the earth than the rain-bearing clouds. It is uniformly dense and regular, without contrast of light and shade or diversity of form ; while even the rain-bearing stratus of the second story it more or less irregular in density with fragmentary variations of light and shade, and in addition to these there will be an entire absence of such changes in the state of the atmosphere in other respects, as usually attend, and are produced by an approaching stormy condition. These facts will enable you to distinguish it.

We come now to one of the most important and most strangely neglected species of cloud, seen only in the surface story,—*the scud.* But before we enter upon the examination of them and the other varieties of cloud, it will be well to look for a moment at the received nomenclature of clouds.

In 1818, Mr. Luke Howard published that nomenclature, and it has since been generally received. He classified the clouds by the following names : Cirrus, Stratus, Cumulus, Cirro-stratus Cumulo-stratus, Cirro-cumulus and Nimbus. This nomenclature was founded upon the peculiarities of *structure alone*, without reference to location, or any other characteristic or quality. It is for that reason exceedingly imperfect, and was so considered by Mr. Howard himself in the later years of his life. The annexed diagrams show the different forms as originally published by Mr. Howard, and copied in all our text-books from that time to this.

FIG. 5.

FIG. 6.

Upon the first diagram, marked by a single bird, are attempted representations of four forms of Cirrus. Three of them represent curls or wisps, sometimes call the " curled clouds " or "mares tails." The *fibres* as represented on the diagram are very *large*, and *coarse*, but as seen in nature they are *fine* and *thready*. On the upper right hand corner there is an attempted representation of the *linear cirrus*, composed of long, straight, white threads or lines. Howard describes this form of cirrus as sometimes extending from southwest to northeast in England, and from one horizon to the other. It is not often so extensive in this country, but it is very often seen as the outlying, advance condensation of a northeasterly storm, approaching from the west and extending in the same general direction. Another form of cirrus is attempted to be represented on the upper part of the second diagram by one bird, and consists of small irregular patches of delicate, thin, white cloud. This also is a very common form in this country, and it may be observed that the cirrus is generally seen taking some one of these forms or some modification of them.

Remembering that the cirrus is always in the upper story—that it is always *white and thin*, and generally fibrous or thready, with slender filaments contrasting with the azure of the sky though sometimes misty and without observable form, and at others in thin sheets or parallel bands, and is the highest form seen, you will soon learn to distinguish it.

The second form, represented upon the first diagram only, by two birds, is the cirro-cumulus. It is represented as having the elevation of the cirrus, but consisting of small separated and isolated masses, like a collection of fleeces of wool arranged for exhibition without touching each other. Howard gave it the name of Cirro-cumulus because it had the elevation of the cirrus and was in small *heaps*. It is, in fact. cirrus broken into small portions. This form of cloud is rarely seen in this country. It does sometimes occur in the upper story during long continued droughts or set fair weather—perhaps half a dozen times a year.

The next form represented upon the first plate is the Cirro-stratus, indicated by three birds, and of this he gives three forms.

These, and modifications of them are very common in this country, occurring in the upper part of the trade or second story, and are especially visible in the advance condensation of approaching storms,—one modification (not represented,) being familiarly known as the " mackerel sky."

The fourth cloud represented upon the first diagram is the Cumulo-stratus, having a stratus base and a heaped-up elevation. It is represented on the first plate by four birds. The same form is seen upon the second diagram as existing in a continuous belt and not as an isolated cloud. This cumulo-stratus is also a common form with us, and usually makes up the eastern portion of our showery conditions while the thunder and lightning prevail, as in the condition which we have described as having passed over Springfield in August, 1859. The following diagram taken from a daguerreotype view, represents the Cumulo-stratus of an approaching belt of showers, which was taken in July of the same year, and which had a stratum of linear-cirrus over it.

Fig. 7.

The cirrus is a very difficult cloud to take by the daguerreotype or ambrotype process, and the artist has imperfectly represented it upon the cut, but it will serve to guide your observation of the cloud as it exists perfectly in nature and may very frequently be seen. It is always white and I have represented it darker than it is, to make it more observable. The stratus is represented at the bottom of the 2d diagram.

As to the Cumulus, Mr. Howard must have applied the term to all the various forms of scud, which exist in masses or heaps, and to the cumulo-stratus when of moderate size. There is no distinct cloud of that kind having a distinct character or use. The last description of cloud mentioned by Howard is the Nimbus, which is nothing but a Cumulo-stratus, dissolving in rain and presenting a smooth appearance. The same appearance precisely,

Fig. 8.

is presented by all our approaching thunder-showers, which have a smooth appearance because the falling rain obscures the inequalities of the cumulo-stratus clouds of which they are composed. The preceding is Howard's representation of this form of cloud, exhibited separately from the others. It is obvious that this is not a distinct form of cloud, and I shall make no further reference to it except to say that the continuous *line* of cumulo-stratus which is presented to view by an approaching belt of showers in summer, will have the appearance of Nimbus. Above, you will see the irregular tops and front of the cloud, below, a smooth, lighter, uniform surface from which the rain is falling, and it is not till that smooth portion gets over us, that the rain descends upon us. That lighter, smoother, seeming surface of the cloud is a deception. It is made by the falling rain, which *obscures* the *irregularities* of the mass of cloud above and beyond.

Having thus referred in a general way to Mr. Howard's nomenclature of clouds and his illustrations of them founded on their structure and form, let us return to a consideration of their qualities and peculiarities, in connection with the stories of the atmosphere, commencing with the scud.

Scud, form and float in all the winds of the surface story, which blow with any force. Light airs and breezes, are felt during the fair weather state, but they are casual, feeble and without scud. Scud form in all the winds which are parts or incidents of conditions. The winds, strictly speaking, which are parts or incidents of the stormy conditions, rarely blow from the south, north, east or west, but from the intermediate points of south east, south-west, north-east or north-west. This results from the fact that the conditions, after they curve in a low latitude, move in a north-east direction. The lateral winds of a condition will therefore be northwesterly and southeasterly. The northeasterly wind moves in under the storm, in opposition to its line of progression. The southwesterly follows after in the same line. The northeasterly wind is very common east of the Alleghanies, and less common west of them But we shall speak of all these winds more particularly hereafter. Our object now is to impress upon you the fact that they form the principal cardinal winds of the differing conditions.

Such being the character of our winds, we may contemplate the scud as they exist and are to be observed in different winds. And first, of the northwest scud.

These scud are seen at different seasons of the year, in every conceivable shape. In the summer season, they assume more or less the cumulus form, in distinct, rounded masses, and are white unless so dense as entirely to intercept the rays of the sun, when their under surfaces will be dark. The following diagram shows approximately the character of this description of scud, as they appear in summer, and as they floated over the heads of The American Association on their way to Amherst, in 1859. The view was taken looking to the north.

FIG. 9.

In the fall of the year, a distinct, rounded form is less prevalent, and they assume more of the stratus form, and float in larger and darker irregular masses. The following diagram is an imperfect copy of a daguerreotype view, taken at 8 o'clock in the morning, looking to the northwest, when a heavy southeaster was clearing off.

Fig. 10.

The rain ceased towards morning, the body of the storm had passed to the eastward, and was still visible, black and gloomy, in the east and southeast. The retreating western edge of the storm, thin and melting, is exhibited on the upper part of the diagram, with an abrupt edge of stratus on its edge. That western port'on with its edge, followed rapidly the body of the storm, and by 10 o'clock it was low down in the southeast. Between that western edge of the storm and the horizon the sky was clear and cloudless, except that near the horizon small masses of northwest scud were forming, which are represented upon the cut. The air was then still, but the northwest wind soon followed, and the scud rose towards the zenith, pursuing the retreating edge of the storm. At 10 o'clock the following view was taken, the wind having fre-hened almost to a gale, and the scud in wild, irregular masses covering all the northern part of the sky, and hurrying on to the southeast.

FIG. 11.

The storm soon disappeared from sight, but the wind continued to blow all day and until toward night in the same direction, filled with the same wild, irregular scud.   Later in the season and early in the spring, these scud are frequently very large and dark and gloomy, coalescing and covering the sky, and dropping for a brief period flurries of snow, but they may always be known by the direction and character of the wind they float in, and their relation to some stormy condition which has passed by and cleared off.   I need not be more particular; you will have fifty opportunities in the year to observe them, and a single glance for a short period whenever the wind blows from that quarter, will make their characteristics as familiar to you as those of the furniture of your room.

The northeast and southeast scud are of a different character. The northeast scud may be seen running under the outlying, advance condensation, towards the southwest, and the body of the storm approaching from that quarter.   They rarely assume a cumulus form, but are almost always in irregular patches, never white but always of an ashy grey. The following is a view of these scud, as they were running from the northeast to the south-west, twelve hours before the rain reached us, and while the pre-cipitating portion of the storm was 150 miles to the west of us.

FIG. 12.

Over these scud, layers of cirrus and cirro-stratus in the second and third stories were visible between the masses of scud, mov-ing to the northeast,—the advance condensation of the storm; but they could not be represented. As the storm approached nearer these masses of scud became larger and denser, and be-fore the rain set in they wholly filled the upper part of the sur-face story, and obscured the storm clouds above from view. The northeast wind and the scud continued to move to the south-west, until after the body of the storm had passed to the east-ward, the rain had ceased, and the layers of cirrus and stratus were again visible through the dissolving scud. I have thus des-cribed somewhat minutely these northeasterly storm scud. I have watched their progress during the continuance of these storms, in hundreds of instances. Those of you who live east of the Alleghanies, will have at least twenty opportunities for observing them in the course of a year. I beg you to observe them carefully, for a knowledge of them and of their movement will be of importance with reference to another branch of our inquiry.

The southeast scud do not differ materially from the northeast, in color or form. Perhaps they are not so uniformly ashy-grey as the northeast, but they are never black or white like the scud which run from the westerly points. The following diagram represents a body of southeast scud, moving during a heavy blow, towards a storm which was then at the northwest. They were very low down, as their appearance indicates, not more than 1500 feet high. And they were very wild in appearance, but that wildness could not be represented.

The view was taken looking northeast, the zenith and whole western sky being obscured by them.

FIG. 13.

Opportunities for observing the southeast scud are less frequent, but they are sufficiently numerous. These southeast scud float in the southeast wind, towards, under, and frequently quite across the long, stormy conditions to which they are incident. I have frequently seen the southeast wind, blowing under and entirely across and beyond the body of a belt of showers or storm, and gradually ceasing its movement, while the northwest wind was blowing in under it and under the retreating edge of the storm.

The southerly and southwesterly scud are less distinct in character than any of the others. When they blow from the south-southwest, towards and under a belt of showers, in the summer season, they very much resemble in form and appearance the scud of the easterly winds. The following cut represents a view of them taken by the daguerreotype as floating in a south wind, in the summer season, towards an approaching belt of showers.

FIG. 14.

But when, after a storm or belt of showers has passed by, and the wind is hauling or veering through the west and southwest, toward the northwest, the scud become whiter and more regular in form.

I have thus given you an outline of the character of the different varieties of scud, and I wish to impress upon you the importance of observing and becoming familiar with them, that you may better understand the organization of the various conditions, the rules for prognostication, and the motive-force of the system.

We come now to a consideration of the clouds of the second or trade story. These are all of the stratus form. They are all rain-bearing clouds, and precipitate the moisture which they bring from the tropics. There are three forms of the stratus. The first is the cirro-stratus, seen in the imperfect condensation which thins out in front and at the sides of the body of the storm. The following is a representation of cirro-stratus as taken by the daguerreotype and seen in the advance condensation of a north-easter. The scud had commenced running under it, but were small, and are omitted. There was a light stratum of cirrus in the third story, which is imperfectly represented.

Fig. 15.

As the storm came on, the bands of cirro-stratus were seen to have coalesced, and before the rain reached us, they had formed a dense, unbroken stratus, that could be seen in spots between the scud, until the latter had become so numerous that a view of the dense mass of stratus could not be obtained. It is from the stratus of the N. E. and S. E. storms, that we obtain most of our rains in the autumn, winter, and spring. In the summer season most of our rain falls from the cumulo-stratus, the remaining form of stratus, and the prevailing form in the eastern portion of our belts of showers. It is among the masses of cumulo-stratus in such a belt, that the lightning plays and the thunder is heard, and it is from the base of some cumulo-stratus which settles down into the surface story, that the lightning descends and strikes upon the

FIG. 16.

earth; and it is from the rounded thunder-head of the same cloud, that the lightning flashes up, to the layer of cirrus in the story above; and in the chamber between these two stories, that the thunder reverberates and rolls, till it dies away in the distance. The preceding diagram will give you some idea of the three strata, when the stratus alone is forming in the middle story. It is made up from observation, for a view of the three strata coëxistent, cannot be taken by daguerreotype. It represents, what I believe is invariably true, that the dense layer of stratus has its irregular rounded projections on the upper surface, though they do not tower up like those of the cumulo-stratus.

In this connection I quote from Mr. Redfield as confirmatory, his observation of the stratus and scud.

In a series of elaborate articles, substantially reviewing the whole subject, published in the American Journal of Science, for 1846, he says:

"In nearly all great storms which are accompanied with rain, there appear two distinct classes of clouds, one of which, comprising the storm scuds in the active portion of the gale, has already been noticed. Above this is an extended stratum of stratus cloud, which is found moving with the general or local current of the lower atmosphere which overlies the storm. It covers not only the area of rain, but often extends greatly beyond this limit, over a part of the dry portion of the storm, partly in a broken or detached state. This stratus cloud is often concealed from view by the nimbus, and scud clouds in the rainy portion of the storm, but by careful observation may be sufficiently noticed to determine the general uniformity of its specific course, and approximately, its general elevation.

"The more usual course of this extended cloud stratum, in the United States, is from some point in the horizon between S. S. W. and W. S. W. Its course and velocity do not appear influenced in any perceptible degree by the activity or direction of the storm-wind which prevails beneath it. On the posterior or dry side of the gale, it often disappears before the arrival of the newly condensed cumuli and cumulo-stratus which not unfrequently float in the colder winds, on this side of the gale."

"The general height of the great stratus cloud which covers a storm, in those parts of the United States which are near the Atlantic, can not differ greatly from one mile; and perhaps is oftener below than above this elevation."

The one great object of Mr. Redfield, in all his observations and investigations, was to obtain evidence to establish his theory, that the surface winds of all storms are revolving winds; and he overlooked many of the important facts we have been considering. But with reference to the stratus and scud of storms his conceptions were clear. Of the theory which he sustained with great research and eminent ability, I shall speak in a more appropriate place.

The remaining form of cloud, and the only form seen in the upper story is the cirrus. I have never been able to get a good daguerreotype or ambrotype view of it. It is uniformly white or whitish, and for that reason exceedingly difficult to take. The description already given will be sufficient to guide you in your observation of it, and I beg you carefully to observe it. It is ordinarily the first seen as the advance condensation of a storm, and bears an important part in the organization of all the conditions.

I have thus given an outline of the stories of the atmosphere and their cloud furniture. Much of it will be new to you. I know of no author who seems to have had any distinct conception of the stories of the atmosphere and their peculiar clouds except Mr. Ruskin, the author of Modern Painters, and his conceptions were limited to the form and appearance, and did not extend to their qualities and uses. We are now prepared to return to the *organization* of the *conditions* in another chapter.

CHAPTER III.

I recur now in the first place, to the condition described as having passed over Springfield in August, 1859. That this was an organized condition is obvious. In the language of the dictionaries, it was " composed of several individual parts, each of which had its proper function and conduced to its existence." Of its body and wings of wind, we have already spoken in a general way. Let us look now at the *parts* which composed its *body*, and their motions and functions. That body was composed of three strata of clouds. There was, in the first place, a layer of cirrus in the upper story. There was, in the next place, a layer of cumulostratus in the eastern portion of the trade story and of dense stratus in the middle portion of that story, extending to the western edge, gradually thinning, breaking up and dissolving. There was in the lower story a stratum or layer of southerly surface wind,

b'owing in under the other strata, humid and moist, and filled
with patches and dense masses of scud; and all these were in
some way, and had been for the days of its existence, and continued to be for at least one day longer, working or acting together,
and traveling, as relative parts, maintaining their relative positions,
to produce and deposit the rain. The thunder and lightning were
incidents, evidential of the intense action which produced the
strong wind and the rain in large quantities, but not essential to the
production of either. The layer of cirrus in the upper story, and
all the forms of stratus in the second story, were moving steadily
to the *E. N. East,* yet the whole body had a *drift to the eastward.*
I shall discuss hereafter the occasion of that drift. At present
I confine myself to the organization of the condition.

Now here we have the several parts which I have named, working harmoniously and continuously together, created and guided
by the Giver of all good, to distribute the blessing of needed
rain over the earth. The condition thus organized, existed and
drifted for several days, passing beyond our ken, out upon the
Atlantic. Some creative and vital force must have organized
and continued it. What was that force? While the condition
was passing over Springfield, Prof. Henry, in a lecture, was urg-
ing the theory that all such conditions, and all conditions and
winds were originated by the heating of the air at the surface of
the earth, and continued by the giving out of latent heat from the
condensing vapor, constituting a vortex which *sucked up* the surface atmosphere, and produced the winds. I shall hereafter de-
monstrate to you that the theory is unsupported by a *single* fact
in nature. But I wish now to allude to some facts then existing.
The lateral winds of that condition were at least 200 miles in
breadth on the eastern side, and 150 on the west, and 2500 feet
in depth, moving at the rate of at least 20 miles an hour. The
whole width of the body of the condition from which rain was
falling, did not exceed at any time, 100 miles, and that space
could not have furnished a vortex sufficiently large to create or
receive the winds. Can any sane or honest minded man believe
it could?

In the second place the N. W. wind did not blow at any time, with any force *into* or *under* the body of the condition, but followed after it as it retreated to the east, visibly piling itself up in the rear, and elevating the Barometer. In the third place, no part of the condition reached above five miles from the earth, and the vision could be extended a long distance over it as it approached from the west, and so over it again as it receded to the east, and no vortex which could have taken up both these winds, could have existed without attaining great elevation and being seen, and nothing of the kind was visible. And in the fourth place, a fact conclusive, after the intense and precipitating portion of the condition, (where the vortex was assumed to be,) had passed by, and the rain had ceased, the layer of cirrus in the upper story and the layer of melting and dissolving stratus in the trade story, were distinctly visible, occupying *undisturbed* and *unperforated* their relative positions, and the southerly wind was moving with its remnants of scud in its relative position under them, toward the western edge of the condition far beyond the place where the creating vortex was assumed to be. If that southerly wind had been created by the suction of a vortex, it would have *obeyed the law of its creation and gone up the vortex.* But there it was, visible from the seat I occupied during the lecture, *moving as I have seen it a hundred times,* AND WITHIN 2,000 FEET OF THE LECTURER'S HEAD, a *conclusive refutation of the fundamental doctrine of his lecture.* In the debate which followed, I alluded to it, and after the lecture was over, I pointed it out to several gentlemen who were staying with me at the same hotel. In that debate it was conceded, and the fact cannot be denied, that if there was such a vortex, the wind and scud must ascend it and could be seen to do so. And I offered Prof. Henry fifty dollars a scud for every one he would prove to have been seen so ascending. Prof. Loomis asked for a renewal of the proposition, saying *he was "up for a speculation."* It was renewed AS A STANDING OFFER and in good faith, although the discovery of a single vortex would sweep away all I possess. It is now ten years since and no vortex has been discovered, or money called for. Nor will there be (although Prof. Henry has more than 100 observers at his service) until the air of his parlor, by being heated, shall be made to ascend through the floors, and carry the furniture with it to his attic.

But it is not my purpose *now* to discuss the theory, or any theory in relation to the organizing and vital force which creates these conditions. I content myself with putting these *facts* on the *record* for consideration hereafter.

There is another important fact in relation to all the conditions of this class which should not be omitted. The edges of the body when abrupt, extend from a southwesterly to a northeasterly point, parallel with the axis of the condition and all the forms of the stratus condensation in the trade-story, lie in *distinct bands* parallel with that axis and conforming to their line of motion. This is frequently the case also with the layer of cirrus in the upper story. And the same fact may not unfrequently be observed in respect to the northwest scud which follow it. They are often arranged in lines, corresponding to the bands of stratus and cirrus. This arrangement of the stratus in bands, is more distinctly observable in the less intense conditions of this class, like that which we have already described. They are also to be seen in all the classes of conditions. In the storms which come up the coast in the fall of the year and move more directly north, the bands of condensation may be observed lying more northerly and southerly, conforming to the line of motion. This fact has an important bearing upon the question to be hereafter considered, relative to the organizing and continuing force of the conditions, and I beg you to observe it. It is visible in most of our conditions when clearing off from the northwest, and in that direction.

There are some other peculiarities attending these belts of showers to which it may be well to refer. They are sometimes composed of imperfectly connected showers or masses of cumulo-stratus, and when the break between them passes over any given point, a shower will pass to the north and another to the south of it, and in such cases, people will say " the showers go round us." They sometimes move too, in the manner indicated in the following diagram:

Fig 17.

And in such cases, the observer must *look in the southwest*, and not at the west or northwest for the *particular portion* or cloud, which is to precipitate rain upon his locality.

This is the general course and manner of progress of these belts of showers, but there is one other peculiarity which should be noticed. The *lateral wind* is always the strongest which blows from the surface that is *most moist*. Thus when a belt is passing south of the Great Lakes or south of a territory which is extensively marshy, or over a dry portion of the country, when there have been heavy rains to the north of it, the N. W. wind will frequently supplant the S. wind and blow entirely across the body of the condition and out upon the southerly side. In some sections of the country, and especially south of the Great Lakes, this is not uncommon.

Before I leave this class of conditions I wish to say that they are the most common everywhere. In the following table taken from Col. Reid's work on "Storms," there is a record of five of them which passed over Bermuda in December, 1839.

Date.	Hour.	Direction of Wind.	Wind's Force.	Weather.	Bar.	Ther.
1839						
Nov. 30	Midnight.	S. S. E.	1	b. c.	30 06	65
Dec. 1	Noon.	S. S. W.	3	b. c.	30 07	71
2	"	S. W.	5	g. m. q.	29 86	70
3	"	S. S. W.	3	g. c.	29 76	"
4	"	S. W.	6	g. m. r.	29 62	68
5	"	W. N. W.	5	p. q.	29 56	"
6	"	N. W.	6	p. q.	*29 55	"
7	"	N. N. W.	5	b. c.	29 78	70
"	Midnight.	N. N. W.	3	b c.	29 89	68
8	Noon.	W. N. W.	2	b. c.	29 82	71
9	"	S. S. W.	5	p. q.	29 84	70
10	"	S. W.	2	b. c.	29 96	"
11	"	W. N. W.	6	b c. m.	*29 88	68
12	"	S. S. W.	"	b v.	29 99	69
13	"	N. N. by W.	"	b. v.	30 01	66
14	"	N. N. W.	5	b. c. v.	30 06	64
"	Midnight.	N. W.	2	b. c. q.	30 05	63
15	Noon.	S. W. by S.	6	g. m. r.	29 72	65
"	P. M. 2	S. S. W.	7	m. q. r.	29 92	64
"	" 4	S. S. W.	"	g. m. q. r.	29 55	"
"	" 6	W. S. W.	"	q. w.	*29 53	"
"	" 8	N. W.	6	b. c. q.	29 54	"
"	" 10	N. N. W.	"	b. c.	29 55	"
16	Noon.	N. W.	7	b. c. m.	29 53	62
17	"	N. W. by N.	"	p. q.	29 67	60
18	"	N. W.	6	c. q.	29 86	"
19	"	N. W. by N.	7	m. q. r.	*29 73	59
20	"	N. N. W.	"	p. q c.	29 89	58
21	"	N. W. by N.	6	c. q.	29 96	56
"	Midnight.	S. W.	1	b. c.	29 95	55
22	Dawn.	——	0			
"	Noon.	S. S. W.	5	g. m.	29 83	56
"	P. M. 4	S.	7	g. m.	29 79	"
"	" 6	S. S. E.	"	g. m. r.	29 61	"
"	" 8	S. S. E.	"	w. r.	29 52	"
"	" 10	S. E.	"	m. w. r.	29 48	"
23	Noon.	S. W.	6	b c. m.	*29 44	57
24	"	W. N. W.	"	b. m.	29 71	59
25	"	W. N. W.	5	b. c.	29 88	56
26	"	N.	3	c.	30 09	62
27	"	S. E.	5	c. q. r.	30 07	61
28	"	S. W.	6	c q.	29 88	66
"	Midnight.	S. S. W.	"	b. c.	29 76	65
29	Noon.	S. W.	7	c. h.	*29 48	64
30	"	W. N. W.	6	b. c. q.	29 83	55
31	"	N. W.	5	b. c.	30 12	58

b. indicates blue sky —c. detached clouds—r. rain—v. visibility of objects—q. squalls—w. wet dew—u. ugly threatening appearance—g. gloomy weather—m. misty—p. passing showers—h. hail.

The foregoing table embraces the weather for a month, on an island where no disturbing element of configuration could inter fere. Examine it carefully, (it is from a standard work and reliable) and answer it to your love of, or desire for truth and knowledge, whether there are not recorded there five successive, passing conditions, with all the elements and changes I have described as belonging to that class. The second one of those five conditions, I beg you to observe, was (like some of those which occur here as I have stated) feeb'e and short and *did not precipitate*. It occurred on the 10th and 11th. As it was late in the season none of them were accompanied by lightning and thunder. The asterisks point to the several periods of low barometer.

So the northers of the Gulf of Mexico are the left hand, lateral winds of this class of conditions as they come up over the Gulf of Mexico in the fall and winter months and curve to the eastward over the United States.

The following description of two winter northers, copied from Colonel Reid's valuable work, will illustrate what has been said.

1843.	Wind.	Force.	Weather	Bar.	Ther.	
Jan. 30.						
A.M. 4.	S.S.W.	2	b. c.	29.90	77	Off Tampico.
Noon.	South.	5	b. c. r.	29.86	76	⎧ Lat. 23° 41′ N., Long. 94°
P.M. 8.	South.	6	b. c. r.	29.84	76	⎩ 50′ W.
Jan. 31.						
A.M. 4.	S. Easterly.	3	b. c.	29.90	74	⎧ Between 6 and 10 A. M., ⎨ wind was variable.
Noon.	N. by W.	9	c. q. w.	29.96	76	Norther com'ced at 10 A M.
P.M. 8.	N.N.W.	9	c.	30.09	73	Lat. 22° 36′ N., Lg. 95° 48′ W.
Feb. 1.						
A.M. 4.	N.N.W.	7	c. g.	36.29	63	Lat. 22° 9′ N., Lg., 94° 50′ W.
Noon.	Westerly.	6	c.	30.30	67	
P.M. 8.	Calm.	0	c.	30.26	67	
Feb. 14.						
A.M. 4.	S.E.	3	b. c. r.	29 66	73	At Sacrificios.
Noon.	S.W.	4	b. c.	29.62		Norther come'd at 5.30 P.M.
P.M. 8.	N. W. by N.	10	c. q. u.	29.72	65	
Feb. 15.						
A.M. 4.	N.W. by N	10	c. q u.	30.10	61	⎧ Gale moderated and again ⎨ freshened about 8 A.M.
Noon.	N.W. by N.	10	c. g. q.	30.19	61	
P.M. 8.	N.W.	4	c. g.	30.20	65	
Feb. 16.						
A.M. 4.	N.W.	5	q.	30.18	62	
P.M. 8.	N.N.W.	2	c. g.	30.21	66	

b. indicates blue sky—c. detached clouds—r. rain—v. visibility of objects—q. squalls—w. wet dew—u. ugly threatening appearance—g. gloomy weather.

Precisely such changes from S.E. rains to N.W. winds, with blue sky and detached dark clouds—N.W. scud—occur every autumn in October and November, in the Eastern and Midd e States, and the falling of the thermometer and rising of the barometer, after rain, and such a change of the wind, are perfectly characteristic. Doubtless the northers are all of the same general character.

In both of these instances there was *rain* with the wind south or southeast, and with a high thermometer and low barometer preceding the northerly wind; and with the N.W. wind the thermometer fell and the barometer rose. The westerly or left hand lateral wind of the autumn *southeaster,* which I have mapped and described to you, was a severe Norther on the western shore of the Gulf of Mexico. If that storm revolved over the Gulf it certainly ceased to do so before it arrived in the position where it was investigated and I have represented it. With a few other instances occurring elsewhere, I will leave this class of conditions.

Take now an instance from Mr. Bassnett's "Mechanical Theory of Storms," as observed by him at Ottowa. The letters in parentheses refer to the wind.

"*June 21st.* Fine clear morning, wind, (S. fresh :) noon very warm, 88°; 4 P.M., plumous *cirri in south;* ends clear.

"22d. Hazy morning (S. very fresh), arch of cirrus in west; 2 P.M., black in W.N.W.; 3 P.M., overcast and rainy; 4 P.M., a heavy gust from south; 4.30 P.M., blowing furiously (S. by W.); 5 P.M., tremendous squall, uprooting trees and scattering chimneys; 6 P.M., more moderate (W.).

"23d. Clearing up (N.W.); 8 A.M., quite clear; 11 A.M., bands of mottled cirri pointed N.E. and S.W.; ends cold (W.N.W.); the cirri seem to rotate from left to right, or with the sun.

"24th. Fine clear, cool day, begins and ends (N.W.).

And another instance in August, less intense, of the same character.

"*August 6th.* Very fine and clear all day; wind from S.W.; a light breeze; 8 P.M., frequent flashes of lightning in the northern sky; 10 P.M., a *low bank of dense clouds in north,* fringed with cirri, visible during the flash of the lightning; 12 P.M., same continues.

4

"7th.   Very fine and clear morning; wind S.W., moderate; noon, clouds accumulating in the northern half of the sky; *wind fresher, S. W.;* 3 P.M., a clap of thunder over head, and black cumuli in west, north, and east; 4 P.M., much thunder and scattered showers; six miles west rained very heavily; 6 P.M., the heavy clouds passing over to the south; 10 P.M., clear again in north.

" 8th.   Clear all day; wind the same (S.W.), a hazy bank visible all along on *southern horizon.*"

The wind for the 9th is not given, but the belt moved very slowly, keeping in sight all day on the 8th, and it was undoubtedly round to N.W. after the belt had moved farther to the eastward on the 9th.

Now follow me into the Southern Hemisphere. But here I must premise that in the south temperate zone the winds vary in direction from ours, but obedient to the same general law.   The conditions move there from Northwest to Southeast. Of course the winds are reversed; the storm-winds there, as here, are on the tropical side of the condition.   The N.E. wind, as a wind lateral to the condition, corresponds to our S.E. wind, and the N.W. wind bears the same relation to the condition that the S.W. does to the conditions in this zone.   So the S.W. wind bears the same relation that the N.W. does here, and is their clearing off wind as the N.W. is here.   And the S.E. wind a'so corresponds to our N.E. and the S. wind there corresponds to the N. wind here. The west is the only wind that bears the same relation to the conditions in both zones.

It is a notorious fact that the difficulties experienced in doubling Cape Horn, arise from the frequency with which the conditions pass at that point, and the almost constant alternation of N.W. storm-winds with S.W. clearing off winds of passing conditions, drifting to the eastward, past the cape.   I have had a strong desire to pass around the Horn and observe the passing conditions there and in both Oceans, but circumstances have prevented it.   An intelligent neighbor, Dr. Samuel Lynes, made the trip from New York to Panama in the Pacific Mail Co.'s Steamer, the "China," commanded by Capt. Geo. H. Bradbury. At my request, he kept a careful record of his observations, which he furnished me, and which I shall hereafter use extensively. I now copy from that record and the log politely furnished me by Capt. Bradbury, in order to show the conditions met with from lat. 24° S. to the Straits of Magellan, a distance of about 2100 miles.

Friday, July 26, 1867. Latitude then at noon was 24° S. They had left the S.E. trades and were standing S.W. by west, and in the afternoon entered the N.E. wind of a passing condition. The record is, " P.M., wind E.N.E. fresh, sent up headyards," and the record for the succeeding days till their arrival at Possession Bay in the Straits of Magellan, is as follows:

"Saturday, July 27, 1867. Begins A.M. with strong wind from N.E. to N.E. by N. Hazy and foggy. 7 A.M. thermometer 66°. *Fresh breeze from N. East,*—nearly calm at noon. Saw a steamer probably bound for Rio Janeiro. Has been very pleasant to-day; sails set to-day. *Lightning in S.E. and S. to-night* from dark to midnight, and likely to be a change. Thermometer highest 67°, Lat. 26° 15 S., Long. 47° 03′ W. Miles run, 240.

Sunday, July 28. Begins A.M. with brisk N. to N.N.W. winds and squally. 7 A.M., ther. 57°, wind round in a light breeze from S.W., sails down this morning, lightning all last night. Fresh S.W. to S.S.W. wind through the day, and fine. Passed St. Catherine's light about midnight. We begin to see many birds now, gulls, etc. Heavy fog at 2 P.M. Saw a brig. Divine service at 4 P.M. Saw " *Magellan's Clouds.*" Lat. 29° 38′ S., Long. 49° 19′ W. Miles run 240.

Monday, July 29. Began A.M. with light N.E. wind till 2 A.M. then fresh and strong from N.E. to N.N.E. 7 A.M. ther. 56°. Fresh breeze from N.E. Thunder and lightning, and heavy showers and hail. Lighted up in S.W. after a deluge of rain with lightning and thunder; at 10 A.M. set sail, saw a brig, wind steady N.N.E. Clouded aga'n at 11 A.M., wind hauled around slowly and got to S.W. at 10 P.M.; heavy showers all day, and heavy E.N.E. swell to midnight. Ther. highest 57°, Lat. 33° 10′ S., Long. 52° 40′ W. Miles run, 250.

Tuesday July 30. Began A.M. with fresh W.S.W. wind and
cloudy with sharp squalls. 7 A.M. ther. 47° cloudy. A S.W.
gale ("Pampero") blowing this morning, Capt. B. says not a
severe one. Let in steam to the saloon to warm us. Off the
river Plata to-day. Passing, sharp squalls with rain and hail in
afternoon and evening. Nearly over a vessel at 8½ P.M.
Stopped engines and backed her and she got by—narrow escape.
Ther. 45°, Lat. 35° 30' S., Long. 54° 37' W. Miles 197.

Wednesday July 31. Began A.M. with wind fresh from W.
S.W. and fine. 7 A.M. ther. 40°, clear, wind and sea not *quite*
so high as yesterday, cloudy at 9 A.M. when the wind veered to
N.W., rainy at 11 A.M. and cold, windy and disagreeable; set jib
and fore spencer, sea smoother. Off Cape Corientes at 4 P.M.
Cleared and wind moderated in P.M. Sun set *most beautiful* to-
night behind Cape Corientes. Ther. highest 48°, Lat. 37° 31' S.,
Long. 56° 40' W. Miles 163.

Thursday Aug 1. Began A.M. moderate to fresh from N. to
N.W. At 4 A.M. wind went to S.W. by W. and blew fresh a
strong wind from S.W. dead ahead, large sea. Wind and sea
moderated in the afternoon and at sundown, only a gentle breeze
from S.W. Has been very clear and bright to-day. Ther. highest 50°, Lat. 39° 48' S., Long. 59° 37' W. Miles 190.

Friday Aug. 2. Began A.M. with wind N.W fresh with
strong intervals. 7 A.M. ther. 51°, cloudy, cool and fresh breeze
from N.W. Sea moderately rough. Set jib and fore spencer at
8 A.M. At 9 A.M. clear arch to the W.S.W. and when overhead wind came out S.W. by W., smart gale. Noon, blows a
gale—a pampero—more severe than any we have had yet. The
waves are very large—ship is very steady. 7 P.M. gale increasing and terrific. Ther. 48°, Lat. 42° 07 S., Long. 63° 00' W.
Miles 208.

Saturday, Aug. 3. Began A.M. with strong S.W. to S.S.W.
gale, and very heavy sea. 7 A.M. ther. 49°, cloudy, wind moderating, heavy sea running. Passing rain squalls through the
night. The gale broke towards morning, ship behaved admirably.
P.M. moderated more and very pleasant, except rough sea and

occasional light passing snow-squalls. Ther. 44°, Lat. 43° 49' S., Long. 63° 44' W. Miles 107.

Sunday Aug. 4. Began A.M. with wind, light from S. by W. till 6 A.M., then light N.N.W. 7 A.M. ther. 40°, cloudy and cool but pleasant, wind freshening from N.N.W., set all sail on foremast, making 10 knots. Off Cape Blanco at noon, in sight; a strong breeze from N.E. continuing through the afternoon; evening rainy and dark. Ther. 41°, Lat. 47° 26' S., Long. 65° 10' W. Miles 226.

Monday Aug. 5. Begins A.M. fresh breeze from N.W. 7 A.M. ther. 38°, clear, pleasant but cold. Fresh breeze from west and rough sea. 1 P.M. snow-squall. 8 P.M. off Cape Virgins, entrance to the Straits of Magellan. Passed in and anchored in Possession Bay at midnight. Blowing fresh S.W. Lat. 51° 3' S., Long. 67° 41' W. Miles 240.

Tuesday, Aug. 6. At anchor in Possession Bay. Began A.M. with fresh S.W. wind and very fine weather."

I have not only examined carefully the record of Dr. Lynes, and the log of Capt. Bradbury, but have conversed with both of them personally on the subject, and the record is entirely reliable. And now remembering that the steamer was running to the S.W. at right angles with and across the path of the conditions, from 8 to 10 miles an hour, and that the conditions which they met were moving to the southeastward at as great and probably greater speed, let us analyze the record and see what the number and character of the conditions were which were met and passed through by the steamer.

It will be observed that at lat. 24° S. they had entered and were running with a fresh N.E. breeze, which was obviously the left-hand lateral wind of a condition moving to the S.E., corresponding to our southeasters. They continued to run in that wind through that night and the next day. As they approached that condition at night, the body of it had crossed their track, but they saw its lightning in the S.E. and S. from dark till midnight. The wind veered in the course of the night through the N.W. and W. with some squalls, and the next morning they entered the S.W.

right-hand, lateral wind of the condition, which continued through the day with fine weather. Thus they entered this condition July 26th, at noon, at lat. 24°, and ran out of it on the 28th, at evening, and the whole condition, body and wings, was about 450 miles wide. On the 29th they entered the N.E. lateral wind of another condition, and in the forenoon had thunder and lightning, and showers and hail. The showers continued through the day. Passing through the condition they entered the S.W wind on the other side of it, at 10 P.M. which freshened through the night and was a S.W. gale or Pampero in the morning. The S.W. wind continued through the 30th and into the morning of the 31st. That condition was about 350 miles wide. At 9 A.M. of the 31st they struck the N.W. wind of another condition, with rain. This N.W. wind lasted until 4 A M. of August 1st, when " it went to S.W. by W.," and so to S.W. and blew strong from that quarter through the day, moderating at sundown to a gentle breeze from the S.W. That condition was about 300 miles wide. On the morning of the second of August, they entered the N.W. wind of another condition, which they ran through in the course of the day, and then encountered the S.W. wind of the condition, which lasted them through the rest of that day and the next, and proved to be a more severe pampero than any they had before had. That too, body and wings, was about 300 miles wide. At 6 A.M. of the 4th, they entered the N.W. wind of still another condition, which veered to the N.E. and continued through the day. In the evening they entered the body of the condition and had the rain and northerly wind through the night. During the next day the wind hauled through the west to the S W. and they anchored in Possession Bay at midnight with a fresh S W. wind and fine weather. That condition was at least 400 miles wide.

This is a most instructive record. It is made up of the usual entries of an intelligent shipmaster and the diary of an intelligent physician, a passenger, posted in regard to such matters, and watching for the changes, and is perfectly reliable.

The steamer was a powerful one, and held her way, night and day, at right angles to the path of the conditions in that locality, and through their lateral wings of wind, and bodies; and taking the speed of the vessel and time of passage through each condition, we could map them out severally and distinctly without the least difficulty. In a run of less than 2100 miles she passed through five of these conditions distinctly and clearly marked, and resembling in every feature, except those incident to the hemisphere, the belts of showers and southeasters known to us. Except in the direction of their progress, and the change of sides for their storm and clearing off winds, they are identical.

There were five of them met with and passed through in the run of less than 2100 miles. Doubtless they are a fair representation of the character of the conditions in that hemisphere. Nothing like our northeasters was met with, nor are they often found elsewhere than on the eastern part of our country and off our coast. They are rare in the Mississippi Valley, and unknown on the western coast. And so, as I have said, they are almost unknown at the Horn, and there they rarely have any easterly winds.

There is other and abundant evidence, derived from the records of travellers and navigators, that the *prevalent condition* which disturbs the normal, fair weather state of the south polar zone, is of the same character. The Sailing Directions of Maury are full of it. Dove collected a large amount of it, to prove his law of Gyration. (See his Law of Storms, p. 107 and onward.) The intelligent man who examines the subject will find it impossible to doubt the fact, if his judgment is not demoralized by false theory. And I leave this class of conditions with the assertion that they are undoubted by the prevailing ones of the Atmospheric System, in both the Polar Zones.

We come now to another atmospheric condition distinctly marked on the east side of the Alleghanies, but not as prevalent or distinctly marked on the west side, called the Northeaster. The distinguishing feature of this atmospheric condition is a thin stratum of N.E. wind which sometimes blows a day or two towards the storm which is appoaching from the southwest before the storm reaches us. From the best information I can obtain, I am satisfied that in such cases, the storm is very wide and presents a wide

front to the eastward. The wind sometimes changes from the
South to the N. East, under a distinct belt of rains, where that
has very considerable width. I am inclined to think that the wind
is always Northeast, where the belts have a width of 500 miles
or more on the east side of the Alleghanies. This wind is very
frequent in the spring of the year, when the focus of the storm
is to the south of us, and its northern edge extended up over us,
and the condensation is not sufficiently dense to precipitate. In
such cases, though we may have N.E. wind for a day or two, while
the storm is passing by us to the south, we may get no rain in New
England, but have what is termed a "dry Northeaster." I think
it is generally true that the hurricanes of the West Indies, which,
when they first commence have the winds N.E., and have their heaviest blow from the S.W., become Northeasters in high latitudes,
where they widen out and lessen in intensity, as shown by Figures
18 and 20.

FIG. 18.

Most of our northeast storms, notwithstanding they have currents from the N.E. and S.W. blowing under them in the direction of the axis, have the lateral winds from the S.E. and N.W. developed on their sides. North of the latitude of 40°, during the winter and spring months, the principal path and focus of the conditions, as we shall hereafter see, is at the southward. When that is so, the N.E. wind backs into the N.W. as the condition passes by, all the surface winds blowing toward the focus or precipitating portion of the storm. When, too, the focus of the storm passes south of the observer, without rain in his locality, the wind soon backs from the N.E. to the N.N.E. or N.,—a sure indication that the focus is to the southward, and this is very common north of 40° in the eastern part of the United States in the early spring months.

When the principal path of the conditions is carried up by the laws of the general system, as will be hereafter explained, the wind hauls around from the N.E. by the South to the S.W., clearing off warm, as it is termed, and then summer breaks upon us. I have frequently seen the lateral southerly wind, with heavy scud, blowing in under the body of a northeaster, while the thin stratum of N.E. wind was running, next the earth, rapidly to the S.W. Such double currents produce our continuous heavy rains. I have seen, too, the northwester, in the rear of a N.E. storm, attempting to wedge itself in between the N.E. current, near the earth, and the S.E. current, in the upper part of the surface story which had blown clear across the storm. It is unnecessary to pursue further the description of the Northeaster. Enough has been said to point out its peculiarities, and guide you to an observation of it.

There are certain irregular conditions which occasionally occur. Thus the whole eastern part of the continent is sometimes covered for days by cloudiness, with here and there irregular patches of snow or rain in winter and spring, or showers and perhaps tornadoes in summer. These long spells of extensive condensation and dampness or irregular rain, and irregular winds, are occasional and exceptional, and of course defy distinct description or classification. But it will be observed that they generally begin with southerly or easterly wind and clear off from the northwest.

The next condition to which we shall advert is the Hurricane. This term is usually applied to violent gales at sea and particularly those which originate in the tropics. They occur most frequently in the Bay of Bengal and China Sea, and upon the Atlantic Ocean, east of the Windward Islands of the West Indies. They are most common in the warm season of the year. Those which occurred in the West Indies between the years 1493 and 1855, of which there is record, comprising a list of about four hundred, are divided between the months as follows:

January,	5	July,	2
February,	7	August,	9(
March,	11	September,	8(
April,	6	October,	6!
May,	5	November,	1;
June,	10	December,	

FIG. 19.

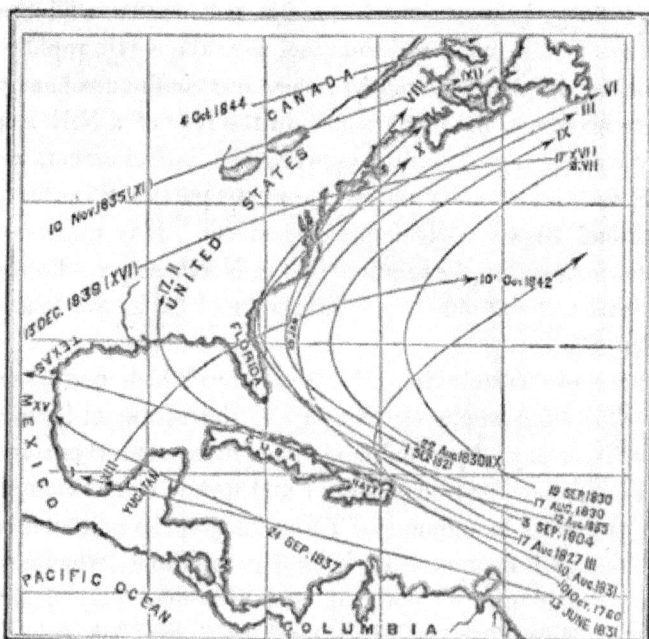

Many of the Hurricanes of the West Indies were investigated and charted by the late Wm. C. Redfield, and published in the American Journal of Science. He considered them revolving gales, and such is now the received opinion. The evidence has never satisfied my mind, and I shall examine the theory hereafter. But whatever my opinion in relation to the theory may be, I can and do most cheerfully concur in the high appreciation generally entertained of the abilily and utility of his investigations. The preceding is a copy of a chart prepared by Mr. Redfield, indicating the tracks pursued by the various storms which he investigated.

The following is also a copy of a map of a hurricane, investigated by Lieutenant Porter of the Navy and National Observatory, from the Logs of vessels, received at that Institution, and published in the " Sailing Directions " of Lieut. Maury. The chart

FIG. 20.

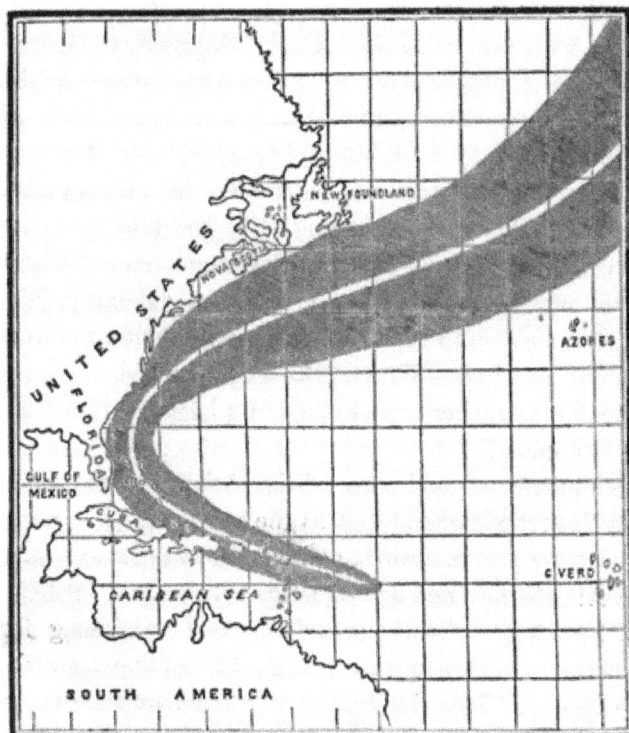

of Mr. Redfield will give you a general idea of the place where
these hurricanes originate and the course which they pursue.
The map of Lieut. Porter will exhibit to you the course and
width of that particular condition, by its shading. The white line
in the centre is intended to indicate the line upon which the barometer stood the lowest.

In that storm the lateral winds were very distinct, and as it
widened out in the North Atlantic, the N.E. wind set in and blew
beneath the lateral winds, precisely as it does under the storms
which pass over the continent when they are widened out to the
same extent. There are many instances on record where others
have observed four currents in a storm, viz.: The N.E., next
the earth, the N.W. over that, the southerly wind over both,—
each indicated by its scud—and the trade current with its stratus
moving steadily to the N.E. over all. The vortex has never been
seen *and never will be.*

These hurricane conditions of the Atlantic, thus originating,
organized and progressive, were frequently traced by Mr. Redfield, as by Lieut. Porter, maintaining their organization and violent action for many days, until they passed far into the North
Atlantic beyond the track of commerce, and all opportunity for
observation. Where their organization was broken up and their
action ceased, we cannot know. The presumption is by no means
a violent one—(curving, as some of them evidently did, to the
westward again on the North Atlantic, in the face of the theory
of rotation—and regardless of converging meridians)—that they
entered the Arctic circle and attained a higher latitude than man
has ever attained.

These hurricane conditions of the Atlantic are of practical as
well as theoretic interest to us, as they generally hug our coast, in
their movement to the north, and extend more or less inland east
of the Alleghanies, and are sometimes destructive in their effects.
But a knowledge of their organizing and continuing force has
an important bearing in the investigation of the general atmospheric system. That interest is increased by the events of the
last two years. During the summer of 1867, a wide belt of coun-

try in the interior of the United States from the Gulf to Canada was afflicted with a destructive drought, while the whole coast-region east of the Alleghanies was suffering with dampness and drench. The extraordinary humidity and precipitation were produced by a succession of these hurricane conditions passing up along the coast and extending inland. They were more numerous and precipitated more than usual, but in other respects their action was less violent. There was obviously a diversion to the eastward of the conditions and trade current, which usually enter over the Gulf of Mexico and upon the Southwestern States, and pass up over the various parts of the country, in the summer months, which diversion gave the coast-states an unusual quantity of rain and left the interior states deficient. I shall allude to this matter again, after we have examined the general organization of the atmospheric system and come to the consideration of Droughts.

I have said that these Atlantic Hurricanes were generally considered revolving Cyclones. Much evidence has been accumulated to prove this, and also with reference to the hurricanes of the China Sea and Bay of Bengal. Whatever may be said or true in relation to the wind blowing in them round a circle, it is conceded and undoubtedly true that their strongest winds are lateral to their line of progress. Their rotary character will be considered when we come to the laws of the conditions.

There is another class of Hurricanes which have no discoverable rotation, and which are the most violent of all. These vary in width from a mile to forty or fifty miles, and under the whole width the lateral winds are masked by a most violent and destructive wind which follows the body of cloud and its line of progress from the westward to the eastward. By the kindness of Col. Lachlan, of Cincinnati, a British officer now resident in that city, who has devoted much attention to Meteorology, and who investigated it with much industry, I am enabled to describe fully that particular condition. It was one of the most extraordinary ones of the class which is known to have passed over the country.

It occurred on the 21st May, 1860. Commencing west of Louisville, Kentucky, it followed the valley of the Ohio to Portsmouth, Ohio, a distance by an air line of 160 miles. East of Portsmouth it does not appear to have been violent. It was from 40 to 50 miles in width, and devastated the country over which it passed. It passed over and injured the residence of Col. Lachlan, and he sent me a full account of it as witnessed by him, and a large collection of other accounts, collected from newspapers and other sources. I regret that my limits will not permit the insertion of much of the material furnished me. I make, however, a few extracts. The following is from one of the Cincinnati Journals published on the succeeding morning :

" About three o'clock yesterday afternoon, a tremendous tornado burst upon the city. We had observed a few minutes before, that the storm rising in the southwest promised to be one of unusual severity. Violent gusts of wind whirled the dust in fierce eddies about the streets, and made the sand and gravel whiz and spin over the boulders and sidewalks. The air suddenly grew dark, and presently a craggy mass of black and ashen clouds were seen, laced and bristling with streams of lightning, rushing up the sky with amazing velocity. So vivid and incessant was the play of electricity, that the storm-cloud seemed like a magazine of rockets exploding and launching volleys of fiery darts in every direction. While the lightnings were flaming above, the tempest was roaring below, and as it advanced, the city was lost to view in the white and hissing spray of the rain.

"The force of the wind was for this latitude almost unexampled.

"In a few minutes immense damage was done. Steeples were prostrated, dwellings overthrown, crushing the occupants in the ruins ; roofs whirled on high, torn into shreds and scattered far and wide ; shade trees uprooted, twisted and broken by hundreds ; signs torn from their fastenings, and shivered into splinters ; chimneys beaten down ; windows burst in ; carriages overturned ; persons hurled from their feet and bruised against the walls. We have indeed a long and distressing catalogue of disaster to present our readers.

"There was universal consternation throughout the city. People lost faith in the roofs over and the walls around them, and ran into the streets, notwithstanding the rain, for safety.

The sun set in a sky of crimson and orange. The western sky was illuminated as if by a vast conflagration. A heavy rain set in early in the evening and continued until after midnight, drenching the habitations made roofless by the tornado, which is conceded to have been the most disastrous known to the present generation."

The following account is from the "Louisville (Ky.) Courier," of the next morning:

"A little after 2 o'clock yesterday the city was swept by one of the most fearful and destructive tornadoes ever known. During the morning it was pleasant, but towards noon it began to cloud up, and about 1 o'clock the heat was oppressive in the extreme. The threatening clouds were gathering in heavy masses in all directions, with not a breath of air stirring in the streets, but that a storm was brewing was apparent to the most casual observer of the elements. The scene soon fearfully changed, with a most singular change in the appearance of the clouds, which varied in color from black to green, and then to straw color, followed by a terrible tornado, which luckily lasted but a few minutes, or the destruction would have been immense. The city in a twinkling was completely enveloped in dust, houses unroofed and demolished, trees uprooted, and the utmost confusion prevailed in the almost cimmerian darkness that encompassed the city. The wind shifted from southwest to the northwest, and amid the clashing of the elements, amid the lightning and loud peals of thunder, the rain poured down in torrents—a miniature deluge, intermixed with a heavy storm of hail."

The following also is from the "Portsmouth Tribune."

"Last evening, about four o'clock, a terrible tornado swept over this city and vicinity, surpassing in violence all storms ever experienced in Ohio, and most disastrous and fatal in its results.

"A dark cloud in the *southwest*, accompanied by muttering thunder, indicated the approach of a storm; but nothing more than an ordinary thundergust was anticipated, until the gale in all its resistless fury burst upon us. Its approach was so sudden that persons were hardly able to close doors and windows, before rain and wind had swept in, carrying all before them. In an instant, buildings in all directions were unroofed—chimneys, gables, fire-walls, steeples, and spires began to fall. The air was filled and literally darkened with timbers, rafters, shingles, bricks and mor-tar, tin roofing, and all manner of wreck, whirling and eddying above and around like feathers in a whirlwind. Crash upon crash —surpassed only by the terrific thunder—followed in succession, as steeples, furnace stacks, and buildings fell, involving men, women, children, furniture, horses and cattle, in one common ruin.

"In its approach to the city it came with overwhelming violence *from a southwestern direction*, sweeping all that remained of the old New York Warehouse from the foundation."

Colonel, then Major Lachlan, also wrote me a graphic account of the effects of the hurricane, remarking that he was necessarily engaged in attending to the preservation of his premises, so sud-denly threatened with destruction, and could not notice partic-ularly all the phenomena as they occurred. He sent me, however, the following letter from Prof. Christy, who happened to be pres-ent and who did notice them. It is an interesting letter, and I insert it entire:

MY DEAR MAJOR—You ask me to state my recollections of the storm of the 21st May, which occurred about half past 2 P.M., while I was visiting you at your residence. You will remember that we had taken our seats in your room, and that I sat near the window opening to the south. My attention was directed to the appearance of the clouds on the Kentucky side of the river. I remarked that I did not like their appearance. They were flying very swiftly and had the peculiar color which characterizes the advance clouds of a hurricane.

A moment later, on looking down upon the city, I found the view obscured by a dense cloud of dust, and could distinctly hear the roar of the storm, as it rushed onward in its devastating course. In the next moment it was upon us, with all its force, and we had to spring to doors and windows to prevent them from being dashed inwards by the force of the wind. Taking my stand by the front door which occupied the point of greatest exposure, I expended my strength in keeping it secure. From that point I could witness the movements of the wind, as indicated by the motion of the objects which were driven along by its power. It seemed to confine itself to *a direct line* in a S.W. and N.E. direction and *not* to move in a curve like an ordinary whirlwind.

The hurricane was *not a continuous* rush of wind, of equal force during its duration, *but swept along with much greater power at some moments than at others ;* nor was its greatest force exerted at the commencement of the storm. The cut hay upon the lawn in front of the dwelling, was partly removed during the first blasts; but it was not until near the middle of the period during which the storm lasted, that the hay was all licked up so as not to leave a stalk of it upon the stubble. It was at this period, too, that portions of the porch were blown down, and the post and rail fence, lifted from the ground, was partly precipitated against the steps leading up to the front door. The roar of the storm was so deafening that the *falling of the chimneys* upon the roof was *unnoticed*, and I must confess that I expected every moment, that the house and all its contents would be overturned and swept away during the elemental commotions which so fiercely beset us.

In reference to the extension of the storm eastward, I can only remark, that a few days since I passed over the Cincinnati, Chillicothe and Marietta Railroad, and had the opportunity of witnessing its effects upon the forests. I learned from the conductor of the freight train, that its force was maintained *from Cincinnati* onwards *to a point some fifteen miles east of Athens*, after which it seemed to lose its power, and at *Marietta amounted only to an ordinary gale.*

The conductor pointed out the spot to me where the storm met him, a few miles beyond Hamden. Hundreds of trees were prostrated, but as the wind was blowing in a line parallel to the road —say from west to east—none of them fell across the track, and thus his train escaped destruction.

Respectfully yours,

DAVID CHRISTY.

I copy next an account of the hurricane as it struck the Covington and Lexington Railroad, 16 miles south of Cincinnati, in Kentucky, where a band of it demolished a railroad train.

"We are indebted to Mr. A. F. Woodall for the following news of the tornado on the line of the C. & L. Railroad:

At 2.25 P.M., the train entered a forest *sixteen miles south of Covington*, and the hurricane struck. Trees fell like reeds. It seemed safer to proceed than to stop, and either seemed destructive. When the train was within fifty yards of a clearing, *twelve trees fell within fifty feet of the cars.* One of them, a soft poplar, *three feet eight inches in diameter*, fell just in front of the engine, *which leaped it, leaving her trucks, cylinders and steam chests behind.* The tree is under the baggage car, and the car on its side. The train stopped within a distance of forty feet, throwing passengers in a heap at the front end. When the train stopped limbs of trees were falling like flails, and two trees came crashing like a thunderbolt, and filling the air with leaves and boughs. The tops of several trees fell within a few feet of the ladies' car. A scene of more utter terror could not be imagined. Death seemed inevitable. *The track of the tornado was about one hundred yards wide*, and the train was in the very center of it. Tom Flood, the fireman, had both hands badly scalded, and was otherwise injured. Isaac Barnet, engineer, stood at his post like a man, and received only a small cut over one eye. He was covered with the branches of trees, and leaves. No passenger car was thrown from the track. The screams of children, the shrieks of terrified women, the pallor of stout men, was a scene to move the soul. It was awful. All the employees of the train were more or less hurt, but the passengers escaped without injury. The fireman was buried under the wood in the tender, and would have been scalded to death but for Frank Lockwood, who worked like a hero to save the poor fellow. *The storm lasted perhaps ten minutes.*"

It may perhaps be well to insert here a copy of the Meteorological Record, kept at College Hill, Cincinnati, for the 20th, 21st, and 22d days of May, 1860.

ABSTRACT OF METEOROLOGICAL OBSERVATIONS AT COLLEGE HILL, DURING
THE 20TH, 21ST, AND 22D OF MAY, 1860.

DATE.	BAROMETER.			THERMOMETER.			CLOUDS.			WIND & DIRECTION.		
	7 AM	2 PM	9 PM	7 AM	2 PM	9 PM	7 A.M.	2 P.M.	9 P.M.	7 A.M.	2 P.M.	9 P.M.
May 20	28.70	28.70	28.60	62	76	70			9 w. 2	w. 2	s. 3	w. 2
" 21	28.57	28.70	28.65	70	70	62	8 s.w. 4	10 w. 8	10 w. 2	s.w. 4	w. 8	w. 3
" 22	28.75	28.70	28.70	64	70	62	10 n.w. 4			n.w. 2	n.w. 2	n.w. 2

GENERAL REMARKS.

"21st. A thunder storm from the west, accompanied with driving rain, lightning and thunder, very violent. Violent gale or hurricane; blew down vigorous trees of large size by scores. Its greatest violence at 3 P.M. By reports from the south and west, it appears to be about 50 miles wide,—direction S.W., W., and N.W., making its bend near the Ohio river, at Cincinnati. Nothing like it here for over 40 years."

Also the following from the Register at Woodward's High School:

MEMORANDUM OF OBSERVATIONS AT WOODWARD HIGH SCHOOL.

DATE	BAROMETER			THERMOMETER.			CLOUDS.	WIND.	RAIN.
	7 AM	2 PM	9 PM	7 AM	2 PM	9 PM			
May 20	28.70	28.70	28.60	62	76	70			
" 21	28 57	28 70	28.70	70	70	62			
" 22	28.75	28 84	28.64	64	70	62			

From these data and a mass of other data collected by Col. Lachlan, the following facts appear:

First, The hurricane was about 40 miles wide. *Second*, It was exceedingly violent and destructive. The injury done between Louisville and Portsmouth, as estimated by a Cincinnati paper, after the accounts were all in from the country, was estimated at $1,300,000. *Third*, The duration of the storm at any place did not exceed 20 minutes, nor was it less than 10 minutes. *Fourth*, The direction of the wind at every place was between

N.W. and S.W., and at no one of the 40 to 50 places was the wind N. of one or S. of the other. At Cincinnati one of the papers described it as from the S.W. and another as from the N. W. The Louisville Courier, as will be seen by the extract, described it as shifting from S.W. to N.W. Nearly all the accounts described it as from the west. One of the Cincinnati papers speaks of it as having blown in bands or separate divisions, and the same thing is mentioned by others, and is clearly deducible from the reports. Undoubtedly the cloud was divided into bands parallel with its line of motion, as I have described them as existing in all storms, and the lines of violent wind corresponded with the bands of intensity and condensation in the clouds. *Fifth*, The barometer rose just before the cloud appeared, and did not fall materially during its passage. *Sixth*, At no spot on the territory which it devastated was there the slightest trace, mark or evidence of rotation or gyration. *Seventh*, It occurred in a passing condition, of considerable extent, which had the usual lateral winds and a subsequent continuous rain.

From these facts it is evident that this was substantially a straight line hurricane, and where the wind varied from its line of progress it was either S.W. or N.W., consisting of a mass of cumulo-stratus clouds settled low down near the earth, moving rapidly over it, whirling and condensing on its front and drawing the air between it and the earth, after it with irresistible force.

I have seen many such hurricanes, but upon a smaller scale, and less violent, and such gusts are very common under parts of intense belts of showers.

It will be observed from the Louisville description that the sky became overcast some hours before the hurricane clouds appeared. Probably there was a dense layer of cirrus in the upper story, with patches of cirro-stratus in the upper part of the trade story, and the bands of cumulo-stratus constituting the hurricane cloud formed in the lower part of the trade story, and gradually settled down into the surface story, drawing after them the intervening surface atmosphere as they passed along, and creating the roaring sound described by the observers. Such has been the manner in which such storms and gusts have formed and moved when I have had opportunity to observe them.

The thermometer, as it appears from the Meteorological reports, stood at about 70°, and as it was in May and a cloudy day, the earth could not have been heated in that locality as high as that.

And now professional men, practical men, and young men, whose minds are uncommitted to theory, answer me, in candor and honesty, is it possible that that organized condition could have been created and impelled over the earth at the rate of 70 or 80 miles an hour, with all that awful and tremendous exhibition of power, by the mere mechanical effect of the air heated by contact with the surface of the earth to 70°? Is not the persistent teaching of such a theory an insult to your understandings?

The remaining form of condition is the tornado, so it is termed when it occurs upon land, but it is called spout or water-spout when it occurs upon the water. The following are its principal peculiarities.

1st. It occurs during a *peculiarly sultry and electric* state of the trade and surface atmosphere, and at a time when thunder showers are prevailing in and around the locality, as an incident of the showery condition, and at any period of the year when such a state of the atmosphere exists. One recently occurred in Brandon, Ohio, in midwinter.

2d. There is always a cloud above, but very near the earth, between which and the earth the tornado forms and rages. It is usually described as a black cloud, ranging about 1000 feet or less above the earth, often with a whitish shaped cone projecting from it, and forming a connection with the earth; at intervals rising and breaking the connection, and again descending and renewing it with devastating energy. Its width at the surface varies from forty to one hundred and eighty rods—the most usual width being from sixty to ninety rods. Sometimes when still wider, they have more the character of thunder-gusts, and are brightly luminous.

3d. Two motions are usually visible, one ascending one near the earth and in the middle, and a gyratory one around the other.

The latter is rarely felt, or its effects observed, near the earth. Occasionally, and at intervals, objects are thrown obliquely backward by it.

4th. It is composed, at the surface of the earth, of *two lateral currents*, a northerly and southerly one, varying in direction, but at right angles in most cases, although not always, with its course of progression, extending from the extreme limits of its track to the axis; which currents are most distinctly defined toward the center, and upward. These currents prostrate trees, or elevate and remove every thing in their way which is detached and movable. There does not seem to be any current in advance of these lateral ones tending toward the tornado, save in rare and excepted cases, and then owing to the make of the ground or the irregular action of the currents; nor any following, except that made by the curving of lateral currents toward the center of the spout as it moves on, and perhaps a tendency of the air to follow and supply the place of that which has been carried upward and forward, like that of water following the stern of a vessel. The south current is always the strongest, and often crosses the axis and curves backward as it rises from the surface, and ascends a little in advance of the other, and covers the greatest area. The proportion of the two currents to each other is much the same that the S.E. trades bear to the N.E. This excess in volume and strength of the southerly current will explain the irregularities in most cases, and the fact that objects are so often *taken up and carried from the south to the north side*, and so rarely from the north and carried south of the axis. These irregularities are such as attend all violent forces, and something can be found which will favor almost any theory; but the two lateral currents appear always to be the principal actors, except when it widens out and assumes more the character of a straightforward gust. See a collection by Professor Loomis, American Journal of Science, vol. xliii. p. 278.

The following diagram is a section of the New Haven tornado, from Professor Olmstead's map accompanying his article in the " American Journal of Science and Arts," vol. xxxvii, p. 340.

The manner in which the main currents flow is shown by their early and unresisted effect in a cornfield, as represented by the dotted lines. The direction in which the fragments of buildings

FIG. 21.

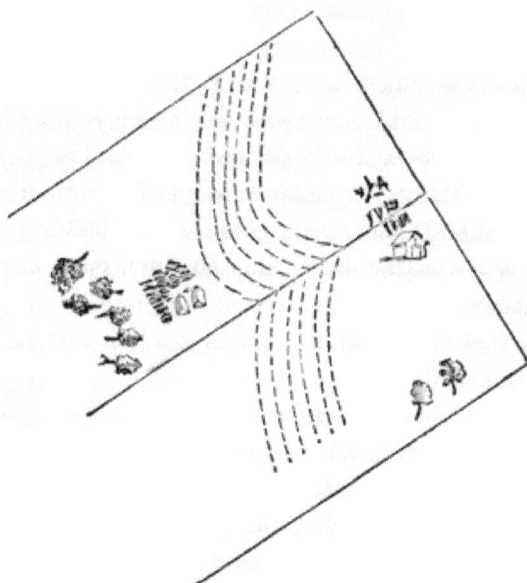

were carried by the greater power of the southerly currents, and by their crossing the axis in advance and curving backward is clearly shown and should be particularly observed and remembered. And so is this irregular action, where a part of the southerly current broke through the northerly one, and prostrated two or three trees backward on the north side of the axis.

5th. This cloud, and its spout, move generally with the course of the counter-trade in the locality—*i. e.*, from some point between S.W. and W., to the eastward, but occasionally a little south of east.

6th. Several exceedingly instructive particulars have been observed and recorded.

a. No wind is felt outside of the track, as those assert who have stood very near it, and its effects show.

b. The track is often as distinctly marked, where it passed through a wood, as if the grubbers had been there with their axes to open a path for a railroad. The branches of the trees, projecting within its limits, are found twisted and broken off, or stripped of their leaves, while not a leaf is disturbed at the distance of a foot or two on the opposite side of the tree, and outside of the track.

c. As the spout passes over water, the latter seems to *boil up* and *rise to meet it,* and *flow up* its trunk in a *continued stream.*

d. As it passes over the land, and over buildings, fences, and other moveable things, they appear to *shoot up,* instantaneously, as it were, into the air, and into fragments. If buildings are not destroyed or removed, the doors may be burst open *on the leeward side,* and gable ends *snatched out,* and roofs taken off on the *same side,* while that portion of the building which is to the windward remains unaffected.

e. Articles of clothing, and other light articles, have been carried out of buildings through open doors, or chimneys, or holes made in the roofs, and to a great distance, without *any opening* being made for the air to *blow in.*

f. If there be a discharge of electricity up the spout from the earth, like that of lightning, the intense action ceases for a time or entirely.

g. Vegetation in the track is often scorched and killed, and so of the leaves on one side of a tree, which is within the track, while those on the other side, and without the track remain unaffected. (Espy's Philosophy of Storms, 359, cited from Peltier.)

h. The active agent, whatever it is, has been known to *seize hold of a chain attached to a plow* and *draw the plow about, turning the stiff sod for a considerable distance.* (See Loomis on the tornado at Stow, Ohio, American Journal of Science, vol. xxxiii, p. 368).

The following is a copy of the diagram prepared by Professor Loomis, of the path of that tornado:

FIG. 22.

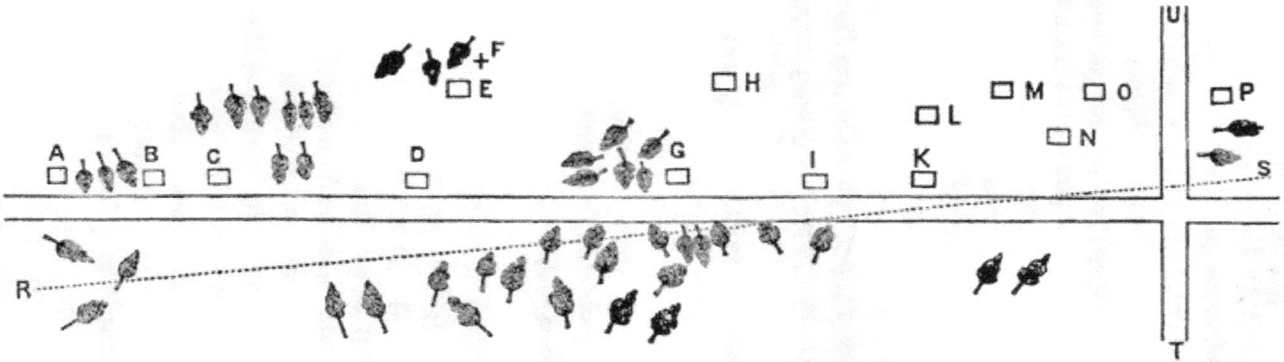

A, represents a house unroofed; B, a barn partly unroofed; C, a house uninjured; D, the house of Mr. Sanford, completely destroyed; E, the barn, somewhat injured; F, the spot where the cart was dropped; G, a house not much injured; H, a barn unroofed; I and K, houses unroofed; L and N, barns unroofed; M, a two story log house with its upper story taken off; O and P, houses uninjured. The horizontal road runs east and west; the perpendicular road T U, runs north to Hudson. R, S, supposed central path of tornado, moving from R to S.—This line is inserted by the Author.

i. In passing over ponds, the spout has taken up all the water and fish, and scattered them in every direction, and to a great distance.

j. The barometer falls very little during the passage of the spout. (See the Natchez hurricane of 1827, Espy, page 337.) Not more than it *frequently* does during gentle showers.

k. Persons have been taken up, carried some distance, and if not projected against some object in the way, or some object against them, have usually been *set down gently and uninjured.*

l. Buildings which stood upon posts, with a free passage for the air under them, although in the path of the tornado, escaped undisturbed. (Olmstead's account of the New Haven tornado, American Journal of Science, vol. xxxvii, p. 340.)

m. A chisel taken from a chest of tools, and stuck fast in the wall of the house. (Ibid.)

n. Fowls have had all their feathers stripped from them in an instant and run about naked but uninjured.

o. Articles of furniture, etc., have been found torn in pieces by antagonistic force.

p. Frames taken from looking-glasses without breaking the glass. Nails drawn from the roofs of houses without disturbing the tiles.

q. Hinges taken from doors—*mud taken from the bed of a stream* (the water being first removed,) and let down on a house covering it completely—a farmer taken up from his wagon and carried thirty rods, his horses carried an equal distance in another direction, *the harness stripped from them,* and the wagon carried off also, *one wheel not found at all.* (American Journal of Science, vol. xxxvii, p. 93.)

Pieces of timber, boards and clapboard, driven into the side of a hill, *as no force of powder could drive them, etc., etc.*

Such are some, but by no means all, the peculiar effects resulting from the action of a tornado. Many others have been surveyed and mapped, but I have not space for them. Conforming to my avowed purpose of avoiding all theoretic discussion until the facts are fully presented to you, I waive inquiry into the charac er of the force which creates this condition. A few words only as to the localities where they are most frequent. Waterspouts are most common in the West Indies and other localities where hurricanes are most frequent, and doubtless for the same reason. Tornadoes are more prevalent in the United States than in Europe. They occur very rarely in the western part of this continent, and indeed thunder is not frequently heard in that section. Tornadoes in the United States are most frequent and most severe between the 35th and 45th parallels of latitude, and between the Alleghany Mountains and the western borders of Iowa and Missouri.

They occur too in much greater numbers and with much greater severity in some years than others. The year 1860 was such a year, and I conclude this notice of them by copying the following article from a western paper, written in the summer of that year:

The Year of Hurricanes.

"The season through which we are passing, will go far to remove the impression that the severest hurricanes are confined to the tropics. All over the Northern and Middle States, tornadoes of unprecedented violence, extent, and destructiveness have occurred. At least three hundred lives have been lost by the force of the elements. There seems to have been a cycle of tornadoes.

Within a fortnight there have been four in the Middle and the Western States, one extending from Louisville up to Central Ohio, one in Cattaraugus county in Western New York, one in Armstrong and the adjoining counties in Western Pennsylvania, and now one in Iowa and Northern Illinois, which, it is said, has surpassed all the others in violence and destructiveness.

Whole villages have been demolished, and their inhabitants either killed outright, or fearfully mangled by the falling of houses, the mad sweep of the winds, or the flying of timbers and trees. The imagination can hardly paint the terror which has overtaken, in stern reality, the inhabitants of the devastated districts. It must have been witnessed to be realized.

Never before has our country been visited by so many, or so widely destructive hurricanes as during the present year. They have rushed through the country with resistless fury, uprooting trees, demolishing fences, houses, and churches, and even taking loaded freight cars from their tracks and dashing them to pieces. The space traversed by the last great tornado, in Illinois and Iowa, was fully one hundred and fifty miles in length. The force of the wind has been almost incredible. Tin from roofs was carried over twenty miles ; large brick houses were lifted completely from their foundations and twisted around in the air, before being dashed to fragments upon the ground. Churches were demolished with a single thunder gust, and the whirling and snapping of whole forests of trees, as they were caught up in the folds of the tremendous whirlwinds, was awfully grand to hear and to see.

In all cases the electric forces appear to have been combined with the atmospheric agitation. The fearful velocity and power of the wind—estimated to have traversed space at a speed of from sixty to seventy miles an hour—was alone sufficiently terrific. But when to this was added the roar of the most startling and tremendous claps of thunder, and the seething glances—flash after flash—of the lightning, it is no wonder that a scene was presented which struck men dumb with fear and alarm, and even made the stoutest hearts to quake. The awful grandeur of the storm, as the whole "artillery of heaven" seemed let loose upon the stricken globe, was indescribable. Add to this the groans of wounded men, and the shrieks and prayers of terror-stricken women and children, ending in a catalogue of dead, lying cold and stiff in every stage of mangled deformity, and we have a picture of human desolation which transcends in horror all the convulsions of nature."

Having thus examined the distinctive characters and peculiarities of the different conditions, we are now prepared to investigate understandingly the systems of conditions which exist upon this continent, and characterize its climatology and its weather, but we will enter upon that branch of our inquiry in a distinct chapter.

CHAPTER IV.

OF THE SYSTEMS OF CONDITIONS ON THE CONTINENT OF NORTH AMERICA AND IN THE UNITED STATES.

Three distinct and diverse systems of conditions in the United States, passing over the continent in three distinct and different paths—The first, the Atlantic system—its conditions originate upon the Atlantic or its connected seas and gulfs, or in the counter trade they furnish—move N.W. and N. and enter upon the southwestern and southern states—curve thence and move to the N.E.—The second originate upon the Pacific, move to the N.E. on to and over the coast and continent—The third, a part of the tropical belt of rain which covers the Gulf coast, Southern Mexico and Central America in midsummer—Direction in which the different systems move—The supply of rain depends upon these conditions—Intermediate between these systems, an arid area—not all or entirely a desert—all parts of it reached by one or the other of the systems, but temporarily—that area scantily supplied with rain—These systems evidence of law, order and organization—All peculiarities in our climatology dependent upon them—Developed thirteen years ago in the "Philosophy of the Weather" —Not since denied, but ignored by scientists—accepted and adopted by practical men—Focal paths of the conditions and what is meant by them— Their situation in winter in the Atlantic system—Rain-fall under and on either side of the focal path—Situation of the Atlantic focal path in February, 1854—Temperature under and on either side of it—Rain-fall for that month, under and on either side of it—Winds under and on either side of it—Weather south of the focal path, warm—The same thing true in respect to the single conditions—Situation of the focal path in March, 1854—Extension N.W. and N.—Situation of the focal path in April—Had extended still further W. and N.W. and N.—Situation in May and June— Rapidly extending W. and N.W. and N.—conforming to its annual progress but more rapid and concentrated than usual, leaving an unusual drought behind it—Situation in July and August—unusual extension and concentration W. and N.W. and N.—Area of the consequent remarkable drought— Descent of the focal path again in autumn and winter—Extension less regular on the Atlantic coast than in the interior —Annual rain-fall upon the eastern states, with chart—Effect of this annual ascent and descent of the focal path on our climatology—causes rainy and dry seasons at the different points—Affects the temperature—Produces sudden changes—Explains the reason why consumptives go to Minnesota—Why storms appear at the southwest in winter and showers at the northwest in summer—Explains the general winds—Prevalent winds in Florida—Prevalent winds in eastern Arkansas—in the Indian Territory—in Tennessee—in the N. west and northern states—in New England—Explains why the winds prevail from the intermediate instead of cardinal points of the compass—Chart of the isothermal lines for the winter—their descent in the Mississippi valley in winter, below the 40th parallel—Their ascent at the same time above the 40th parallel—Explanation of anomalies—Owing to the situation of the focal paths of the two systems at that time—Climatology of the Pacific states—more simple, but has the same elements—Its conditions move from the Pacific in a northeasterly direction—Have a focal path like those of the Atlantic system—that path has a similar transit to the north in summer—that transit produces a like effect upon the temperature—produces like rainy and dry seasons—has similar attendant and prevalent winds—The Pacific conditions are less intense—the phenomena less violent and more uniform—Description of the third system of conditions—Reaches the Gulf coast and Central America in summer only—movement of the conditions in that system, and their character—Summer rain-fall under that system.

   There are three distinct and diverse systems of atmospheric conditions passing over this continent in distinct and different paths.

   The first is the ATLANTIC SYSTEM, which consists of *conditions* that originate upon the Atlantic Ocean within the tropics, the Caribbean Sea, the West Indies, and the Gulf of Mexico, or form in the Equatorial current which comes from that part of the tropics, and moving N. and N.W. enter upon the Southern and Southwestern states of the Union, curving and moving to the northeast, supplying the Eastern and Central States with rain.

   The conditions of the second system originate upon the Pacific Ocean, and move in upon the Western coast, and supply California and the Northwestern states and territories, the British territories and Alaska, and the country northeastward of them.

   The third system is a part of the tropical, central belt of rain which surrounds the earth, and which moves up in summer far enough north to cover some portion of Florida and the Gulf Coast, the West India Islands, Southern Mexico and Central America. The path of the conditions in this central tropical belt, is from the eastward to the westward, across Southern Mexico and Central America and out into the Pacific.

The path of the Atlantic conditions is northward from the Gulf States, curving to the N.E. and passing off on to the North Atlantic.

The path of the Pacific system of conditions is northeastwardly from the Pacific, across the northwestern part of the continent, into the Arctic Circle.

To these different systems of conditions, and their diverse paths, we owe fundamentally the diverse character of the climates of our country. Thus, the eastern portion of the United States is largely supplied with moisture by the Atlantic conditions. The western coast from San Diego to the Arctic Circle is supplied—California moderately and Oregon, Alaska, and the country east of them abundantly—by the Pacific system. And Southern Mexico and Central America are abundantly supplied during their rainy season by the central belt which moves up over them in summer; intermediate between these three systems, Lower California and Northern Mexico, the valley of the Gila, Western New Mexico, the Staky Plain, the valley of the Colorado and Utah, and the territory east of the Rocky Mountains, and west of the 100th meridian, are in some places nearly, and in all comparatively dry or desert. Both the Atlantic and Pacific systems reach them by an extension of the paths of their conditions at particular seasons of the year, but those extensions are for brief periods, temporary and exceptional—the Atlantic extending up upon a part of them in summer, and the Pacific system reaching down on a part in winter, as we shall see. In this diverse system of conditions and in their paths, we shall find law, order, and organization, and we shall also find an explanation of *all the phenomena and peculiarities in the climatology of our country.* No savan or meteorologist, whatever his position or pretensions, who ignores the existence of the systems, and their transits, has ever explained or can ever explain them. When I was writing the "Philosophy of the weather," the differences in these systems became very apparent to me, but it was not until the "Army Meteorological Register" was published, that I was enabled to trace and develop them. In an appendix to that book, I traced the systems and their transits by tables and diagrams, and showed their effect in producing our varieties of climate and differences of fertility and sterility. Prof. Henry, with the book before him, and the facts *demonstrated by mathematical expression*, while publishing his series of compilations in the Patent Office Report, at the expense of the country, professedly for the people of the country, and professing by implication at least to give them, and induce them to believe he had given them *all the important facts* pertaining to the subject, ignored also that most important discovery.

But the Agricultural Department at Washington, or some writer connected with it, (whom I do not know,) appreciated and adopted the substance of that discovery and appendix, in an elaborate article on American Cotton, Wool, and Tobacco, and Climatology with reference to them, with the following recommendation, in the monthly report of November and December, 1864:

" These maps are taken from Butler's Philosophy of the Weather, one of the most practical books on Meteorology, and one that should be studied by every farmer desiring to learn the character of that atmosphere which rules the productiveness of the earth."

I have spoken of the focal paths of the conditions; by that I mean the paths in which the *greatest number* of conditions, or the most *intense* conditions, or the *focus of precipitation* in the conditions, pass for the time being, over the country. Thus, to speak generally, the path of the Atlantic conditions is upon the southeastern portion of the United States, and there the greater number of those conditions, or the most intense of them, or the focal precipitating parts of those which spread all over the Eastern states, are found. This appears quite obvious upon a bare inspection of the rain-chart for the winter season of the year. The following is an abridgement of one (probably as accurate as any), which accompanied the " Army Meteorological Register." The map is without territorial subdivisions, but the great lakes, great rivers and coast outline are given, and the reader, holding it right side up, cannot mistake the location of the lines.

FIG. 23.

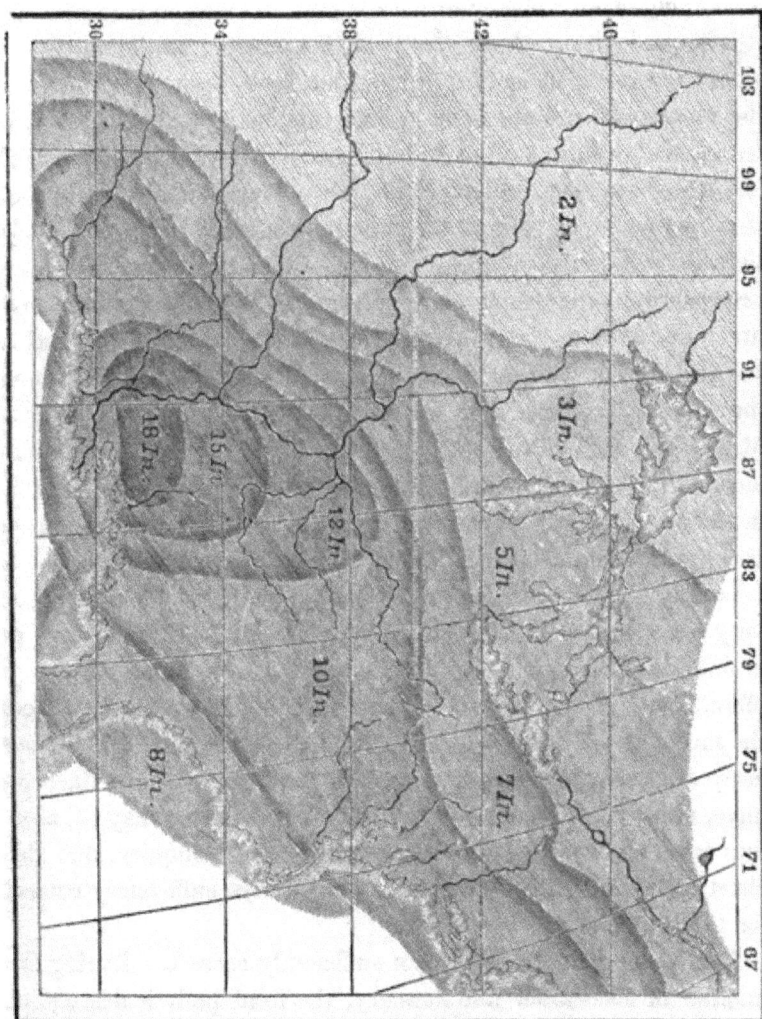

Upon an examination of this chart, it will be seen that west of a line drawn from Lake Superior to the mouth of the Rio Grande, the fall of rain and melted snow is but two inches for the three winter months. Upon a space nearly parallel with that, on the east of it, the fall is but three inches. On still another space farther east, extending from the Gulf of Mexico to the Gulf of

St. Lawrence, the fall is but five inches. Upon still another space of like character, further to the east, and curving to the northeast it is seven inches. And on still another, curving to the northwest where the conditions entering upon the Gulf Coast move up over the southwestern states, and curving again to the northeast, it is ten inches, while upon the Atlantic coast of the Carolinas, the fall is but eight inches. That portion over which the fall is ten inches is the winter focal path of the Atlantic conditions—for obviously the rain was carried there by those conditions, and they must have been more numerous or more intense, or their focal portions of precipitation must have passed over that portion of the states. It will be observed that a still greater fall was experienced where the conditions were met by the mountain chains of Alabama, Mississippi, and Tennessee, where the fall reaches to twelve, fifteen, and eighteen inches. I have not implicit faith as a rule in the accuracy of the various registries of the weather, and especially in reference to the fall of rain and direction of the wind. I have in my investigations quite too often found them differing very much where kept by different persons in the *same place* or at different places *very near* together, and I have found them very often, from my own personal knowledge, differing materially from the facts; but I think the registers kept by the officers of our army are quite as reliable as any we have had, or are likely to have, under our present system. But however that may be, absolute truth is not essential to our purpose in this inquiry, and this chart founded on Army Registers is doubtless sufficiently correct for that purpose.

But in one particular it is not sufficiently correct. During the months of December and January, the focal path is descending rapidly to the southeast and east. It is not until after the first of February, and not always so early as the middle of that month, that the conditions begin to extend their paths to the west, over the Gulf coast and States, and to the northwest and north, as they curve to the northeast. The chart, therefore, does not show the lines and rain-fall when the focal path is at its greatest descent in February. I deemed it important, therefore, to ascertain the location of that focal path as near as could be on the 1st of February, and trace it thence, as it moved to the west and north, month by month, until it attained its highest elevation about the 1st of August. It seemed important also to take a specific year, and I selected the year 1854, because the last one embraced in the publication, and because it was characterized by a memorable drouth. The following diagram shows the position in which I found that focal path on the 1st of February, 1854, and also the path of the Pacific conditions, as shown by the shading and arrows, and the comparative extent to which they deposit their moisture on the mountains which lie in their path:

FIG. 24.

The arrows indicate the direction in which the conditions moved. The intermediate space between these two systems of conditions is left unshaded, not because no rain at all fell upon it, but because there did not fall upon any portion of it, with here and there an exception, more than two inches of rain for the three months, and where only two inches or less of rain falls, even during a single month, drought may be said to exist. The rule is arbitrary, but convenient for illustration. The following table exhibits the

5

amount of rain which fell at the various posts situated *under* this
focal path, with the mean of previous years, in February.

| 1854. | FEBRUARY. | |
	1854.	Mean.
Key West, ...	2.55	1.38
Fort Myers,.............	4.70	2.16
" Brooke,..........	6.89	3.01
" Mead,............	2.21	1.01
" Pierce,..........	3.40	2.72
" Barrancas,.......	5.55	4.95
Mt. Vernon Arsenal,.....	12.83	6.04
Baton Rouge,............	5.50	4.91
Fort Moultrie,..........	2.84	2.33

It will be seen that there was an *excess* at *every station.* The
conditions were not more numerous probably, but they were more
concentrated—folded down, so to speak—and confined to the
focal path, and did not spread out to the west and north as much
as usual, and for that reason the fall of rain was greater and temperature higher on the focal path, and at the stations enumerated.

If now we look at stations to the west and north of the then
depressed focal path, and adjoining it, we shall find the fall was
less than the mean, during that month and also in January.

	JANUARY.	FEBRUARY.
Fort Towson, Indian Territory,.....	1.01	2.00
Mean,.........	3.13	2.97
Fort Gibson, Indian Territory,.. ...	0.30	1.43
Mean,......................	1.33	2.26
Fort Smith, Arkansas,..............	1.37	2.05
Mean,......................	1.96	2.17
St. Louis Arsenal, ...	0.65	2.40
Mean,......................	1.93	3.37

The following table will also show the actual temperature at
the same places under the focal path during the month of January:

	LAT.	LON.	JAN.
Fort Moultrie,......................	32.45	79.51	**50.83**
Mean of 28 years,............			50.36
Fort Pierce,...................	27.30	80.20	**67.91**
Mean of 5 yea s,............			62.75
Fort Meade, ...	28.01	82.00	**63.75**
Mean of 3 years,............			58.40
Fort Brooke,.........................	28.00	82.28	**62.94**
Mean of 25 years,............			61.53
Fort Myers,................	26.38	82.00	**67.56**
Mean of 4 years,............			63.39
Key West,........................	24.32	81.48	**71.75**
Mean of 14 years,............			66.68
Fort Barrancas,....................	30.18	87.27	**54.71**
Mean of 17 years,............			53.61
Mt. Vernon Arsenal,............... ...	31.12	88.02	**51.52**
Mean of 14 years,............			50.44
Baton Rouge,......................	30.26	91.18	53.43
Mean of 24 years,..........			53.47

It will be seen that the temperature, also, was above the mean in January at every post under the concentrated focal path, except Baton Rouge, and there it was at the mean. We shall see hereafter that Baton Rouge was near its western line.

Now by the following table, we see that at the same time, at the posts surrounding this path, west and north, the temperature was below the mean:

	JANUARY.	
	1854.	Mean.
Western Texas.		
Fort Brown,.......................	59.34	60.41
" Ewell,...........	50.47	52.92
" Inge,......................	47.24	49.46
Indian Territory.		
Fort Towson,......................	36.32	43.14
Forts Gibson, Washita, and Arbuckle, in much the same proportions,		
Arkansas.		
Fort Smith......................	33.92	40.18

	JANUARY.	
	1854.	Mean.
Missouri. St. Louis Arsenal,............ ...	25.47	31 44
Kentucky. Newport Barracks,.........	31.75	34.04
Pennsylvania. Alleghany Arsenal,...	29.08	29.25
Delaware. Fort Delaware,....	32.38	33.67
New York Harbor. Fort Columbus,................. ...	28.71	30.18

And this was in obedience to a general law. It is generally
cool or cold west or north of the focal path, *when concentrated,* as
then, and always *cool* or *cold* when the *focus* of a storm passes to
the *south* or southeast of a place, and *warm* when it passes to
the *north* or northwest. In the first case the storm *clears off,* as
it is expressed, *cold,*—the wind hauling *through the north* to the
northwest, as the focus of the storm passes by to the south. In
the second case, the wind hauls round by the south, as the focus
of the storm passes by to the north of the place, and it clears off
from the S.W., warm. When the latter becomes the *rule,* "*summer sets in,*" as it is said. Please remember and observe these
facts, for they are of great importance in several respects, as will
appear hereafter, more fully and clearly.

The following cut and table are inserted to show the situation
of the focal path in March, of the same year. when it was rapidly extending west along the Gulf coast, northwest on to the Indian
Territory, and north on to Missouri and over Kentucky. Its progress west and northwest was greater than its progress north.
This will be alluded to hereafter. In this table I give not only

the rain-fall for March, but also that for January and February, with the annual means, that you may continue to compare them as we proceed. I have remarked that Baton Rouge was near the western line of the focal path, and you will observe that the extension was felt at that post and at Fort Barrancas as soon as the movement of the focal path to the west and north commenced in February.

FIG. 25.

1854.	JANUARY.	FEBRUARY.	MARCH.
Fort Barrancas, Pensacola Bay,....	3 45	5.55	**7.21**
Mean,...................	8.87	4.95	5.87
Baton Rouge, Louisiana,....,..........	2.85	5.50	**6.15**
Mean,.................	5.26	4.91	4.68
Fort Towson, Indian Territory,......	1.01	2.00	**5.10**
Mean,.................	3 13	2.97	4.38
Fort Gibson, Indian Territory,......	0.30	1.43	**7.83**
Mean,.................	1.33	2.26	2.54
Fort Smith, Arkansas,............	1.37	2.05	**7.05**
Mean,...,................	1.96	2.17	2.92
St. Louis, Arsenal,...............	0.65	2.40	**7.10**
Mean,.................	1.93	3.37	3.82
Newport Barracks, Kentucky,........	3.20	5.30	**8.10**
(No Mean given.)			

Now run your eye over the diagram and table and see how the conditions extended their path to the west and north in March, on to the western part of Florida and the eastern and northern part of Louisiana, curving on the Indian Territory, and covering Missouri and Kentucky. The evidence that they were following their annual course and obeying a general law, although in advance of their usual mean progress, and the precipitation was in excess of the annual mean, is perfectly satisfactory. The movement was a general one. It occurred at all the stations, and although somewhat early and excessive, was in precise conformity with an annual, unfailing progress which can be traced in all the records known to the country. We shall see clearly hereafter what occasioned that more rapid progress and greater precipitation, when we come to examine the laws of the general system which created the conditions and held them to their annual path.

Following up this most interesting development, we find by the record that in April it reached a curving line of posts, still further removed to the W., N.W., and N., and that it was still in advance of its mean annual progress and excessive in its precipitation. I give the fall in January, February, and March, at the same places, that you may contrast them also ; and you may begin in this table to compare the rain-fall in June and July with the means, to see the *commencement of the drouth*, which will be fully developed as we go on. I wish also that you would carefully compare and watch the means—for the object of this exposition is to show you by those *means*, the manner in which the extension of the focal paths takes place *every year*, as well as their peculiar early progress and greater precipitation in that particular year.

We begin now to reach the northwestern states. Take for instance the mean of Fort Leavenworth in Kansas, and look at it separately, and see how the paths gradually spread up over it. Or take any other of the western or northwestern stations.

The following cut and table show the positions reached by the focal path in April :

FIG. 26.

TABLE V.

APRIL, 1854.	JANUARY.	FEBRUARY.	MARCH.	APRIL.	MAY.	JUNE.	JULY.
Fort Riley, Kansas,....................	0.00	0.94	1.86	**4.55**	4 35	1.10	0.00
Fort Leavenworth, Kansas,.............	0.04	1.78	1.33	**3.35**	**5.55**	4.5ᵗ	0 18
Mean,......................	0.72	1.01	1.61	2.74	3.62	5.80	3.15
Alleghany Arsenal, Pittsburgh,........	2.23	2.33	2.82	**4.21**	2 24	2.06	1.45
Mean, ...	2.18	2.17	2.70	3.10	3.58	3.56	2.97
Fort Columbus, New York Harbor,.....	2 60	4.00	0.70	**8.80**	7.70	2.20	1.90
Mean,......................	2 78	2.92	3.44	3.33	4.78	3.46	3.17
Fort Independence, Boston,...........	2.50	3 36	2.55	**5.40**	4 28	2.00	
West Point	3.52	5.04	2.81	**10.53**	5.08	1.62	
Mean, ...	3.50	3.44	3.71	4.55	6.18	4.79	

And here you get not only evidence of the continued extension
to the northwest, but of a fact alluded to which could not be
proved without anticipating, that the extension to the *northwest*
commences *earlier* and *progresses more rapidly* than that to the
north *upon the Atlantic coast.* The focal path in conformity with
the winter chart, covered Boston and New York in February, probably the early part of it, but had descended below them in March.
And we all know that March and early April are comparatively
dry in the northeastern states. But there were heavy floods

there in April, 1854, from the earliest extension of the focal path
and the concentration of the conditions, peculiar to that year.

Let me call your attention to the appearance of the drouth at
a part of the stations in that table in June, and the rest in July,
for the same stations will not appear in the next table.

Fig. 27.   May, 1854.

Fig. 28.   June, 1854.

The preceding diagrams show the position of the path in May and June. And in connection I give a full table of the stations reached in those months, so that not only the earlier and more excessive progress of that year, and the drouth which followed behind, may appear, but also the mean annual progress in other years. The table is full and instructive on each of these *three points*, and I ask you to study it well. At most of the stations the return of the focal path on its descent to the south ended the drouth in September.

TABLE VI.

JUNE, 1854.	JANUARY.	FEBRUARY.	MARCH.	APRIL.	MAY.	JUNE.	JULY.	AUGUST.	SEPT.
Fort Brown,............	0.45	1.50	1 15	0.05	4 10	**7.65**	4 25	5.00	11.81
Mean,............	1.61	2.25	1.20	0.56	2.21	4 55	1.95	2.76	6 73
Ringgold Barracks,.......	0.70	1.69	0.22	0.00	2.83	**10.98**	4.06	1.58	3.02
Mean,............	1.24	1.18	0.72	1 08	2.09	3.47	3.18	1.50	3.22
Fort Merrill,............	0.11	1.99	0.05	1.16	**7.66**	4.70	5 44	3.13	5.01
Mean,............	0.23	2.09	0.09	1.62	3.43	4.19	6.13	3.40	4.60
Fort Duncan,............	0.05	0.69	1.50	0.00	2.53	**6.83**	0.83	0.90	4.81
Mean,............	0.26	1.27	1.34	0.71	1.50	5.63	3.35	0.93	3.28
Fort McKavet,............	0.01	0.77	2.10	0.28	**3.72**	0.15	2 91	0.04	3.86
" Belknap,............	0.11	1.10	1.42	1.75	4 97	**8.33**	0.00	0.75	1.53
" Kearney,............	0.23	1.33	1.87	2.56	4.15	**5.40**	3.51	1.18	4.60
Mean,............	0.50	0.48	1.55	2 68	6.57	4.36	5.07	2.62	1.83
Fort Laramie,............	0 18	0.40	0.80	3 98	**4.46**	3.67	3.26	1.27	1.60
Mean,............	0 27	0.71	1.87	1.93	5.39	2.95	1.83	0.92	1.33
Fort Ridgley,............	1.20	0.01	1.18	2.83	**6.84**	2.70	2.49	2.28	2.58
" Snelling,............	0.72	0.03	1.03	2.51	**4.30**	3 31	3.92	1.75	6.55
Mean,............	0.73	0.52	1.30	2.14	3.17	3.63	4.11	3.18	3.82
Fort Ripley,............	0.67	0.03	0 79	0.97	**4.34**	3.68	0.62	1.69	4.40
Mean,............	0.68	0.37	1.80	1.42	3.09	5.15	5.20	2 27	4.92
Fort Mackinac,............	2.59	1.23	1.56	1.04	2 65	**6 35**	5.67	4.26	3.22
Mean,............	1.25	0.82	1.14	1.21	2.32	2.81	3.20	2.87	2.97
Fort Brady,............	2.49	1.18	1.34	2.14	**3.61**	1.23	3.21	3.86	3.18
Mean,............	1.84	1.13	).37	1.83	2 24	2.83	3.73	3.39	4.33

Observe first, its progress to the *west* over Louisiana and Texas. Fort Brown is near the mouth of the Rio Grande, opposite Matamoras. It was reached in May, and had its greatest excess from the Atlantic conditions in June. But it was also reached, like the rest of the Gulf coast, later by the *central conditions*, as shown on the diagram, and had no drouth. Ringgold Barracks are on the Rio Grande, further west and north, and were reached in June and had the drouth in August. Fort Merrill is on the Neuces, further east, and was reached in May by the Atlantic conditions, and by the central ones afterwards. Forts Duncan, McKavet, and Belknap are in northwestern Texas, and were reached by the Atlantic conditions in May and June, and then had the drouth. Then observe secondly, its progress to the *northwest* and *north.* Fort Kearney is in Western Kansas. Fort Laramie is in Nebraska. Forts Ridgely, Ripley, and Snelling are in Minnesota, and Forts Brady, and Mackinac in Michigan. Run your eye along the columns opposite to those places for 1854, until you come to the excessive fall of the then focal path, and continue it on to August, and contrast the actual fall of that year with the mean which is given immediately below it, and you will realize the extent and severity of the drouth, even on those western and northwestern states. Follow on until you come to Forts Brady and Mackinac, high up in Northern Michigan, which were beyond the drouth. As this table contains the means also, you can see again its *average annual progress* at these other posts, and upon the northwestern states, as well as the continued development and extension of the drouth. That extended drouth, it will be seen, reached Nebraska and Minnesota in July, but its full effect was not felt there until August, when it reached beyond Fort Kearney and almost to the base of the Rocky Mountains. It reached Forts Snelling and Ripley in Minnesota, but did not reach Forts Brady and Mackinac in Michigan, which are situated in a somewhat higher latitude and above its northern limit.

The conditions continued to extend their paths to the west and north, till they reached the position shown by the following cut and table in July and August. Glancing at the table you will perceive that their extreme western edge reached Fort Yuma and San Diego on the Pacific, gave New Mexico an unusual quantity of rain and spent themselves upon Northern Minnesota, Northern Michigan, and Southeastern British America.

FIG. 29.

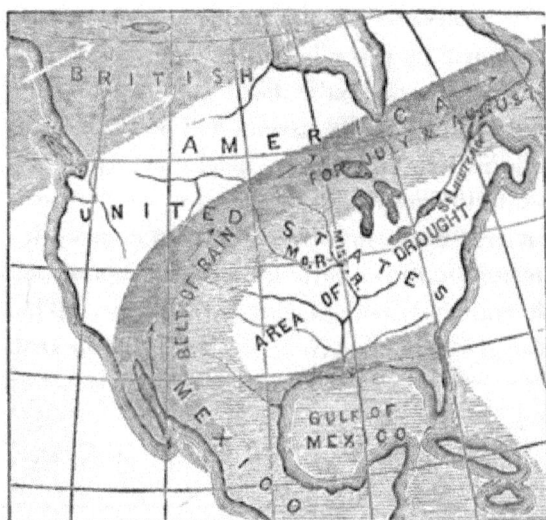

TABLE VII.

Situation of the focus of Precipitation in July and August.

	JUNE.	JULY.	AUG	SEPT.	OCT.
New Mexico.					
Fort Thorne, ...	0.08	2.23	6.01	3.50	0.00
Albuquerque,.....	0.28	2.50	1.19	2.67	1.37
Santa Fe,.............	0.32	4.11	3.86	4.06	2.50
Fort Defiance,........	1.24	3.94	5 24	3.47	0.62
" Yuma,...........	0.00	0.01	2.37	0.17	0 30
San Diego,.....	0.42	0.07	1.35	0.13	0.01
Fort Snelling, Minnesota.	3.31	3 92	1.75	6.35	1 23
" Brady,...........	1.23	3.21	3.86	3.18	3.40
" Mackinac, ...	6 35	5.67	4.26	3.22	2.28

I will now add a table showing the extension of the tropical belt of rains from the south up on to the Gulf coast, forming the southern limit of drought as shown on the diagram.

1854.	JULY.	AUG.
Fort Moultrie,...	5 09	3.82
Key West,...	3.45	5.83
Fort Myers, Fla., ...	9.70	9.90
" Brooke,..	15.53	11.23
" Meade, ...	8.55	10.20
Baton Rouge,...	6.55	7.41

I have added to the foregoing cut the situation of the focal path of the Pacific conditions in August. The table which will verify its correctness, will be given hereafter.

The descent of the Atlantic focal path in the fall and winter is but the counterpart of its ascent, and it reaches its southern position again in February, having made the transit up and down, in obedience to the laws of the general system. I give a single table with a few representative stations on a curve in the south-western, western, northwestern and northern states, for the entire year, in illustration of both ascent and descent. The focal path is not so sharply defined in Autumn, but of that hereafter.

	JANUARY.	FEBRUARY.	MARCH.	APRIL.	MAY.	JUNE.	JULY.	AUGUST.	SEPTEMBER.	OCTOBER.	NOVEMBER.	DECEMBER.
Fort Duncan, Texas,	0.26	1.27	1.34	.71	1.50	5.63	3.35	0.93	3.78	1.43	1.61	0.89
" Smith, Arkansas,	1.66	2.17	2.92	5.10	4.46	4.74	3.82	4.47	3.01	3.43	3.49	2.73
Washita, Indian Territory,	1.65	2.88	3.27	3.94	5.98	5.04	3.57	2.66	3.87	3.06	3.85	1.89
Jefferson Barracks,	1.91	2.04	3.32	3.06	4.18	5.07	3.67	4.14	2.88	2.76	2.38	2.42
Fort Scott, Kansas,	1.92	1.18	1.79	3.70	7.08	8.13	4.55	3.69	2.30	2.66	3.43	1.69
" Winnebago, Wisconsin,	0.91	0.82	1.07	2.26	2.25	4.24	4.21	3.01	3.62	2.00	2.01	1.09
" Snelling, Minnesota,	0.73	0.52	1.30	2.14	3.17	3.63	4.11	3.18	2.32	1.35	1.31	0.67
" Kearney, Nebraska,	0.50	0.48	1.55	2.68	6.57	4.36	5.07	2.62	1.83	0.88	1.11	0.33
" Mackinac, Michigan,	1.25	0.82	1.14	1.21	2.32	2.81	3.20	2.87	2.97	2.12	1.92	1.24
" Niagara, N. Y ,	2.25	1.89	2.12	2.20	2.55	3.28	3.49	3.04	3.95	2.37	2.36	2.27

I have not included any of the eastern states because the focal path covers them in December and January, and sometimes even in the early part of February. This will be seen by referring to the diagram of winter rains. (Fig. 23.)

But when the ascent commences in February, it first spreads out over the states to the west and northwest, and continues to extend in that direction much faster than over the Atlantic states. In a class of seasons, this is so much the case that serious spring drouths are occasioned in the northeastern states. I shall recur to this when I come to classify and explain our drouths, but I mention the fact here and wish it remembered. It will be found hereafter not only an important fact in relation to drouths, but an important element in prognostication.

And here I also insert, rather as a matter of interest than as

showing the movement of the system of conditions, a chart of the
rain-fall for the year. It conforms of course generally, to the
chart, tables, and facts already given, but the movements of the
focal path are masked in the general result. The chart is copied
from the "Army Meteorological Register." It may not be entirely accurate, but I think it is substantially so. It shows doubtless, more clearly than the winter chart I have given, the effect
of configuration on the quantity of rain, and also where the system of Atlantic conditions is focal for the longest portion of the year.

FIG. 30.

And now professional men, practical men, and especially young

men, I have given you the key to the climatology of your continent, in each and all of its various localities—*your locality*, reader,
wherever it is.   I have given you the key to your RAINY AND
DRY SEASONS—and they occur and alternate more or less perfectly everywhere.   I have given you a key to understand the
TEMPERATURE of your locality as it is affected by the situation
of the focal path of the conditions in its relation to that locality.
I have given you a key to the SUDDEN CHANGES you all experience so often and talk so much about, and you can now understand them, and forecast them, as the conditions which produce
them, pass over or within influential distance of you.   You will
understand why the *consumptive* leaves the Atlantic coast, where
the conditions pass in winter with their sudden and extreme
changes, to seek restoration in the *colder* but *dry and comparatively changeless* atmosphere of *Minnesota*, where the conditions
rarely pass or extend their influence at that season of the year.
You will understand also, why, if in the central or eastern states,
you *do* look or *may* look in the *southwest* in *winter*, and in the
northwest in *summer* for the first appearance of the advance clouds
of a stormy condition, the focal path being in the one case at the
south, and the other at the north ; and why, when an approaching
storm is seen in the northwest, you may expect rain, and in the
southwest snow, east of the Alleghany mountains.   And you will
also understand the PREVALENT WINDS of your country, alternating east of the Rocky Mountains, from southerly to northerly
and from northerly to southerly as the focal path of the conditions
is west or north of you, or east or south of you, or over you.
And you will also understand the SPECIAL WINDS of your climate
and their changes, remembering that these special winds are *parts
of conditions*, and that you are enveloped in one wind, a *warm*
one, when the condition is approaching you, and in another and
cool one when the condition is passing away from you.

But on this general subject of the winds it may be well to be
more particular, and I copy the following tables from Prof. Coffin's " Winds of the Northern Hemisphere." The first is in relation to the winds at Tampa Bay on the west coast of the penin-

sula of Florida. Here it will be seen that the prevalent winds are southerly, during all the time that the focal path of the conditions is ascending to the northwest, but that during the winter months they have the northeast trade.

No. 50.—Tampa Bay, Florida.

11 YEARS.

MONTHS.	MEAN DIRECTION OF WIND.	RATE OF PROGRESS.
January, ...	N. 9° 17' E.	10
February, ...	S. 86 14 E.	12
March, ...	S. 51 17 W.	12
April, ...	S. 30 23 W.	10
May, ...	S. 3 56 E.	14
June, ...	S. 18 33 E.	26
July, ...	S. 8 21 E.	35
August, ...	S. 19 58 E.	29
September, ...	S. 80 46 E	25
October, ...	N. 66 14 E.	22
November, ...	N. 55 2 E.	15
December, ...	N. 27 20 E.	13

No. 67.—Forts Gibson, Smith, and Wayne.
3 stations.

8 YEARS.

MONTHS.	MEAN DIRECTION OF WIND.	RATE OF PROGRESS.
January, ...	N. 61° 8' E.	10
February, ...	S. 46 13 E.	20
March, ...	S. 46 45 E.	18
April, ...	S. 24 52 E.	23
May, ...	S. 29 20 E.	60
June, ...	S. 31 31 E.	66½
July, ...	S. 33 44 E.	37
August, ...	S. 40 28 E.	34
September, ...	S. 79 40 E.	31½
October, ...	S. 70 43 E.	18
November, ...	N. 2 6 W.	19
December, ...	S. 8 57 W.	2

Following the focal path to the northwest on to Arkansas, we have at Forts Gibson Smith, and Wayne the preceding table, the winds conforming to the focal path which covers the eastern part of Arkansas in winter, and the curvature of the conditions over that state as shown in the diagrams.

Turning now to Fort Towson on the Red river, in the Ind'an Territory, southwest of Fort Smith, we have the winds ranging through the year from the south, bearing more or less to the west. Fort Towson as we have seen lies southwesterly of the focal path of the conditions as situated in winter.

No. 65.—Fort Towson, on Red River.

8 YEARS.

MONTHS.	MEAN DIRECTION OF WIND.	RATE OF PROGRESS.
January,..........................	S. 34° 1' W.	17
February,..........................	S. 65 42 W.	22
March, ...	S. 31 11 W.	26
April,..........................	S. 6 40 W.	31
May,..........................	S. 5 33 W.	47
June, ...	S. 4 53 W.	49
July,..........................	S. 6 21 W.	56
August, ...	S. 13 38 W.	40
September, ...	S. 5 13 W.	17
October,..........................	S. 47 37 W.	16
November, ...	S. 44 46 W.	28
December, ...	S. 27 5 W.	21
The year,..........................	S. 17 48 W.	29

Turning now to Nashville, Tenn., which is covered by the focal path even in winter, and where the conditions are always moving to the northeast, we have the winds conforming to them, and prevailing between the south and west during every month of the year. Observe with what substantial regularity the winds grow more southerly from February to September, as the focal path is carried up by its annual transit above the place to the north and northwest, and how they turn to the west again when it return in October.

Nashville, Tennessee.

5 YEARS.

MONTHS.	MEAN DIRECTION OF WIND.	RATE OF PROGRESS.
January,................................	S. 39 41' W.	30
February,...............................	S. 65 22 W.	22
March,.................................	S. 70 35 W.	21
April,..................................	S. 57 38 W.	41
May,...................................	S. 57 29 W.	38½
June,..................................	S. 45 1 W.	49
July,...................................	S. 39 18 W.	27
August, ...	S. 20 31 W.	25
September,..............................	S. 34 30 W.	18
October,................................	S. 81 13 W.	27
November,..............................	S. 62 42 W.	23
December,..............................	S. 60 59 W.	39½
The year,...............................	S. 57 20 W.	30

If now we ascend to the northwest, to the stations from Iowa to Maine inclusive, between the parallels of 45° and 50°, we have the following as the mean monthly direction of the winds, during a period of 17 5-12 years.

MONTHS.	MEAN DIRECTION.	RATE OF PROGRESS.	NO. OF DAYS.
January,................................	N. 58° 40' W.	14	31.00
February,...............................	N. 47 43 W.	19	28.24
March,.................................	N. 24 37 W.	9	31.00
April, ...	S. 69 34 W.	15½	30.00
May,...................................	S. 12 27 W.	3	31.00
June,..................................	S. 51 31 W.	18	30.00
July,...................................	S. 64 3 W.	37	31.00
August,................................	S. 56 43 W.	28	31.00
September,..............................	S. 69 58 W.	20	30.00
October,................................	S. 70 59 W.	19½	31.00 -
November,..............................	N. 50 56 W.	11	30.00
December,..............................	N. 69 50 W.	10	31.00

And at Amherst, which is a fair representative of New Eng
land, we have the following:

Amherst, Massachusetts.

5 YEARS.

MONTHS.	MEAN DIRECTION OF WIND.			RATE OF PROGRESS.
January,............................	N.	69	42′ W.	36
February,.....................	N.	63	34 W.	35
March,.............................	N.	53	39 W.	41
April,.............................	N.	55	2 W.	33
May,..............................	N.	85	9 W.	22
June, ...	S.	67	5 W.	22
July,.............................	S.	70	47 W.	37
August, ...	S.	88	34 W.	26
September,.........................	S.	76	54 W.	16
October,...........................	N.	78	53 W.	30
November,.........................	N.	55	19 W.	41
December, ...	N.	57	2 W.	47
The year,..........	N.	73	13 W.	30

And the explanation of this prevalence of S.E. winds in the
eastern Gulf states; of southerly winds in the western Gulf
states; of S.W. winds in the central states; and alternation of
N.W. and S.W. winds in the northwestern, northern, and eastern
states, is found in the location and extension of the focal path of
the Atlantic conditions, and the *alternation* of the *special winds*
of the *conditions*, as they pass over the northern and eastern portion of the country.

And in the foregoing facts the reader will find the reason why
the winds prevail from the northwest, northeast, southeast, and
southwest more than from any other points of the compass. It
is because the line of progress of the conditions, after they curve
over the eastern part of our country, is to the northeast, and their
winds are either lateral or conform in their direction, or blow in
opposition to their line of progress. Southerly winds prevail on
the south side of the focal path, northwesterly and northeasterly
on the north side. Where the path is central, west of the Alle-

ghany mountains, the winds are southwesterly, and east of them northeasterly. If the conditions moved from one cardinal point to another, the winds would prevail from the cardinal points.

And now let us see how perfectly this discovery explains the heretofore unexplained peculiarities of temperature.

The following is a reduced copy of the chart of isothermal lines (lines of equal temperature) *for the winter*, which accompanies the Army Meteorological Register. It has been questioned in some respects and may not be entirely accurate, but it is substantially so, and sufficiently so for illustration.

Fig. 31.

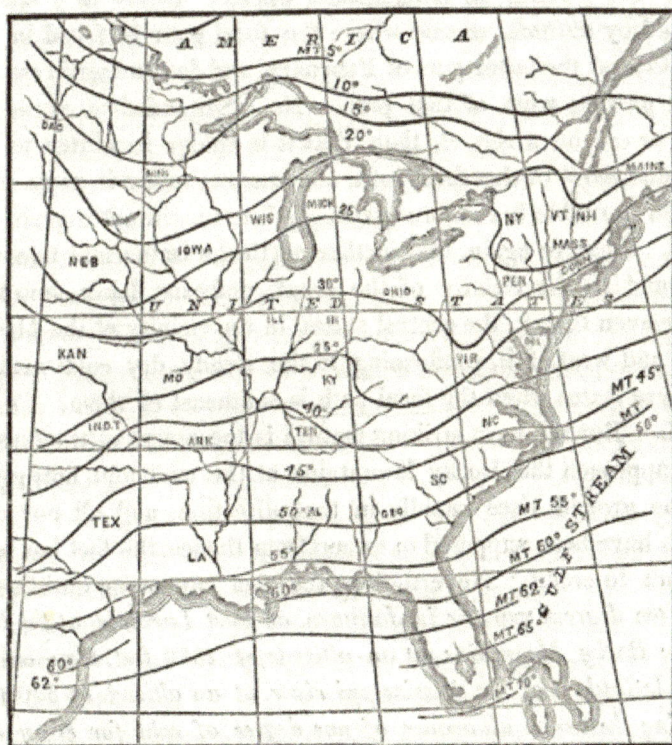

If this chart could have been drawn for the first half of February, when the focal path is at its extreme southeastern depression, its peculiar features would have been much more obvious, but they strikingly appear as it is.

Take then, 1st, the isothermal of 50°. It commences off the coast, on the Gulf Stream, where the ocean is much warmer than the land, a little above the parallel of 35°, and descends till it meets the focal path of the conditions at eastern Georgia, and there its descent is arrested until it gets to the *west of that path* upon Texas, where it descends with great rapidity, as it should do, in accordance with the tables I have given you. Observe the sudden angle made by the descent, even on the coast and upon the Gulf. It is as cold at 30° in Texas, as at 35° in N. Carolina.

2d. Observe again, that over the Mississippi Valley all the lines, after passing the Alleghanies, where of course they are depressed by *altitude*, ascend where the focal path is found in December, and the latter part of February, and descend again rapidly after passing west of that path. No other sufficient cause has been or can be assigned, than that it is colder in winter to the *west* and *north* of that focal path, and warmer *under* it, as we have seen in the table heretofore given.—Altitude will not explain it.

3d. Observe again, that further north the isothermal lines rise over and in the vicinity of the great, unfrozen lakes, but fall, lower even than in the central states, in the vicinity of the Mississippi and west of it, conforming to the steady, dry cold weather of these states when the focal path is southeast of them.

4th. But the most striking feature is the ascent of the lines as they approach the Rocky Mountains at the west and northwest. As the ground rises rapidly in that direction, and all our cold winds have been supposed to sweep from thence, the fact has been difficult to credit. Nevertheless, the fact cannot be questioned. *It is ten degrees warmer in January, at Fort Laramie, at the base of the Rocky Mountains, at an altitude of 4519 feet, than on the same latitude near the Mississippi river, at an altitude of 500 feet. Making the usual allowance of one degree of cold for every 300 feet of ascent, it should be 13° colder at Laramie.*

There is then the very great difference of twenty-three degrees, to be explained. It has not been explained. But our key ex plains it. *Fort Laramie is reached or influentially approached in winter by the southern edge of the focal path of the Pacific winter conditions, and has not only the greater warmth which is found under that path, but somewhat of the warmer s uthe ly and easterly winds instead of northerly winds, which are found on the immediate southerly and easterly borders of both focal paths.* The following tables will show this. And first as to temperature.

	JAN.	FEB.	MARCH.	APRIL.
Fort Laramie, mean of six years, Lat. 42° 12′, Long. 104° 47′, Altitude 4519 feet,..........................	31.03	32.60	36.81	47.60
Fort Kearney, 6½ years, Lat. 40° 38′, Long. 98° 57′, Altitude 2360 feet,....	21.14	26.11	34.50	47.18
Council Bluffs, mean of 7 years, Lat. 41° 30′, Long. 95° 48′, Altitude 1250 feet,	19.36	25.23	33.77	51.84

No station having precisely the same latitude as Laramie exists. Council Bluffs is the nearest to it, and sufficiently so for illustration. Now observe two facts shown by the foregoing table. The first is that there is at least ten degrees of difference in January as stated. The second is that the stations further east feel the extension of the Atlantic conditions in advance of Laramie, and the Pacific having moved beyond influential distance, the temperature increases more rapidly in March and April at those stations than at Laramie.

In the following table we compare the winds for the same months in the same year, and take a year about the middle of the decade, 1856, giving at the same time the mean temperature and range of the thermometer. Our comparison here is with Fort Kearney only, Council Bluffs being then broken up.

1856, JANUARY. TEMPERATURE.				WINDS.							
	Mean.	Max	Min.	N.	N.E.	E.	S.E.	S.	S.W.	W.	N.W.
Laramie....	19.98	42	− 7	0	1	13	1	0	14	50	8
Kearney......	6.05	38	−20	24	0	0	2	2	3	18	38

This table does not differ materially in its results from the mean of ten years. Looking to the fact that Laramie is much more elevated, and nearly a degree farther north, where northerly winds should be more prevalent, the absence of northerly winds at Laramie, and their presence at Kearney, together with the presence of easterly and southerly winds at Laramie and their absence at Kearney, in the same month of the same year, indicate decisively the influence of the Pacific system upon the weather at Laramie. There is nothing else in either locality to affect the result. The exposure of Laramie is open, and the valley of the Platte river where Kearney is situated is broad, and the bluffs distant and moderate. There is much other evidence which might be adduced on this point, but it does not seem necessary to adduce it. The facts given are sufficient until impugned.

Follow me now to the Pacific states, and we will analyze their climatology. We shall find it much more simple, but substantially the same in its elements.

In the first place, we find that the conditions all enter upon the Pacific coast from the S.W. and move northeasterly. We have but few data on this subject, and those are mostly from California and Oregon. All these data prove that the rain-bearing clouds move in from the southwest, and this is as it should be. The Pacific is broad, and all the conditions which reach the western coast of the continent form upon its surface and curve to the N.E. before they reach the coast. They necessarily, therefore, reach it from a southwesterly direction and pass on to the northeastward precisely as the Atlantic conditions do, above 35° in summer and 30° in winter. There is therefore precise conformity in respect to the direction in which the conditions move.

In the second place, there is a focal path in which the conditions move, precisely similar to that found in the Atlantic system. In the winter season, when the southern limit of the path is at or near San Diego, the focus of the path is at Astoria in Oregon, and the northern limit in the vicinity of Sitka. The following table will illustrate this, and it will be seen that in January, when there was very little rain at San Diego, or at Sitka, there was 27 inches at Astoria, and 11.8 inches at Puget's sound, which is nearly in the same latitude, but lies in the interior, east of the coast range of mountains which affect the rain-fall there. In August and September, when the focal path has moved up to the north, we find that 10 and 14 inches fell at Sitka, while there was none in California and very little in Oregon. There is other evidence on this point, but it is unnecessary to adduce it.

	Lat.	Jan.	Feb.	March	April	May	June	July	Aug.	Sept.	Oct.	Nov.	Dec.	Year.
San Diego, Cal.,	32 41	0.8	1.7	1.1	0.9	0.5	0.0	0.0	0.2	0.0	0.1	1.5	3.4	9.6
San Francisco,	37 48	1.7	0.5	.4	2.1	0.4	0.0	0.0	0.0	0.4	0.6	3.0	5.5	18.8
Cant., Far. W., Cal.,	39 02	3.3	0.6	6.4	2.2	0.9	0.0	0.0	0.0	0.3	0.1	3.5	4.6	21.9
Astoria, Oregon,	46 11	27.0	10.9	6.1	4.4	5.9	2.6	0.0	2.8	1.9	6.7	13.2	6.2	87.2
Puget's Sound, Ore.,	47 07	11.8	3.9	4.7	4.1	0.8	0.6	0.5	1.3	1.6	3.6	5.9	6.1	44.8
Sitka, Alaska,	57 3	2.5	9.6	3.5	3.3	1.9	5.9	3.7	10.1	14.8	12.7	7.4	4.2	79.5

The figures are for inches and tenths of an inch of rain

In the third place, we find that there is a transit of this focal path to the north in summer, and that all or nearly all the conditions enter upon the coast in August above California and Oregon. From evidence which I shall have occasion to introduce in another connection, it will appear that during certain years of the decade, the transit to the north is so great as to leave Oregon as rainless as California. For the present, the foregoing table is sufficient evidence of the *fact*, and the *extent*, of the *transit*, and that it is in precise conformity with that of the Atlantic system in the eastern states, in its material features.

In the fourth place, there is satisfactory evidence that the winds attendant upon the focal path are the same as those in the eastern states. This is so well expressed by Mr. Blodgett in his Climatology that I copy from him.

" The rains of this best known portion of the Pacific coast, (California,) are, as has been said, peculiar in regard to the attending winds, which, from San Diego to Puget's Sound, are in nearly all cases from the southeast and south with a strong and steady force. These are also simply *attendant* winds, and not those which may be said to bring the rains,—the course of clouds above the local or surface wind, being quite regular from the west. But no sooner is precipitation begun than the attendant southeast wind sets in, to be continued steadily to the end of the rain in most cases. And at the northernmost stations it begins always earlier than at the next southward,—in fact beginning and ending with the rain in all cases, and as they begin earlier at the northerly points it has more days of duration there." As the centre of the focal path is never south of Oregon, there are of course none other than the southerly *lateral* winds which form a part of the conditions south of that point.

If now we go north of the focal path to Sitka, we find the easterly and westerly winds are prevalent, and that northerly winds are prevalent when the focal path is at the south. The east wind is not known south of the focus. At Sitka it is more prevalent than any other, as the following table of the relative number of winds taken from Prof. Coffin's work will show:

WINDS IN BRITISH AND RUSSIAN AMERICA.

SITKA, RUSSIAN AMERICA.

Course.	Jan.	Feb.	March.	April.	May.	June.	July.	Aug.	Sept.	Oct.	Nov.	Dec.	Total.
North,	11	6	78	42	12	36	42	18	18	30	12	6	311
N.E.	127	7	198	90	24	42	78	42	18	30	18		704
East,	48	14	222	233	156	162	90	156	246	240	270	210	2047
S.E.	167	330	48	59	114	60	6	90	102	174	90	204	1444
South,	3	48	48	52	78	60	66	90	60	90	30	78	703
S.W.	5	31	48	113	156	66	120	54	72	18	42	36	42
West,	57	13	66	65	144	186	168	108	42	72	72	48	1041
N.W.	41	10	36	53	30	72	36	24	36	6	24	48	416
Calm,	279	237	0	14	18	36	120	162	120	66	150	98	1298

But it should be stated that all the conditions are less intense on that coast than in the eastern states. Thunder storms are exceedingly rare,—thunder is not heard more than two or three times a year in California. Gales are uncommon and the conditions seem to partake of the *pacific* character of the ocean on which they originate. Why the ocean and the conditions are thus pacific, I am not yet at liberty to suggest, for I have undertaken to unfold the system to you, *as matter of fact*, leaving the consideration of its motive and controlling forces for a concluding chapter.

I turn now to the third system of conditions, which in certain years and seasons covers the southern portion of the Gulf States, Southern Mexico, and Central America. Here, as we have said, the conditions originate upon the Atlantic or the Gulf of Mexico and pass to the westward out upon the Pacific. They are not directly connected with the other two systems, and a further consideration of them is scarcely necessary. They consist almost wholly of limited and isolated thunder showers, passing frequently and rapidly over the track, giving dashes of rain in large drops and pouring masses for a brief period, and are gone. But the aggregate amount of rain which they deposit during the rainy season, even during the brief period they are over the Gulf coast, is very large, as the following table will show:

Fort Myers, Florida, mean annual rain,	-		-		-	62.26
" " " mean for summer,		-		-	-	30.91
Fort Brooke, Florida, mean annual rain,			-		-	55.47
" " " mean for summer,		-		-	-	28.44
Baton Rouge, Louisiana, annual mean,	-		-		-	62.10
" " " mean for summer,		-		-	-	19.14

With this table, which requires no comment, I leave this branch of the subject. The importance of the foregoing development of the systems of conditions and the transits of their focal paths, cannot be doubted. Nor can the truth of the development be honestly denied. There is much other evidence to support it for which I have not space. Some of it will hereafter appear in another connection. Let not the practical reader doubt the truth of that, and other developments, made and to be made, because they have been in substance thirteen years before the professed Meteorologists of the country, and have been ignored and suppressed, because adverse to a baseless and pernicious theory.

> " Truth crushed to earth will rise again,
> The eternal years of God are hers;
> But error wounded writhes in pain,
> And dies amid her worshipers."

And now if the reader has attentively followed me in the foregoing exposition of the manner in which the conditions produce the states of the atmosphere constituting the weather, the character of those conditions as they enter upon or form over our continent, and their various systems and the paths they pursue, he is prepared to follow me as I trace them back to their sources, and examine the *organization* of the GENERAL SYSTEM, and the *operation* of the *laws* of that system, by which the conditions are produced. Upon that inquiry we will enter in another chapter.

CHAPTER V.

THE GENERAL ORGANIZATION OF THE ATMOSPHERIC SYSTEM.

Origin of a class of conditions in the central zone—Central belt of precipitating clouds—encircles the earth—Mean width about 500 miles—chart of it and its connections—Situation in August, north of the equator—Polar zones of rain connected with it—Situation of polar zones in August—On each side of the central belt an area covered by dry trade-wind, called N.E. and S.E. trades—these areas also surround the earth—The central belt of cloud and rain, the two areas of dry trade-wind, and the two polar zones of rain, make five permanent and connected parts of a general system—All as a connected whole have a northern and southern annual transit—All commence their movement together to the south about the first of August, and reach their southern position about the first of February—Chart showing the position in February of the five parts, to wit: the central belt, the two trade-wind zones, and the two polar zones—The connected whole commences its northern transit early in February, and reaches its northern position in August—This organized whole to be particularly examined—Some preliminary facts to be considered—Transit more extended north and south some years than others—Some sections which are not covered by either zone of rains during the transit—such sections constitute a class of deserts—Such are New Holland and Kalahari in the Southern Hemisphere—Arabia, Egypt, Sahara, Colorado, &c., in the Northern Hemisphere—There are also arid areas in the polar zones—Principal rivers of Africa and South America rise under the central belt of rains—Central belt passes over some places twice a year—Monsoons—Examination of the great central condition—Constituted by or composed of a central belt of condensation and two wings of winds, the trades—Analagous in this respect to the conditions of the polar zones—Analagous also in all its essential elements—Critical examination of its elements—Examination of the trade-winds and their character—Examination of the central belt of condensation and its character—Contains no vortex—Theory that it does a mere assumption—Originally made by Halley in 1686—never proved by any direct evidence, nor capable of such proof—Every fact in nature bearing upon it adverse to it—Earth, air and water under the belt of rain colder than in the trades everywhere—Fundamental base of the

theory therefore untrue—Review of evidence on the point—Fact undeniable, undenied, yet not regarded—Coast wind of California, not an exception—That wind does not reach the valleys which are assumed to cause it
—The air as heated in the atmosphere has not the ascensiv force attributed to it—Review of the evidence on this point—Impossible therefore that
it should produce a vortex—Description of a violent West India hurricane
—Originated where the air and water were only 84°. Ascensive force of
unconfined air at that temperature does not exceed a quarter of an ounce
to the square foot—Utterly impossible that it should produce storms of
such violence, or any storms—All attempts to prove the existence of such
a vortex, directly or analogically, failures—Review of evidence on that
point—Examination of the nature of the central condition, as disclosed by
actual and positive observation—Review of evidence on that point—Observations of Squier in Nicaragua—Fendler in Venezuela—Herndon in the
valley of the Amazon—Gibbon upon the Andes—Livingston in South
Africa—Du Chaillu in Equatorial Africa—Barth in North Africa—All
concur to prove the theory of Halley untrue—Critical examination of the
belt continued—Has an upper layer of cirrus—a second layer of stratus
or cumulo-stratus, beneath them the scud of the trade-winds—These three
elements with the squalls and slant winds at the surface constitute the
organized central belt—Trades pass each other in this belt—Review of the
evidence of it and of the objections made to it—After passing, continue in
the same general direction—Facts pointing in a different direction are exceptions—Review of evidence on this point—Resume of the facts contained
in the chapter.

Taking now one of the distinct and intense conditions whose
path is along our coast, in the early part of August, and following it back to the place of its organization, as shown upon the
map of Mr. Redfield; or taking an intense belt of showers which
has come to us from the westward, and tracing it back on its curving path to where it entered upon the continent over Western
Texas, and thence to the south and southeast to its place of origin
in the West Indies or the Caribbean Sea, we find that place to be
covered by a belt or zone of precipitating clouds, averaging about
500 miles in width and encircling the earth. That belt or zone
is the centre of the atmospheric system, the Atmospheric Equator. Always existent, always active, and permanent.

Taking a general view of it and its connections, we find order,
arrangement, and organization. The following diagram shows that

central belt or zone of rain, substantially as it is found in normal
years on the 1st of August, and its connected polar zones.

FIG. 33.

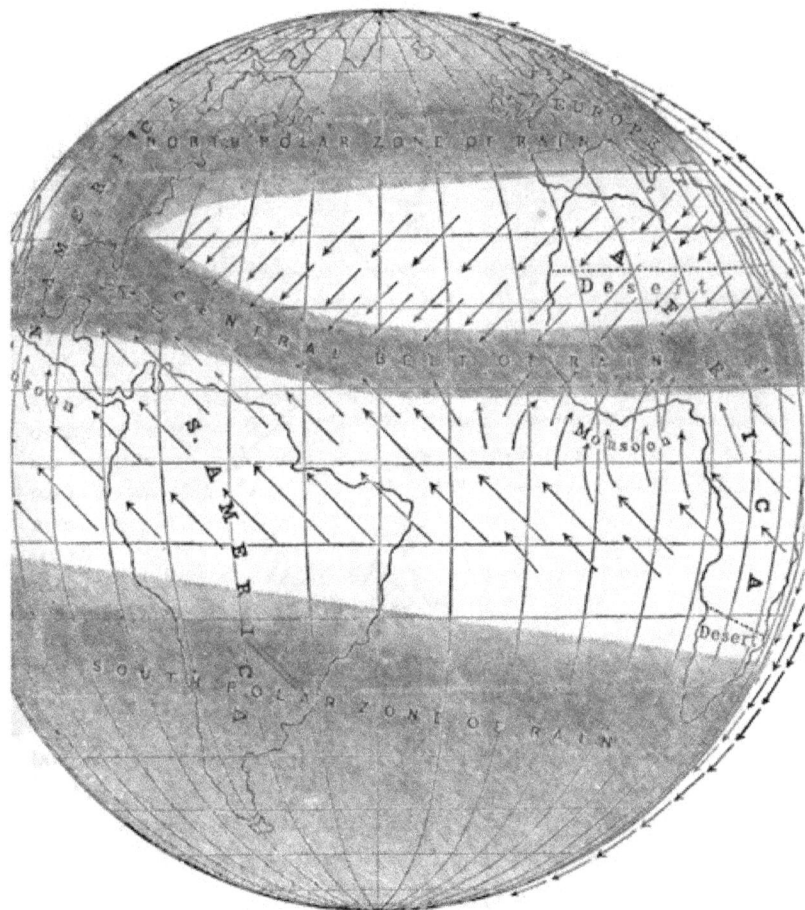

You see that it is situated above the geographical equator, on
the eastern part of the Atlantic, ranging up on to Florida and
the southern part of the Gulf states, and on to the Gulf of Mexico, Central America, and Southern Mexico. This is the belt to
which I alluded as the path of our third system of conditions.
It is generally felt in July and August on the peninsula of Flor-

ida, and more or less on the Atlantic coast as high as Charleston, but not as high on the Gulf coast except in occasional years in

FIG. 32.

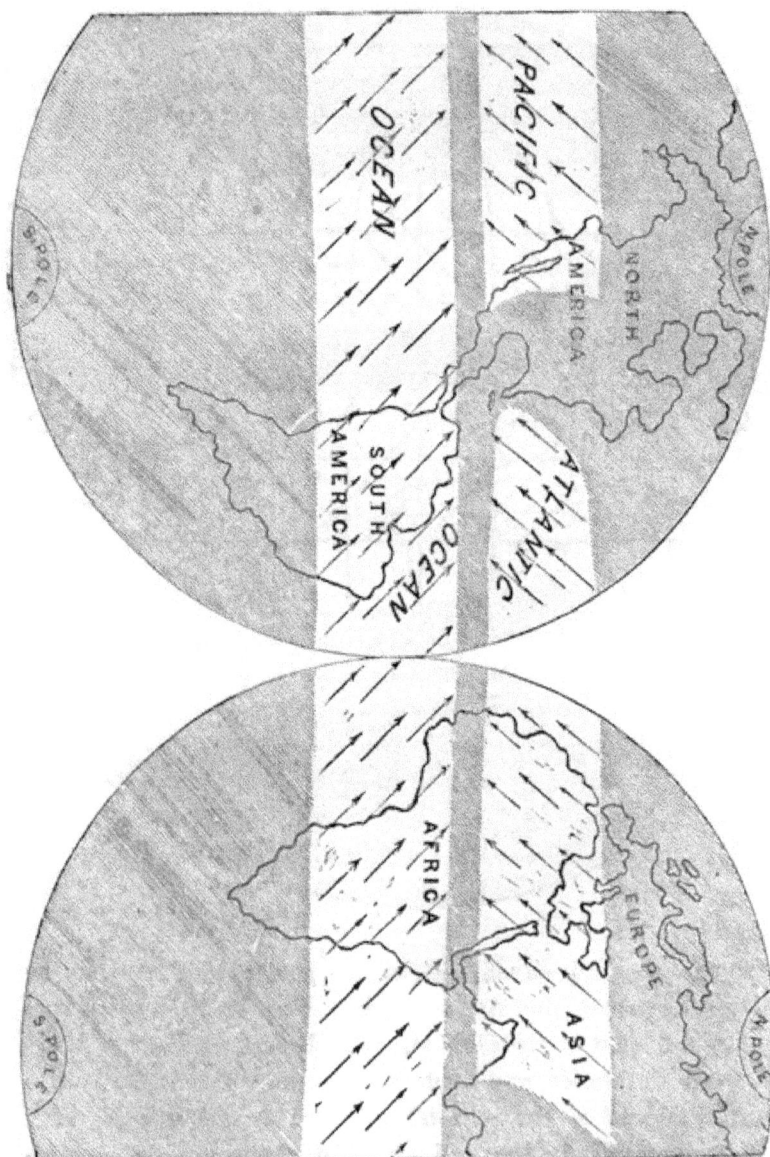

particular parts of each decade. The shading on the diagram represents it with substantial accuracy.

Connected with this central belt of rains on the north, there is a polar zone of rains, exhibited by the shading of the map. In looking at it you will see that that polar zone of rains has moved far to the north in **August**, uncovering the southern part of Spain and Italy on the east, and California and Oregon on the west. Another view of its situation, not quite as accurate, but sufficiently so for our present purpose, is exhibited upon the preceding diagram, and will show its situation also upon Africa and Asia.

It will be seen by the shading that upon the S.E. part of the North American continent, and the S.E. part of the continent of Asia, the zones are represented as connected. Casting your eye upon the southern part of the first diagram, you will see another zone of rains, termed the south polar zone, which about the 1st of August, extends north, particularly upon South America, almost to the Equator. These zones on the north and in the south are of precisely the same character, having similar atmospheric conditions to which they owe their rains, those conditions curving in the south polar zone to the S.E., instead of the N.E., but in both drifting to the eastward, and furnishing both alike with rain. You must remember that the shading in all the zones is only to show the territory upon which the distinct and separate conditions drop their rain during a given period. It is not intended to represent that it is constantly raining upon those portions of the earth. Intermediate between the central zone of rains and each of the polar zones, there are two belts of drouth, and constant wind blowing toward the central zone of rain. These winds are called the *trade winds*, and as they blow most frequently from the N.E. on the north side, and the S.E. on the south side, as indicated by the arrows, they are known as the N.E. and S.E. trades. The dry belt on the south, with its S.E. trades, surrounds the earth, unbroken except as it is crossed by an occasional storm. The dry belt on the N. side, with its N.E. trades, also surrounds the earth unbroken save by an occasional storm, and save also as the central zone is connected with the north polar zone on Southeast-

ern America and Southeastern Asia. It will be observed that
the central zone of rains does not extend over Africa upon the
Desert of Sahara, nor does it extend on to Northern Mexico or
Southern California. Thus then, on the 1st of August, in each
year, we have these five parts of a permanent, atmospheric arrangement. Three zones of clouds and precipitation, and two of
drouth and constant wind, and the five embracing the whole atmosphere, and each in its appropriate place, encircling the earth.

FIG. 34.

About the first of August, every year, these five belts commence as a connected whole, a movement to the south. The northern edge of the south polar zone of rains descends to the south towards its pole, the zone of S.E. trades and drouths, the central belt or zone of rains, and the belt of N.E. trades and drouths, all three descend also, and so does the southern edge of the north polar zone of rains. All together and all as one, they thus continue to descend until about the 1st of February, when they reach substantially the position shown in the preceding diagram.

Strictly speaking, (for in an investigation of this kind we must aim to be accurate,) the northern edge of the south polar zone has not, as a permanent edge, retreated to the south; nor has the southern edge of the north polar zone, as a permanent edge, extended to the south; but the central zone of rain, the *prima mobile* and *causa causans* of all, the central organized body, has moved down, carrying with it its wings of wind, withdrawing its north wind from the north polar zone, and permitting the northern conditions, its creations, to commence curving and precipitating further to the southward. So it has moved its southern wing gradually farther and farther down, causing its S.E. dry winds to commence blowing still farther and farther from the southward, and compelling the southerly conditions of its creation to commence curving and precipitating farther and farther to the southward until its annual southern limit was reached.

Generally by the middle of February, in some *particular year* of the decade a little earlier, and in others still later, the central belt of precipitation, with its wings of winds and connections, following the sun, commences its northern transit. The ascent to the north is but a counterpart of the descent to the south. As the S.E. trades cease to blow at their extreme limits, the southern conditions commence curving and precipitating farther and farther north upon the surface from which they had been excluded by the trades and drouth, or to speak more correctly, by the influence which caused the trades. So, on the other hand, as the central belt moves to the north, the N.E. trades commence blowing farther and farther from the north, compelling the northern conditious to commence curving and precipitating still further and further northward. And so the process goes on, until about the 1st of August, the central belt and the belts of trade and drouth, and the edges of the polar zones, reach the positions in which we found them on the first of August before.

Before we examine critically the character of the organization of which we have thus taken a general view, it will be well to look at a few facts, which will be useful in a critical inquiry. The first is, that although the transit of the central belt of rains is substantially the same every year, at intervals of 10 or 11 years, its transit both north and south is perceptibly extended; and there is *irregularity* and *periodicity* connected with it, which we shall see has an important influence on the weather of our zone.

Another fact is, that over extensive portions of the earth in the dry belt covered by the trade-winds, the central zone of rain and the polar zones of rain do not lap, so to speak, during the transits, and such portions are rainless deserts. Thus upon New Holland, under the southern belt of trade, the central zone of rains, in its transit to the south, covers a part, but only a small part of it on the north. And so, in its movement to the north, the south polar zone of rains moves far enough north to cover a part only on the south, leaving an intervening, extensive, rainless, central desert not reached by either. Following the same belt of trade to the west, we find a desert, though not as perfectly arid in South Africa—that of Kalahari. On the diagram I have marked the extent to which the south polar zone of rain extends up beyond Orange river, during its northern transit, in normal years, and also, by a dotted line, the extent to which the central belt of rains in such seasons reaches in its southern transit. The space which is not covered by either, is the Kalahari Desert, so fully described to us by Livingstone. One fact may here be noted, mentioned by Livingstone, viz.: that occasionally the central zone of rains does extend down far enough to cover the desert, and that it is a tradition among the natives that this occurs once in 10 or 11 years. We shall advert to it again when we come to consider the occasional extreme transits to which we have alluded. Going west to

South America, south of the equator, we find that the central and south polar zones of rains do lap, though imperfectly, and there is no place which is not reached by either of them, and no absolutely rainless desert. South America is peculiarly situated for the reception of moisture from the trades, and peculiarly elevated, and is therefore more abundantly supplied with moisture than any other territory south of the Equator. If the country was flat, and its supply of an ordinary character, the imperfectly watered Salinas and Pampas would be rainless deserts.

There is no other continent covered by the S.E. trades, where, for the reason stated, a desert can exist. The desert of Peru is partly owing to the cause mentioned, and in part also, to the barrier which the Andes present against the passage of either the upper or lower trades over it.

Coming now to the belt of N.E. trades, we find under it extensive tracts of rainless desert. The central zone of rains, and the north polar zone of rains, connect together, as represented on the diagram, on Southeastern Asia, as they do upon Southeastern North America. Commencing now at the western edge of that connection on Northwestern Hindostan, and going west, we find an unbroken rainless desert reaching through Beloochistan, Afghanistan, Arabia, and Egypt, and across northern Africa to the Atlantic. The north polar zone of rains reaches down, for a few weeks in winter, on to the northern part of Egypt, and upon the Barbary States, but the central portion of that entire and extensive desert has no rain, except now and then an occasional brief shower. Except upon Southeastern Asia and Southeastern North America, the central zone of rains does not often extend north of 20°, nor the polar zone of rains south of 25°.

Coming to this continent, the belt of deserts is broken on the eastern part, by the extension of the central belt of rain up to 30°, but the line of desert appears again on Northern Mexico, and in the valleys of the Gila and Colorado. Thus you may understand that there is a general law of the atmospheric system which produces the great central deserts of the earth, and that it is the law which regulates the *transits* of the zones of rains.

There is another important fact to be noticed here, bearing upon future inquiry. In Western Asia, westward of its system of Pacific conditions, and northward of the belt of trade-wind desert which I have described, there is an extensive tract of country, including Beloochistan, Afghanistan, Persia, the Punjaub, Northwestern India, Toorkistan and the northwestern part of the Chinese Empire and Desert of Gobi, that receives but little rain, and is comparatively infertile, and very much of it desert. To what is it owing? Directly south lies the Arabian Sea, and over that for 20° or more, in summer, the central belt of rain is situated. Why should not the equatorial current in its assumed uprising and overflowing, flow from *that* over on to this vast arid country? And if not in summer then why not in winter, when that central belt lies over the Indian Ocean, an extent of more than 50°? There are the facts, what is the explanation? We shall find it in the law of the system which carries up the central belt to meet the polar zone of rain upon Southeastern Asia, and establishes limits for the paths of the conditions, which no "heating of the land" ever disturbs or affects.

Again, upon the southwestern portion of this continent we have a similar state of things. Between the Atlantic and Pacific systems of conditions we have the arid areas before described, of Southern California, Northern Mexico, the Deserts of the Gila and Colorado and the Stakey plain, and Utah, while upon the south of them lie the Gulf of Mexico, the Caribbean Sea, and the Pacific, on which the equatorial belt of rains rests in summer and from whose "*overflow*," if there was such, as assumed by Sir John Tyndall and Prof. Henry, they too might be watered. Again, over all central and eastern Hindostan, even where the central belt of rains passes up over it in the spring, it precipitates but very little, but on the western coast and over the Ghauts the rain falls in large quantities, while east of them the country is dry. But when the central belt returns from its transit to the north, in the fall, it supplies central and eastern Hindostan abundantly with rain. Why, over that great country should it precipitate during its transit to the north, on the west coast and not on the east coast and in the interior, if precipitation depends simply on overflow? And why, in its transit to the south, should it precipitate on the east coast and in the interior and not on the west coast? The questions cannot be answered by alleging a vortex and overflow, but there is a satisfactory explanation for all these apparent anomalies which will appear hereafter. I cannot now give it without anticipating.

There is another matter which it may be well to glance at, as a matter of interest, rather than as furnishing important evidence in our inquiry, viz. : The manner in which all the great rivers in Africa and South America originate in the track of the central belt of rains. Thus, the immense river system of South America is fed mainly by that central belt. On the south is a system which has its trunk in the La Plata, and whose branches are the Uruguay, Paraguay, the Parana, and its affluents the Salado the Vermejo, and the Pilcomayo, and all rise to the northward, supplied by the central belt of rains. So is the vast system of the Amazon supplied by the same belt. The same is true of the Orinoco and its tributaries, and of all Northern South America.

Turning to Africa, we find all its great rivers rising in the centre of the continent, fed by the central belt of rains, and running northward or southward to the sea. Thus the Niger on the west, rising in the western part of Soudan, on the north side of mountains, curves northward and eastward to Timbuctoo, and thence running to the southward, receives the Benuwe and other small affluents, and enters the Gulf of Guinea. So the Nile, rising near the equator, fed by the central belt of rains, runs north through more than 30° of latitude, and much of the way through a sterile country, emptying into the Mediterranean Sea. And so of the Zambesi, and the Kingami, which flow southeasterly into the Indian Ocean. There are a few inconsiderable streams entering the Atlantic on the northwest, and the Mediterranean on the north, which derive their water from the north polar zone of rains. And there is one considerable stream, the Orange river, which drains extreme South Africa of water supplied by the south polar zone of rains, during its extension to the north.

There is still another feature of this general system which it is important to notice. In its transit north and south over South America, Africa, and Hindostan, there are central portions over which it passes twice each year. When it is at the north of such a place the winds are southerly, and when south of it they are northerly. This change in the direction of wind at different seasons of the year gave rise originally to the term Monsoons, but it is obvious that they are not a distinct system of winds. They are simply trade-winds. There is a different class of winds, which I have represented upon the diagrams to which I have applied the term monsoons, although it is not very descriptive of them. I allude to a class of winds—eddies, perhaps they may be called—which blow in under the central belt of rains, on the western coasts of the continents and islands where that coast is elevated, but not too elevated. Such winds are found on the west coast of Hindostan, blowing from the S.W. in upon the Ghauts; and again upon Senegambia on the west coast of Africa; and still again on central America, and the northwestern coast of South America. The same thing is measurably true of Burmah and Siam, and the western coast of Madagascar. These are *deflected S.E. trades*. Of the force which deflects them, I cannot yet speak without anticipating.

Having thus taken a general view of the organization of the general system, as a connected whole, let us now attempt a *close* and *careful analysis* of its path and character,—commencing with the great permanent central condition, consisting of the central belt of rains and its wings, the trades. In this inquiry we omit all notice of the polar zones of rains and variable winds. We have already examined their character, and find that they are constituted by conditions which are originated and controlled directly or indirectly by the cause or force which organizes and sustains the central belt and its wings, the trades. Our inquiry then narrows to the organization of that belt and its trades.

One thought may perhaps already have entered your minds. On a general view, that arrangement of clouds and winds constituting the *great central condition*, seems to resemble precisely the arrangement of clouds and winds which constitute our intense a:.d distinct summer conditions or belts of showers, and we shall find, upon analyzing it, that *all its elements* are precisely like those of the condition that I have described as having passed unregarded, and unnoticed, over the heads of the assembled and associated scientists of the country, at Springfield, in August, 1859.

If now you cut out a section of the central belt and its winds, from Africa to South America, as represented upon the diagram, or make a *fac simile* of it, and slide it endwise to the northwest to the latitude of 30°, and there curve it gradually so as to give it a northeast direction at the parallel of 35°, and then as you pass it up to the northeastward, give it a sliding drift to the east, you will have the appearance and apparent character and movement of an intense belt of summer showers. And in fact they are, as we shall see, precisely alike in their essential elements. But to the evidence.

Passing from that general appearance and similarity, we find in its composition the following state of things. The trade-winds which blow into it, or are drawn into it on the north side, vary considerably in width. On the eastern side of the Atlantic, the northern limit of the central belt of rains ranges in summer from 12° to 15° N. Lat., and the northern limit of the N.E. trades ranges from 32° to 35°, both varying in different years. On the western side of the Atlantic, the northern limit of the central belt is carried much higher, curving to the northwest, passing north of the Antilles and Windward Isles. The breadth of the N.E. trade is therefore much less on the western than on the eastern side. On the eastern side of the Atlantic the breadth is therefore about 20°, and substantially uniform. On the west the breadth is much diminished, the S.E. trades being very strong, blowing across the central belt and following it up on to the southeastern portion of the United States.

The S.E. trades all round the world, are stronger and cover a larger area than the N.E. They are often found on the South Atlantic and South Pacific as low down as 37°, when the southern limit of the belt of rains is not much south of the equator. They must therefore sometimes have a breadth of more than 30°, especially when they blow across the equator up over the West India Islands even to Florida.

What is the character of these winds? I have not been in them and cannot speak from my own knowledge. But I have taken great pains to ascertain that character, by reading and inquiry. They are *substantially constant*, blowing *night* and *day*. They are *attended* by their *scud*, like the winds which blow into our summer belts of showers. These scud frequently enlarge and make dashing showers, and sometimes gusts, constituting what the sailors term " white squalls." Coalescing and gathering together in large masses, they make the cumulo-stratus and heavy thunder showers, and in larger masses still the hurricanes of the surface. From all I can learn of them, from the treatises of meteorologists and the logs of navigators, they *differ* in *no essential respect* from the *lateral storm winds* of our *distinct conditions*. What then is the character of this cloud belt? Beneath are the scud and showers and squalls of the trade-winds and over all a layer constituted of cirrus and cirro stratus, precisely as over the precipitating cumulo-strati of our belts of showers and of the extensive stratus of our less intense southeasters. To the evidence of all this I will advert hereafter.

This statement is adverse to the prevailing assumption in relation to this cloud-belt, inasmuch as it is adverse to the idea that there is an *ascending vortex there*, which creates the belt and produces the trade-winds, and we must inquire into the evidence on which it is rested and dispose of it. Whether there is a vortex in its composition or structure is a *matter of fact* and pertinent and important here.

What then is the prevailing assumption and theory? It is thus concisely stated by Sir John Tyndall in his treatise on "Heat. as a mode of motion," as late as 1863. " From the heat of the sun our winds are all derived. We live at the bottom of an aerial ocean, which is to a remarkable degree permeable to the sun's rays, and is but little disturbed by their direct action. But those rays, when they fall upon the earth, heat its surface; the air in contact with the surface shares its heat, is expanded and ascends into the upper regions of the atmosphere. Where the rays fall vertically on the earth, the heating of the surface is greatest, that is to say, between the tropics. Here aerial currents ascend and flow laterally north and south toward the poles, the heavier air of the polar regions streaming in to supply the place vacated by the light and warm air. Thus we have an incessant circulation. Yesterday I made the following experiment in the hot room of a Turkish bath. I opened wide the door and held a lighted taper in the doorway, midway between top and bottom. The flame rose straight from the taper. I placed the taper at the bottom and it was blown violently inward; I placed it at the top, it was blown violently outward. Here we had two currents or winds, sliding over each other in opposite directions. Thus, also, as regards our hemisphere, we have a current from the equator setting in toward the north and flowing in the higher regions of the atmosphere, and another flowing toward the equator, in the lower regions of the atmosphere. These are the upper and the lower Trade Winds." The same theory, in substantially similar terms, was promulgated by Prof. Henry, in his patent office compilations, and his Springfield lecture.

This is but a repetition of an assumption, and a mere assumption, made by Halley in 1686, nearly two centuries ago. The assumption seemed plausible then, for little was known of the atmosphere, there were none to question it, and it has been handed down, *like a tradition*, from that time to this; and he has been the most distinguished meteorologist who could discover, or group, or assume the most facts to sustain it, or invent the most plausible assumption or incidental theory to reconcile inconsistent and indisputable facts as they have been constantly brought to light by the progress of knowledge. I stated, and I think demonstrated satisfactorily to the minds of all intelligent readers not committed to the theory of Halley, by essays or lectures, in the " Philosophy of the Weather," published in 1856, that the assumption was unsupported by evidence, and was a fallacy. And now after 13 years more of continued and careful investigation, with the fullest conviction of which my mind is capable, I reassert that the theory in its fundamental and auxiliary assumptions—is baseless as the fabric of a vision. THERE IS NOT A FACT IN NATURE, RIGHTLY UNDERSTOOD, WHICH IS NOT OPPOSED TO IT. And I also include the modification involved in the Espyan theory.

Now in the first place, *there is not a spot upon the land, covered at any season by the central belt of rains, which is hotter than the dry surface on either side of it, from which the trade-winds blow into it.* Let us test this.

Commencing on the west coast of Hindostan, we find from a table prepared by Kaemtz, and taken from his treatise on Meteorology, that at Anjarakandy, between 12° and 13° N. Lat., the thermometer rises during the month of April, and while the belt of rains is south of it, to a mean of 29° 8′ Centigrade, and 85° Fahrenheit. As the belt of rains moves north and covers it, it falls rapidly although the sun is vertical, until July, when it is but 25° 8′ Centigrade and 77° Fahrenheit. From this it rises again so that in November, after the belt of rains has receded to the south, it reaches 26° 9′ Centigrade and 80° Fahrenheit.

I give the table of Kaemtz, with the rain-fall, to show the situation in the respective months, of the belt of rains.

	Temp'ture Fahrenheit.	Rain in inches
January,.................................	79	00
February,................................	81	00
March,...................................	83	00
April,....................................	85	1.00
May,.....................................	83	6.00
June,....................................	79	31.00
July, ...	77	31.00
August,..................................	79	22.00
September,...............................	79	12.00
October, ...	80	6.00
November,...............................	80	2.00
December,................................	79	1.00

Fractions of degrees of temperature and inches of rain are omitted.

Between April, when the belt is not over Anjarakandy, and July, when its centre is over it, and the sun also is vertical or nearly so, there is a difference of 8 degrees. In India then, on either side of the belt of rains, the surface from which the trades are blowing, is hotter, materially, than the surface and the air under the belt of rains to which they blow.

Passing now to Northern Africa, we find from the observations of Dr. Barth, at Kukuwa, in latitude 12°, that in April and May and early June, the thermometer ranges during one half of the days above 100° while in August, when the central belt has reached up over the place, and the sun is vertical, the thermometer rose above 90° on two recorded days only. We also find that in October, after the rains had retreated to the southward, the thermometer again rose on a majority of days, and indeed upon every day but two, above 100°. The mean difference in Soudan, according to the observations of Dr. Barth, between May and October, when the sun was *not* vertical, and the earth was *not* covered by the belt of rains, and July and August, when the sun *was* vertical and the rains *were over it*, is at least 15°.

Passing now to the North Atlantic, we find that in summer, when the northern limit of the belt of rains is 14° North in August, the mean temperature of the atmosphere for 5° north of it is at least 2° higher. This is shown by the tables of Dové and so many others that it is not necessary to insert any evidence of it. At this place let me introduce a reduced map of the trade-winds prepared by Prof. Coffin, and published by the Smithsonian Institution.

It is a correct reduction, but as a representation of the trade-winds, it is imperfect. It represents their mean location for the year merely.

But it also represents an indisputable fact which the false assumption we are considering cannot explain or obscure, having, as I think, a conclusive bearing on the question in hand. It shows that from 600 miles of flat coast, adjoining the great Desert of Sahara, on which the sand is heated every day to an elevation of from 130° to 160° Fahrenheit, and where the same thermometer stands in the shade during the summer months as high as 112°, the wind draws off to the S.W., on to a surface and into a belt of rain, where the water never rises above 82° or the air above 84°.

FIG. 35.

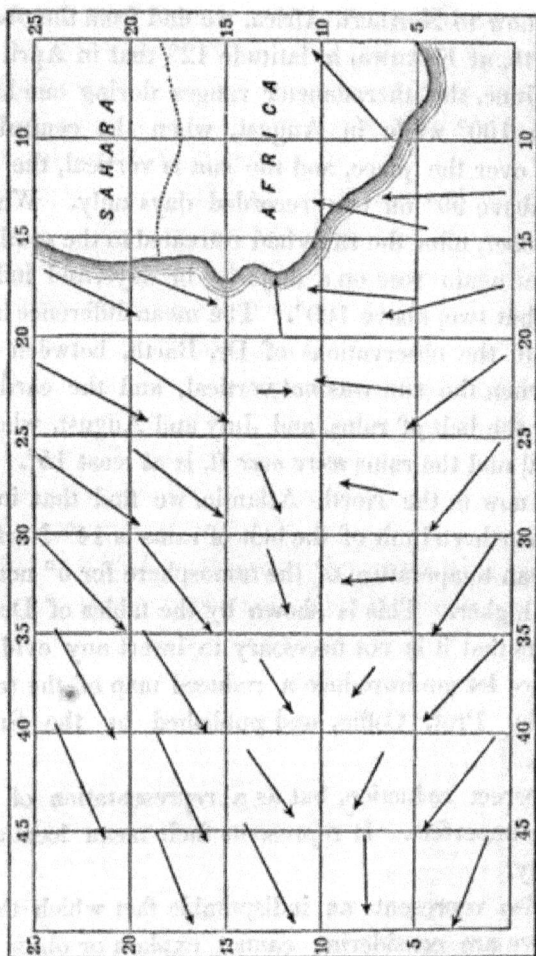

WINDS OFF THE COAST OF AFRICA

In this connection I also introduce the following tables from Prof. Coffin's " Winds of the Northern Hemisphere," showing the direction of the winds in summer along the coast of North Africa, below and opposite Sahara.

LATITUDE 0° TO 5°, LONGITUDE FROM GREENWICH 10° TO 55°.

Course.	July.	August.	Course.	July.	August.
North.	0	0	S.S.W.	54	111
N.N.E.	8	2	S.W.	1	29
N.E.	6	2	W.S.W.	6	19
E.N.E.	27	16	West.	2	9
East.	31	20	W.N.W.	1	6
E.S.E.	120	96	N.W.	1	0
S.E.	216	276	N.N.W.	0	2
S.S.E.	218	443	Calm.	8	4
South.	69	279	Total,	768	1,314

Here it is evident from the foregoing table that the S.E. trades are the prevailing winds, but their course is variable at that point south of the rains.

Ascending to the region between 5° and 10° north latitude, and 10° to 55° west longitude, the northern part of which at this season is covered by the rains, we find by the following table the monsoon, the S., S.S.W., and S.W. winds, the prevailing ones in August, although the winds are variable, as usual under the rainy belt.

Course.	July.	August.	Course.	July.	August.
North.	19	6	S.S W.	188	368
N.N.E.	26	11	S.W.	63	94
N.E.	104	32	W.S.W.	73	93
E.N.E.	30	16	West.	33	48
East.	45	29	W.N.W.	30	18
E.S.E.	36	40	N.W.	21	9
S.E.	93	53	N.N.W.	17	13
S.S.E.	225	307	Calm.	109	74
South.	239	514	Total,	1,351	1,725

Ascending to the region of 10° to 15° north latitude, and 15° to 45° west longitude, still along the coast, we find the winds, as shown in the following table, exceedingly variable, and the monsoons diminished remarkably. If Professor Coffin's theory was correct, they should increase as they approach the desert; but they in fact diminish, and the N.E. trades are found at the north portion. The peculiar variable character of the trade, under the rains on the west coasts of Africa and South America, has been much commented on by seamen. They call the location "*the doldrums.*'

Course.	July.	August.	Course.	July.	August.
North.	17	55	S S.W.	30	71
N.N.E.	64	74	S.W.	33	63
N.E.	155	149	W.S.W.	19	43
E.N E.	91	71	West.	12	25
East.	83	60	W.N.W.	17	21
E.S.E.	25	26	N.W.	13	24
S.E.	17	26	N.N.W.	24	56
S.S.E.	13	33	Calm.	62	78
South.	9	44	Total,	684	919

Ascending to the region between 15° and 20° north latitude, and 15° to 45° west longitude, we get north of the belt of rains, *and lose the monsoons entirely although still below the desert ;* and find the regular N.E. trades, with less variable winds than are found in almost any other part of the ocean, as shown by the next table.

Course.	July.	August.	Course.	July.	August.
North.	39	20	S.S.W.	0	5
N.N.E.	210	185	S.W.	0	5
N.E.	112	87	W.S.W.	8	3
E.N.E.	114	104	West.	0	1
East.	20	36	W.N.W.	0	4
E.S.E.	21	17	N.W.	3	4
S.E.	0	2	N.N.W.	3	31
S.S.E.	2	11	Calm.	20	8
South.	5	1	Total,	557	526

Ascending still further to the region between 20° and 25° north latitude, and 15° and 45° west longitude, which borders in part on the S.W. corner of the desert, and we have not, during the month of August, a single wind between S.S.E. and W.N.W. which blows in upon the land; and *only twelve instances out of three hundred and ninety-four in this hottest month in the year, and on the southern portion of the desert, when the wind blows on shore from any quarter.* This is demonstration. The monsoon winds are confined to the rainy belt; they do not reach the desert, nor does the desert attract the winds from the ocean, or reverse, hold back, or disturb the trades. They blow uninterruptedly from land heated up to 130° every day, on to water which is never heated above 82°.

Course.	July.	August.	Course.	July.	August
North.	25	20	S.S.W.	3	0
N N.E.	210	153	S.W.	2	0
N.E.	129	77	W.S.W.	13	0
E.N.E.	110	86	West.	0	0
East.	8	20	W.N.W.	0	3
E.S.E.	4	11	N.W.	2	1
S.E.	0	3	N.N.W.	5	8
S.S.E.	1	7	Calm.	2	5
South.	1	0	Total,	515	394

Passing on to Northern South America, we find abundant evidence of the fact. Humboldt has given us a graphic description of the heat of the earth and air during the dry season upon the plains of the Orinoco, when the trade-winds are blowing over it and into the belt of rains visibly approaching from the south.

" When, beneath the vertical rays of the bright and cloudless sun of the tropics, the parched sward crumbles into dust, then the indurated soil cracks and bursts, as if rent asunder by some mighty earthquake. The hot and dusty earth forms a cloudy vail, which shrouds the heavens from view, and increases the stifling oppression of the atmosphere; while the east wind (*i. e.* trade-wind), when it blows over the long heated soil, instead of cooling, adds to the burning glow.

" Gradually, too, the pools of water, which had been protected
from evaporation by the now seared foliage of the fan-palm, disappear. As in the icy north, animals become torpid from cold,
so here the crocodile and the boa-constrictor lie wrapped in unbroken sleep, deeply buried in the dried soil. Everywhere the
drouth announces death, yet everywhere the thirsty wanderer is
deluded by the phantom of a moving, undulating, watery surface,
created by the deceptive play of the reflected rays of light (the
mirage.) A narrow stratum separates the ground from the distant palm-trees, which seem to hover aloft, owing to the contact
of currents of air having different degrees of heat, and therefore
of density. Shrouded in dark clouds of dust, and tortured by
hunger and burning thirst, oxen and horses scour the plain, the
one bellowing dismally, the other with outstretched necks snuffing
the wind, in the endeavor to detect, by the moisture in the air,
the vicinity of some pool of water not yet wholly evaporated.

" Even if the burning heat of day be succeeded by the cool
freshness of the night, here always of equal length, the wearied
ox and horse enjoy no repose. Huge bats now attack the animals
during sleep, and vampire-like suck their blood ; or, fastening on
their backs, raise festering wounds, in which mosquitoes, hippo-
bosces, and a host of other stinging insects, burrow and nestle.
Such is the miserable existence of these poor animals, when the
heat of the sun has absorbed the waters from the surface of the
earth.

" When, after a long drouth, the genial season of rain arrives,
the scene suddenly changes. The deep azure of the hitherto
cloudless sky assumes a lighter hue. Scarcely can the dark space
in the constellation of the Southern Cross be distinguished at
night. The mild phosphorescence of the Magellanic clouds fades
away. Even the vertical stars of the constellations Aquila and
Ophiuchus, shine with a flickering and less planetary light. Like
some distant mountain, a single cloud is seen rising perpendicularly
on the southern horizon. Misty vapors collect and gradually
overspread the heavens, while distant thunder proclaims the approach of the vivifying rain. Scarcely is the surface of the earth

moistened, before the teeming steppe becomes covered with Killingiæ, with the many-panicled Paspalum, and a variety of grasses. Excited by the power of light, the herbaceous Mimosa unfolds its dormant, drooping leaves, hailing, as it were, the rising sun in chorus with the matin song of the birds, and the opening flowers of aquatics. Horses and oxen, buoyant with life and enjoyment, roam over and crop the plains. The luxuriant grass hides the beautiful and spotted jaguar, who, lurking in safe concealment, and carefully measuring the extent of the leap, darts, like the Asiatic tiger, with a cat-like bound on his passing prey."

Such is Humboldt's description of the dry season on the Orinoco, and the return of the belt of rains from the south.

There is other and abundant evidence that over all Northern South America, upon the mountains and upon the plains, varying in degree according to the elevation, in the month of April, before the arrival of the belt of rains from the south, the temperature is higher than it is after they have arrived, and that it increases again while the belt moves north of them in summer, or where it continues and returns again in the fall. And so of May in the West Indies.

And now from these and other data, it may be safely asserted, that from every spot upon the circumference of the globe the trade-winds blow, on the north of the central belt, during its northern transit, from extended surfaces where the earth and water and air are not less than 5°, and in some places 40° hotter than under that belt as it lies to the south of them.

Turning now to the south side of the belt, upon South America, we have the evidence of Lieut. Gibbon, who was sent in connection with Lieut. Herndon to explore the sources of the Amazon. Lieut. Gibbon reached Cochabamba in December, Lat. 18° S., and remained until May. The central belt of rains reached Cochabamba in the middle of January.

In December, before the arrival of the belt, the thermometer rose frequently at midday to 80°, and after its arrival, and while it remained, it did not rise except in two or three instances above 70°. When the belt moved to the north again in February and

March, the thermometer rose again to 80° at 3 P.M. Making allowance for the altitude of the sun, and the altitude of Cochabamba among the mountains, the difference in temperature, both before the arrival and departure of the rains, was at least 10°; and if Cochabamba had been situated at the level of the plains of Orinoco, it would have been from 20° to 30°, as Humboldt found it there.

From the log of an intelligent shipmaster, found in the wind and current charts of Lieut. Maury, I abridge the following, which will illustrate the state of things on the South Atlantic. Capt. Young, in February, found the N.E. trades at about 17° north latitude, with the water at 75° and air at 76°, trade-wind N.E.

At	12° 16′	the water was 75°		the air 76°	wind	N.E.
Feb. 22d.	9° 49′	"	"	76½°	"	77°	"	N.E.
" 23d.	7° 13′	"	"	78	"	78°	"	N.E.
" 24th.	no obs.	"	"	79½°	"	79°	"	N.E., E.S.E., rain.
" 25th.	3° 10′	"	"	81°	"	83°	"	E.S.E. rain.
" 26th.	no obs.	"	"	82°	"	82°	"	S.E. to E.S.E. hazy, rain and squalls.
" 27th.	2° 24′	"	"	82°	"	82°	"	calm with rain.
" 28th.	no obs.	"	"	82°	"	82°	"	calm rain.
March 1st.	0° 29′	"	'	82°	"	82°	"	E.S.E., sqs., rain.
" 2d.	1° 27′ S. L.	"	"	82°	"	82°	"	S E., sqs., rain
" 3d.	2° 44′	"	"	82°	"	83°	"	S.E. and S.S.E. weather settled.
" 4th.	4° 17′	"	"	82°	"	83°	"	S.S.E. and S.E., fair weather.
" 5th.	6° 08′	"	"	82°	"	84°	"	S.E., fair weather.
" 6th.	8° 08′	"	"	82°	"	84°	"	S.E and E.S.E., fair weather.

Here the air was seven degrees colder at the extreme limit of the N.E. trades than in the *center* of the belt of rains, as it is, usually, in mid-winter, but not in summer. On the other hand, *after he left the region of calms and rains*, where the water and air stood with almost entire uniformity at 82°, on the 3d of March, and for three days thereafter, during which he was in the S.E. trades with fair weather, the water was the same as under the supposed vortex, *viz.*, 82°, and *the air rose to* 83° *and* 84°. *This is demonstration*, that on the summer side of the central belt, the air is hotter than under it.

Passing east to South Africa, we have the evidence of Livingstone. He had the misfortune to lose his records in his journey to Loanda, and he does not give us tabular observations, but he furnishes other evidence which is perfectly satisfactory.

On his return from Loanda to Linyanti, he made preparations for his journey to the eastern coast, and was ready to start in September, but the natives with whom he was staying, opposed it, and of that opposition he thus speaks: "Being near the end of September, the rains were expected daily; the *clouds were collecting* and *the wind blew strongly from the east*, but it was *excessively hot.* All the Makololo urged me strongly to remain till the ground should be cooled by the rains; and as it was probable that I should get fever if I commenced my journey now, I resolved to wait. The parts of the country about 17° and 18° (S. L.) suffer from drouth, and become dusty. It is but the commencement of the humid region to the north, and partakes occasionally of the character of both the wet and dry regions. Some idea may be formed of the heat in *October* by the fact that the thermometer (protected) stood, in the shade of my wagon at 100° during the day. It rose to 110° if unprotected from the wind; at dark it showed 89°; at 10 o'clock, 80°, and then gradually sunk till sunrise, when it was 70°."

"During the whole of my stay with the Makololo, Sekelutu, supplied my wants abundantly, appointing some cows to furnish me with milk, and, when he went out to hunt, sent home orders for slaughtered oxen to be given. That the food was not given in a niggardly spirit may be inferred from the fact that, when I proposed to depart on the 20th of October, he protested against my going off in such a hot sun. "Only wait," said he, "for the first shower, and then I will let you go." This was reasonable, for the thermometer, placed upon a deal box in the sun, rose to 138°. It stood at 108° in the shade by day and 96° at sunset."

Under date of 22d October, the rains set in, and he made ready to go, and on the 3d of November departed, meeting a tremendous thunder-storm on the night of his journey. That the state of things on Southern Africa is substantially the same as upon Northern Africa is sufficiently shown by these extracts.

The only remaining land surfaces in the Southern Hemisphere from which evidence could be derived, are Australia and the Islands of the South Pacific. Of these there is little to be learned, except from the general statements of navigators. Northern and Central Australia have been but little explored, and the southern part of Australia is watered by the south polar zone of rains.

Those who desire to look further at this matter, will find much interesting evidence in the 3d chapter of Hopkins on " Atmospheric changes." I shall have room for a few extracts only.

" It is true that regular alternating sea and land breezes are found on coasts of continents and islands in many parts of the world ; but, to those who think that they are produced by sun-heated land, it must seem a singular fact, that such breezes should not be found in those parts where the causes which are supposed to produce them exist in the strongest degree. The sun must heat the northwestern part of the Sahara, or North African Desert, every day up to a very high temperature, yet no regular sea-breeze prevails on this coast. Now if sun-heated land caused the sea-breeze in the way described, how could this be? It is known also that the temperature of the land in this part sinks greatly at night, and there is, consequently, that difference of temperature between day and night which it is presumed always produces the sea and land breezes, but no such breezes are found here. The northeast trade-wind, in a gentle form, prevails during a great part of the year off this coast, and the atmosphere is clear. To use the language of Malte Brun—" The earth beneath is scorching, the sky above is on fire on the desert land, while the trade-wind near to it is comparatively cool, and yet there is no sea breeze. It may be said that the trade-wind, blowing away from the coast, is sufficiently strong to overcome the tendency of the air to flow from the sea to the hot desert. That, however, is giving up, in this instance, the theory in question."

" But the northwest coast of Africa is only one out of a number which might be named. In the neighborhood of the Arabian

Sea, land is found which in our summers is heated up to a high temperature. The southern part of Arabia is very hot, compared with the temperature of the adjoining sea, but does the air flow from that sea toward and over the heated land? It certainly does not. On the contrary, when the summer monsoon is the strongest, and the land of Arabia the hottest, the wind blows from the west, and consequently it blows from the greatly heated land towards and over the cool sea, or just in the opposite direction to what would be found if the common theory respecting the cause of sea-breezes were true."

" And to the south of Arabia, along the eastern coast of Africa, from Cape Guardafui, by the dry desert of Ajan, Zanguebar, and Mozambique, extending from 10° N. to 20° S., there is land so greatly heated as to make it some of the hottest in the world; yet the comparatively cool air of the Indian Ocean does not flow toward this land, but as on the Arabian coast, the wind blows from the heated land over the cool sea."

" If any one part of the broad expanse of the continent of Asia could be heated so as to draw air from the Arabian Sea and the Indian Ocean in summer, it would be that part which lies between Hindostan and the Lake of Aral, including the region between the valley of the Oxus and Persia, and the land of this part, un-like Hindostan, is not screened from the sun by thick vapors. But what says Burnes respecting the winds of this part? Why, that about the latter end of June, though the thermometer was at 103° in the day, ' In this country a steady wind generally blows from the north.' And on the 23d of August, after having passed the Oxus, ' The heat of the sand rose to 150°, and that of the atmosphere exceeded 100°, but the wind blew steadily, nor do I believe that it would be possible to traverse this track in summer, if it ceased to blow. The steady manner in which it comes from one direction is remarkable in this inland country.' Again, ' the air itself was not disturbed but by the usual north wind that blows steadily in this desert,' and he has many other similar passages."

" Indeed, from Madeira, by the Canary Islands, along the coast to Cape Blanco, and the Cape Verd Islands, either calms or dry winds from the desert are generally met with. The harmattan, or land wind of the desert, is sometimes encountered, and it is this wind which takes the fine dust so far out to sea. The whole of the coast is remarkable for the want of water, in that particular resembling the south coast of Arabia and the eastern coast of tropical Africa, and in none of those parts does the wind blow from the cool sea to the heated land."

With these extracts, and with the assurance that during nearly fifty years attention to this subject, and the devotion to it, as an incidental speciality, of very much of my miscellaneous reading, I have not been able to discover within the zone in question a single spot upon the land adjoining the central belt of rain upon the summer side of it, where the earth and the air are not hotter than under the belt toward which the air is moving, nor any upon the water, except where a cold, antarctic polar current is interposed, as upon the western coast of South America. Allusion is often made to the cold ocean wind which blows in upon the western coast of California, particularly at San Francisco from about 11 A.M. to 5 P.M., from the W. or W.N.W , and the cause assigned, is the rising of the heated air in the valleys of the San Joaquin and Sacramento which lie east of it. The entire valley east of the Coast Range is unquestionably hot, but the cold ocean wind that blows in over the western coast *never reaches it.* Mr. George Bartlett, in the " American Journal of Science and Arts," for September, 1856, thus speaks of it:

" The most wonderful phenomenon of the California climates is the marked manner in which they are *cut in two* by no higher chain of mountains than the coast range. This range extends along the coast of California from latitude $34\frac{1}{2}°$ to $41\frac{1}{2}°$ and is so low that snow collects during the winter only on a few of the highest peaks. Now while the western side of this range has the cold summer above described, the valley on the east side is one of the hottest portions of the earth. This valley, through which flows in opposite directions the waters of the Sacramento and the San Joaquin, extends about 400 miles from north to south, with

an average breadth of perhaps 60 miles from the Coast Range
on the west to the Sierra Nevada on the east. It is a very flat
valley, much more level than the western prairies, and occupies
the great portion of the interior of California. It has been quite
difficult to obtain exposures of a thermometer which were unob-
jectionable. In the cloth tents and stores which were in use in
1849 and '50, the temperature would range in the warm days
from 115° to 120°. On the north side of a large tree, also in a
wooden cabin covered with earth, a friend of the writer observed
the mercury at 110° and 112° during many of the days of 1850.
On the north side of a large two-story frame house, with but one
other house near, and that one several rods distant, the writer has
observed the mercury at 109°. But Dr. Haille of Marysville,
by hanging his thermometer in a draught of air, in the back part
of his office, where it was shaded by high buildings around, succeeded in keeping the mercury down to 102° during the summer
of 1852. The sun rises clear in the east, rolls up over the heads
of the inhabitants, drying and scorching everything in sight, and
sinks into the west. " One unclouded blaze of living light."
And this is repeated day after day, and month after month. *The
hottest time of day is about 5½ in the afternoon.* The nights are
cool; you need two or three blankets to sleep comfortably even in
the hottest part of the summer. A plate of butter, set in a common wooden house, will be perfectly liquid at night, and entirely
hard in the morning, and these changes will occur every 24 hours
for months in succession.

" The change from the cold climate of the coast, to the heat of
the valley is marvellous. You go on board a steamboat at San
Francisco at 4 o'clock, in the afternoon, and find the passengers
all dressed in winter clothing, flannels and overcoats, huddled
around the stove in the cabin with its hot anthracite fire. The
next morning at sunrise, you find yourself going up the Sacramento river, and as your state-room is insufferably hot, you put on
the thinnest summer clothing, and go out on the guards of the boat,
oppressed with the heat, and the prespiration starting from your
pores."

Mr. Hittell, whose work on the " Resources of California" is just published, says, that at Stockton, which is nearly east of San Francisco, the thermometer may be found every noon at 100° nearly every summer, and that the county assessor of Fresno County stated in his annual report for 1857, that the mean temperature at Millerton, during the summer months was 106°. To this temperature of Millerton, I have already referred.

It is clear that the valley of the San Joaquin is hot enough to create wind of any degree of force which could ever be created by sun-heated land, and equally clear that it creates no wind at all, for the heat goes on increasing every day, until 5 o'clock, according to Mr. Bartlett, unaffected by the cold coast wind which has then been blowing 6 hours at the rate of from 12 to 20 miles per hour at San Francisco, within 40 miles of it. *Wind created by suction begins at the place of suction*, but where wind is not known, suction must necessarily be presumed to be absent.

Before I leave the consideration of this peculiar coast wind, I want to call your attention to the fact, that there is combined upon the west side of the coast range of California, all the elements required by closet-theorists to get up a daily rain; there is a cold ocean wind, saturated with moisture, rushing up the sides of the mountains from 2,000 to 3,000 feet high, at the rate of from 12 to 20 miles an hour, but not producing rain, an observable vortex, or even a cloud. Why one or the other is not produced, as theory requires should be, has not been, and cannot be satisfactorily explained by them. What the true explanation of this ocean wind is, will appear hereafter.

I have thus shown conclusively, that the theory of a *vortex* in the central belt of rains, producing the trade-winds, is not supported by the claim that the air and water are there materially hotter, but is disproved by the fact that everywhere they are materially colder. Such evidence will be sufficient with professional and practical men, to dispose of any theory in relation to any matter of knowledge or science, but it has been and will be otherwise, with the class of men who claim to be meteorologists. The false assumption of Halley, made at the time as only an assumption, supported since by *assumptions* and unsupported by any *facts* then or since known, has been so long received without question, and so many other theories to which such men have committed themselves, in lectures and essays, are dependent upon it, and they so persistently ignore and repudiate all facts or discoveries inconsistent with it, that it seems a duty to go farther and show, that rightly understood, there is not a *fact in nature* that is consistent with it. I proceed then in the second place, to examine the question whether *heat*, as a mechanical agent, operating upon portions of the atmosphere, *can* produce such effects.

Sir John Tyndall found that the hot air of a Turkish bath, when the door of the room is opened, will flow out near the top of the door with sufficient force to flare a taper, and that the cold air flows in with a like force at the bottom. How hot the room was he does not say; the general practice is to heat such rooms up to about the temperature of 150° or 160°. Neither of the currents blew out the candle; now a candle cannot be carried in the open air without being blown out, where the wind is blowing at the rate of 12 miles an hour. The force of the current must then have been considerably less than that, and judging from experiment, it was not probably half of it. Again, the air of this room was confined, and heated air not only tends to expand horizontally as well as perpendicularly, but it is subject to the law of diffusion, and if it had not been confined by the room, would have been influenced by that law of lateral diffusion, and the tendency to rise perpendicularly would have been lessened. The experiment of Sir John then showed the concentrated action of heat as a mode of motion, upon the atmosphere, which might have been attended with very different results if made in the open air. All the most violent hurricanes, in which it is conceded that the wind must blow 80 or 90 miles an hour, originate where the temperature of the air and water do not exceed 84°. Now if Sir John could not get motion enough by heat, as a mode of motion, when applied to air in a *confined* room, until its temperature was raised to 150°, to blow out a taper, how perfectly absurd to claim that by heat as a mode of motion, raising the atmosphere to a temperature of 84°, an uprising suction is created, which produces a current of 90 miles per hour, whose effects are so awfully destructive? The experiment of Sir John with the Turkish bath would not have varied materially if his door had opened through the floor above, instead of at the side. Suppose now we state it as a mathematical proposition. Assuming that the current that creates the vacuum, and the current that fills it are of the same velocity, and assuming that heated air at a temperature of 150° will rise at the rate of 10 miles an hour, what must be its temperature to give it an ascensive velocity of 90 miles per hour? Answer, 84°. And just such a problem as that is being taught every day in the colleges of the country. But let me suggest a few experiments. Many of you have furnaces in your houses. Take now some occasion when there is a strong fire in your furnace and when the air outside is still, and the wind does not blow into the cold-air box, to shut all the registers except the one most directly over the furnace. Hold a thermometer in the current, until satisfied that the air in the hot air chamber beneath has attained a temperature of 130°, which is a greater temperature than is ever given to the earth's surface by the heat of the sun at midday, except upon some arid, rainless desert. Now fold three double sheet newspapers, weighing 5 ounces, and lay lightly upon the register, and you will find there is not ascensive force enough in the heated air of 130° to raise them. Lift them up and hold them one foot above the register, then let them drop, and they will fall, notwithstanding the air is rising with nearly the same rapidity as if there were no current. Then take two papers and you will find they fall in the same way, though not as rapidly, and substantially prevent the ascent of the heated air. Even a single paper will not be blown away from over the register, with the air at 130°, although it may be moved somewhat. At 84° and even at 100°, the ascensive force of a column of heated and confined air, 10 inches in diameter, and 8 feet high, will not exceed 1 ounce. In the open air, where the law of diffusion can operate, it is very much less.

Or, if you have not a furnace in the house, try the experiment in another way. Nail together four boards of any convenient length, say 7 or 8 feet, so as to make a tight, square box one foot in diameter, in the clear, and raise it perpendicularly a convenient distance, say 10 or 12 inches from the ground or floor, place in it a thermometer with an angular tube, if you have one, or adopt any other method of ascertaining the temperature of the atmosphere within the box, cover it with a piece of pasteboard of ordinary thickness or weighing not more than two ounces. Then put inside the box, resting upon the floor, two or three kerosene lamps, which will heat the air within the box, and see what temperature is required to raise the pasteboard from the top. Aid the experiment if you will, by opening the windows and letting the cold air in, or try it by any other means, and you will find that the ascensive force of a column of heated air 7 or 8 feet long, and 1 foot square, does not exceed 2 ounces. Then reduce the temperature to 84°, the temperature which is said to produce hurricanes, and which is higher by 14° than that which existed before and during the Cincinnati hurricane, or than that which existed where some of the most violent tornadoes have originated, and you will find that the ascensive force of a column of air 7 feet long and 1 foot square, does not exceed $\frac{1}{4}$ of an ounce, when the outside air is at zero.

There is a strong draught undoubtedly, up a chimney, where there is a strong fire, and the air is heated to a great heat, but the strongest draught that ever was created by mere heat, without a blower, in chimney or furnace stack, can be stopped by a less weight than that of an inch pine board, on the top. I have passed round a burning block in a city, in which there was a large piano-factory, containing a very large amount of seasoned lumber, and the whole interior part, and much of the rest of the block on fire; the air was rising, full of sparks, heated to a temperature of 300 or 400 degrees, at a velocity not exceeding 20 miles an hour, and the inward draught in the most favorable spots would not have extinguished a taper. And there was an obvious settling down of the air *adjoining the vortex* to supply the place of that which was ascending, clearly visible by the eddies of smoke, so that no wind was created.

The air which ascends from the chimney of a farmer's hard wood fire, over which the tea-kettle and pot are boiling, goes up into the cold winter air, after leaving the chimney, (though charged with vapor and smoke and heat,) but a little way before it is diffused into the atmosphere. It has little ascensive force or velocity. A moderate breeze will turn it almost at right angles.

The smoke of the most extensive conflagrations which occur in times of drouth, has never been carried above the surface story of the atmosphere, but it is diffused and expanded near the surface, and floats near the earth to the leeward, sometimes hundreds of miles. When the railroad was first built upon Long Island, the locomotive frequently set the combustible material upon its scrub oak plains on fire, occasioning extensive conflagrations, in times of drouth. At such times I have seen across the level surface that intervenes, and at a distance of 15 to 20 miles, and, in a still time, for several successive days, a column of smoke rising from the extensive fire, gradually diffusing, dissipating, and becoming invisible. I never saw such a column of smoke, or *any* column from *any* fire, however extensive, rising above 1500 feet; I doubt whether there ever was a column which had ascensive force enough to carry it a greater distance, however extensive the fire or great the heat. By the time it reaches that elevation, its ascensive force is overcome by the laws of diffusion and expansion. Nor have I seen a stove heated hot enough to hoist off one of the revolving toys made of paper, and which are hung on a point above it by their weight merely.

In the early history of ballooning, the Montgolfiers attempted it with balloons filled with confined heated air. They succeeded, by building a balloon 35 feet in diameter, containing 23,000 cubic feet of air, heated by the burning of chopped straw and wool to a high temperature, in raising a weight of 500 pounds. To what degree of heat the air in the balloon was raised, to enable it to raise 500 pounds, I do not know, but it must have been far above any created by the sun in the atmosphere, in any part of the earth, and yet the ascensive force was but $\frac{1}{4}$ of an ounce to the foot. Their second experiment was a success, and they attained the elevation of 3,000 feet, with a balloon 74 feet high and 48 feet in diameter, but *by keeping the air of the balloon heated to a high temperature by constant fire at its mouth.* Balloons are still made and sent up, as matter of amusement, filled with heated air, and provided with a lamp at the mouth, to keep the air heated. But attempts to carry any considerable weight into the atmosphere, by means of balloons filled with heated air, ended with the Montgolfiers, and no man now would think of acquiring for a balloon or anything else, any considerable ascensive force or power, with the temperature known to exist in the central belt of rains, or anywhere else upon the earth where hurricanes or storms of any description originate, without a *very cold* surrounding atmosphere. The rule for rarified air balloons, as given by Mr. Wise, is one cubic foot of air heated *one hundred degrees above the surrounding air*, to attain an ascending power of *one-third of an ounce !* but I give it in his own words.

" In making small fire balloons, we must not calculate on more than one-third of an ounce of ascending power, for every cubic foot of capacity in the machine. Thus, if we look at the table of capacities and dimensions of balloons, we find a three feet diameter one to contain fourteen cubic feet. Hence one of this size, with highly rarified air would raise about five ounces, and hence if the balloon and sponge weigh more than five ounces, it *will not rise*, at least not by raising the temperature of air in it *one hundred degrees above its surrounding atmosphere.*"

And now by way of contrast to this infinitesimal power, let me copy a description of a West India hurricane which occurred at Barbadoes, 10th August, 1831, as described by Gen. Reid. It was violent, but not the most violent of its class.

" At 7 P.M. the sky was clear and the air calm ; tranquility reigned until a little after 9, when the wind blew again from the north ; distant lightning was observed at half past 10 in the N.N.E. and N.W. Squalls of wind and rain from the N.N.E. with intermediate calms succeeded each other until midnight ; the thermometer fell to 83° Fahrenheit, and during the calms it rose as high as 86° ; after midnight the continued flashing of the light-

ning was awfully grand, and a gale blew fiercely from the north and
north-east; but at 1 A.M. on August 11th, the tempestuous rage
of the wind increased, the storm which at one time blew from
the N.E., suddenly shifted from that quarter and burst from the
N.W. and intermediate points. The upper regions were from
this time illuminated by incessant lightning, but the quivering
sheet of blaze was surpassed in brilliancy by the darts of electric
fire, which were exploded in every direction. At a little after 2,
the astounding roar of the hurricane, which rushed from the
N.N.W. and N.W., cannot be described by language. Lieut. Col.
Nickle, commanding the 36th regiment, who had sought protec-
tion by getting under an arch of a lower window, outside his house,
did not hear the roof and upper story fall, and was only assured
this had occurred by the dust caused by the falling ruins. About
3 the wind occasionally abated, but intervening gusts proceeded
from the S.W., the W. and W.N.W., with accumulated fury.

"The lightning also having ceased for a few moments only at
a time, the blackness in which the town was enveloped, was inex-
pressibly awful; fiery meteors were presently seen falling from
the heavens; one in particular, of a globular form, and a deep,
red hue, was observed to descend perpendicularly from a vast
height; it evidently fell by its specific gravity and was not shot
or propelled by any extraneous force. On approaching the earth
with accelerated motion, it assumed a dazzling whiteness and an
elongated form, and, dashing to the ground, in Beckwith Square,
it splashed around in the same manner as melted wax would have
done, and was instantly extinct. In shape and size it appeared
much like a common barrel-shade; its brilliancy and the spatter-
ing of its particles on meeting the earth, gave it the resemblance
of a body of quicksilver of equal bulk. A few minutes after the
appearance of this phenomenon, the deafening noise of the wind
sank to a solemn murmur, or more correctly speaking, a distant
roar, and the lightning which from midnight had flashed and dart-
ed forkedly, with few and but momentary intermissions, now for
a space of nearly half a minute, played frightfully between the
clouds and the earth, with novel and surprising action. The vast

bo ly of vapor appeared to touch the houses, and issued downward flaming blazes, which were nimbly returned from the earth upward.

"The moment after this singular alternation of lightning, the hurricane again burst from the western points, with violence prodigious beyond description, hurling before it thousands of missiles, the fragments of every unsheltered structure of human art. The strongest houses were caused to vibrate to their foundations, and the surface of the very earth trembled, as the destroyer raged over it. No thunder was at any time distinctly heard. The horrible roar and yelling of the wind, the noise of the ocean, whose frightful waves threatened the town with destruction of all that the other elements might spare, the clattering of tiles, the falling of roofs and walls, and the combination of a thousand other sounds, formed a hideous and appalling din. No adequate idea of the sensations which then distracted and confounded the faculties, can possibly be conveyed to those who were distant from the scene of terror.

" After 5 o'clock, the storm, now and then for a few moments abating, made clearly audible the falling of tiles and building materials, which by the last gust had probably been carried to a lofty height. At 6 A.M. the wind was at S., and at 7, S.E., at 8, E.S.E., and at 9, there was again clear weather.

As soon as dawn rendered outward objects visible, the writer proceeded, but with difficulty to the wharf. The rain, at the time, was driven with such force, as to injure the skin, and was so thick as to prevent a view of any object much beyond the head of the pier. The prospect was majestic beyond description—the gigantic waves rolling onward, seemed as if they would defy all obstruction, yet as they broke over the careenage they seemed to be lost, the surface of it being entirely covered with floating wrecks of every description. It was an undulating body of lumber, shingles, staves, barrels, trusses of hay, and every kind of merchandise of a buoyant nature. Two vessels only were afloat within the pier, but numbers could be seen which had been capsized or thrown on their beam-ends in shallow water.

8

"On reaching the summit of the cathedral tower, to whichever point of the compass the eye was directed, a grand but distressing picture of ruins presented itself; the whole face of the country was laid waste, no sign of vegetation was apparent, except here and there small patches of a sickly green. The surface of the ground appeared as if fire had run through the land, scorching and burning up the productions of the earth; the few remaining trees stripped of their boughs and foliage, wore a cold and wintery aspect, and the numerous seats in the environs of Bridgetown, formerly concealed amid thick groves, were now exposed and in ruins. From the direction in which the cocoanut and other trees were prostrated next to the earth, the first that fell must have been blown down by a N.N.E. wind; but far the greater number were rooted up, by the blast from the N.W."

And here your minds will naturally recur to the hurricane which I have described, from the account furnished me by Major Lachlan, which originated in the valley of the Ohio, in May, when the temperature of the air was but 78°, and as the sky was overcast for some hours previously, the temperature of the earth must have been much less. And there, as in the central belt, there was no contrast of temperature, *absolutely* NONE, between the assumed vortex and the surrounding air. These examples might be multiplied indefinitely. It is incomprehensible to me how any mind which is *honest with itself*, can continue to assert that such effects are attributable to uprising air, moved by the mechanical effect of heat at 84° and 78° Fahrenheit, *with a surrounding atmosphere of the same temperature.*

And in the third place, I assert that all attempts to prove directly, or analogically, that there is such a vortex, have been utter failures. No man has ever discovered one. In 1834, Prof. Espy advanced the theory that an ascensive force was given to the air, in which the vapor of water was condensed, by the escape of the latent heat of the vapor. This theory assumes, first, that the vapor of the atmosphere exists in the form of diffused steam. That is a fallacy. Vapor exists in the atmosphere in a state of combination with one or more of its elements. Cold will

condense it into mist, but not into cloud. Electricity, passed
through air saturated with vapor, will produce cloud. Human
ingenuity may produce water and mist by cold, but a vesicular
cloud, reflecting light, has been produced by no agent but electricity. Second, the heat assumed by Prof. Espy to be liberated by
condensation has never been found in a cloud or its vicinity, although in *numberless* instances the clouds have been penetrated
by aeronauts, and have surrounded and enveloped travelers upon
mountains. In the third place, no cloud, however dense, or however suddenly formed, has ever been seen to rise above the story
in which it formed or to which it belonged. I have watched them
ever since the theory of Mr. Espy was started, and have never
been able to discover it in a single instance. It is within the
observation of every man, that the densest fogs form suddenly in
the morning hours of night, to the height of one or two hundred
feet from the earth, and *no heat* is evolved and *no ascent* takes
place. The high fog forms at the same hour, visibly—with a
density and depth sufficient to obscure the sun as perfectly as
any layer of cloud can do, and there it remains without being elevated by any "tremendous power," which the Espyan theory and
Prof. Henry assume must be developed in it. So of the scud
which form and float in the surface wind; you may see them
forming and en'arging till they occupy very considerable space,
but you discover no ascent. And so of an immense cumulo-stratus
towering high as a thunder-head, its base settling down into the
first story, and sometimes near the earth, black, dense, dropping
rain in sheets, but never ascending. Prof. Henry amused his
Springfield audience by a large picture of a cumulus thunder-
cloud, with arrows representing an upward current in the cloud.
The picture was borrowed from Prof. Espy. The Prof. thought
he had established the fact that there was an upward current in
such a cloud, but he did not attempt to tell his hearers *why* the
whole cloud did not go up. The theory he was advocating, assumes that the latent heat is given out as the result of condensation. It must then be given out wherever there *is* condensation,
and therefore in every part of the cloud and according to the com-

pleteness of the condensation. Why then is not the entire dense, black mass shot up by the "*tremendous power*," which it is claimed must then and there be developed? Why does not the hurricane-cloud, which in its incipiency is but an extensive thunder shower, and which forms in the very focus of the central belt, shoot up, impelled by its own developed tremendous power, aided by the suction of the assumed vortex above, also acting with the same supposed "tremendous power?" How does it get away from the vortex and its own ascensive force developed by condensation? To these questions, and many others of equal pertinency which might be put, no answer, consistent with the theory, ever has been or ever can be given. We SEE and we KNOW, or MAY SEE and KNOW, that clouds, whether in dense masses or in extensive, deep strata, are never affected by any such tremendous power, and that the assumption is false. Mr. Redfield was a constant, careful, and close observer, and he always admitted as the result of his observation—although the theory of a vortex would have favored his views, that all considerable storms were overlaid by an extensive field of stratus cloud. The following is one of his admissions, and I cite it again here:

" In nearly all great storms which are accompanied with rain, there appear two distinct classes of clouds, one of which, comprising the storm scuds in the active portion of the gale, has already been noticed. Above this is an extended stratum of stratus cloud, which is found moving with the general or local current of the lower atmosphere which overlies the storm. It covers not only the area of rain, but often extends greatly beyond this limit, over a part of the dry portion of the storm, partly in a broken or detached state. This stratus cloud is often concealed from view by the nimbus and scud clouds, in the rainy portion of the storm, but by careful observations may be sufficiently noticed to determine the general uniformity of its specific course, and approximately, its general elevation.".

Mr. Espy was a laborious and pains-taking investigator, and did good service by his investigations, although he could not establish his theory. He well knew that the scud which run in the surface wind of a storm, must ascend the vortex, if there was one, with the wind in which they float, and by their means the existence of the vortex could be discovered. He doubtless watched for such vortex and found it not. If one had ever been discovered, we should have heard of it. No man, living or dead, has ever seen or discovered an existing fact indicating the truth of the theory. Mr. Wise, the aeronaut, on one occasion, found an upward current in a thunder cloud, but there was a downward current in the same cloud, and *the cloud did not rise.* He, and all other aeronauts, have found the strata of clouds moving horizontally in their places, and have passed through and above them.

Again, in the fourth place, it is settled by experiment at the Peak of Teneriffe, that the N.E. trade-wind at that point, upon the Atlantic, is 9,000 feet deep, when the north line of the central belt is 14° to the south. The depth of the current cannot be supposed to be less, and may be more, when it reaches that belt. Indeed, if it blows obliquely upward, as is claimed, it must be much more. And this vast current, it is claimed, is created and continued by suction, and enters an atmospheric throat in common with a still larger current from the south, and both ascend into the upper regions of the atmosphere. This is undoubtedly a "tremendous" operation, requiring "tremendous power," but what power, equally tremendous holds up the superincumbent atmosphere *at the throat* of the vortex, and to the extremities of this current, and above it *all the way*, while it can thus be created and sustained by the suction of a vortex? Why does not the air rush in at the sides of the vortex, or press down at its throat, or upon some of the thousands of miles where the influence of this suction is assumed to be felt? Can such a phenomenon be imitated by man in water or in air? *I answer*, NO, *emphatically*, NO—it never has been done and it never can be done. It is impossible and incredible. Moreover, the throat is not there, nor is the towering column of the ascending trades there. The whole theory is a false assumption and a "tremendous" DELUSION; and of this there is still other evidence, equally and perfectly satisfactory.

I come now, in the fifth place, to another species of evidence, that of actual and positive observation. It has long been my wish to pass through this central region, and observe for myself the phenomena we are examining, but it has not been convenient to do so. In the summer of 1867, a neighbor and personal friend, Dr. Samuel Lynes, took passage for his health in the Pacific Mail ship, steamer "China," from New York, and went down through the South Atlantic, and the Straits of Magellan, and up the Pacific to Panama, as I have already stated. At my request, he observed carefully the character of this great central condition, and on his return gave me the following concise abridgement of his recorded observations. I was particularly desirous that he should observe the character of the *trade-winds* and *their scud*, and on that point the record is very clear. To parts of his extended record I have referred in another place, and shall hereafter refer to others.

" Struck the N.E. trade-winds July 7, in Latitude 27°, N. Longitude 57° W. They blow steady day and night, a little fresher at times, seemingly they freshened as the moon rose. Scud flying almost constantly, sometimes so thickly that one might call it cloudy. A good deal of the scud resembles that flying in *our northwest winds* after a storm. It flies mostly not very high. Occasionally there would be very slight showers from the passing scud. Temperature very uniform, about 80° day and night. Temperature of the water about the same. Struck the belt of tropical rains July 14th, in Lat. 10° N., Long. 45° W. Capt. Bradbury estimated them about 5 or 6 degrees wide. He says they are very narrow on the coast of South America, but wider as they approach the coast of Africa. It rained heavily in frequent showers for two days, wi h warm and light breezes. Temperature about 82°. We ran through this belt and struck the S.E. trades July 16th, in Lat. 4° 10' N., Long. 41° W. These S.E. trades differ not at all from the N.E. in the phenomena exhibited by them. They blow equally day and night, at times a little more freshly than others. Scud constantly flying of the same character and appearance as that of the N.E. trades.

"Temperature of course, as we got down lower, grew less than it had been in the N E. trades, this being the winter season here now. We ran out of the S.E. trades about July 26th, in Lat. 25° S., Long. 45° W. After getting into the south polar zone of rains, it corresponded with your diagram, that is, we did not get much or heavy rains until about Lat. 32° 30' S., then the rains were often and heavy, with frequent lightning and thunder, until we reached Lat. 41° S. From thence to Cape Virgins and through the Straits of Magellan, the weather was unpleasant with frequent rain and snow squalls, and a more of less overcast condition of the sky. Thermometer was at the lowest in the Straits at 30°. The prevailing winds about the end of South America are northwest and southwest, and on leaving the Straits at the west end, N.W. winds are very frequent and strong. We struck the S.E. trades on the west coast of South America, August 15th, at about Lat. 33° S., Long. 77° W. The characteristics of the S.E. trades in the Pacific very much resemble those of the Atlantic. I saw no special difference. They continued steady day and night from light to fresh, rather fresher at night. Scud flying constantly, and of the same appearance as on the east side. We ran out of the S.E. trades in about Lat. 3° S., and struck the belt of S.W. monsoons with frequent showers, which followed us to Panama, Lat. 9° N. We had not much rain till we were in Lat. 3° N. Temperature of air and water about 80°. The S.E. trades were very cool all the way up, ranging at no time above 65°."

The foregoing statement conforms to the record kept by Dr. Lynes and that of Capt. Bradbury, and is undoubtedly a true statement. It shows the trades to be *continual*, and that fact calls for a *continual vortex* and *uninterrupted rain* to answer the requirements of the Halley and Espyan theories. There is abundant other evidence that the rain of the central belt is not continuous, but in showers, mainly in the afternoon and night. Let us look at this also as we proceed.

Commencing with Nicaragua, in Central America, which lies near the northern limit of the central belt in summer, we have the evidence of Mr. Squier, in his work upon that country, published in 1863. There, according to Mr. Squier, the belt arrives from the south, and the rainy season commences in May. From page 30th I take the following extract: "The year is divided, rather anomalously to the stranger, into two seasons, the wet and the dry; the first of which is called Winter and the latter Summer. The wet season commences in May, and lasts until November, during which time, but usually near the commencement or close, rains of some days duration are of occasional occurrence, and showers are common, but do not often happen except late in the afternoon, (commencing about 4 o'clock,) or in the night."

All the rains described by Mr. Squier in his book were from showers. These showers occur occasionally on the east side of the Coast Range, and in the valley of the San Juan, at all seasons of the year, when the scud of the N.E. trades are enlarged or coalesce in passing over the mountains. All the rains described by Mr. Squier were from thunder-showers formed and moving in the N.E. trades. A few extracts will illustrate this. Speaking of a night on Lake Nicaragua, he says:

"The night was wonderfully still, we could distinctly hear the tinkling of guitars at the fort, at least three miles distant, interrupted by bursts of gay laughter, until a late hour. Before I slunk under the chopa, however, clouds began to gather in the N.E., lighted up momentarily by flashes of lightning, while fitful gusts of wind veering in every quarter, betokened the approach of a thunder storm. A little past midnight we were all roused in a summary manner, by a dash of water full in our faces, followed the next instant by the lurching of the boat, which tumbled passengers, arms, books, and whatever was moveable, all in a heap together." (page 123.)

"It was just sunset when we entered the streets of Chichigalpa. A heavy thunder-storm was piling up its black volumes behind the volcanoes in the east, and the calm and silence which precede the tempest rested upon the plain. The winds were still, and the leaves hung motionless on the trees. It came down from the mountains with the majesty of an ocean, poured along their trembling sides." (page 351.)

He also speaks of the passage of the showers on either side of him and says:

" All that night, thunder-storms, like invading columns, swept over the lake around us, but we fell in the course of none of them. They all seemed to linger against the high volcanes, on the opposite (southwestern) shores of the lake, as if they would level, in their wrath, the daring rocks which opposed their progress."

Again, in speaking of a thunder-storm which was coming on from the northeast, he says:

" The wind had sprung up and carried the impending storm to the southward."

Mr. Squier gives no register, but these extracts will suffice to show you the manner in which the rains fall, under the central belt of Nicaragua.

I turn now to some intelligent, particular, and instructive observations by Mr. A. Fendler, in about the same latitude, at Colonia Tovar, in Caracas, a province of Venezuela. Colonia Tovar is situated among the mountains, at an elevation of 6,500 feet above the sea, in Lat. $10° 26'$, about 30 miles west of Laguayra, and 12 miles south of the Caribbean Sea. The observations of Mr. Fendler are remarkably particular, and are tabulated with great care. They may be found in the Smithsonian report for 1857. The rainy season commences there as in Nicaragua, in May, but continues longer, until December. All their rain is in showers. Mr. Fendler, during the rainy season of 1857, noted down the time of beginning and ending of these showers each day. Nothing like long-continued rain appears in his tables. There were sometimes three or four successive showers of short duration in a day. As illustrating the manner in which these tropical rains fall, I copy some of his tables. I know of none more perfect or instructive.

Look as a sample, and a fair one, at a page of the record for the first part of August, the middle of the rainy season, and observe the beginning and ending of the rain. Remember this was under

DATE	RAIN		CLOUDS 7 A.M.				CLOUDS 2 P.M.				CLOUDS 9 P.M.			
1857 Aug.	Time of beginning	Time of ending	Amount	Course	Velocity	Kinds	Amount	Course	Velocity	Kinds	Amount	Course	Velocity	Kinds
1	2.45 p.m.	4.5 p.m.	9	S.E.	2		7				7			
	4.30 p.m.	6.35 p.m.										E.	1	Mist.
2	12.45 p.m.	1.30 p.m.	3	E.	1	Low	6	S.E.	2		10			
3	4 p.m.	5.35 p.m.	8	E.	2		10		2		10	N.E.	1	
4	1.45 p.m.	4.25 p.m.	9	S.E.	1		10	E.–From W.	Top.	{ of lower clouds. }	5	E.–Stationary	1	
	6.45 p.m.	8.30 p.m.						From E.	Base.					
	10.30 a.m.	10.50 a.m.		E.			10	E.S.E.	2		9			
5	6.15 a.m.	7.10 a.m.	10		1									
	9.20 a.m.	12.15 p.m.												
	1.30 p.m.	1.45 p.m.												
	2.5 p.m.	4.15 p.m.												
6	5.10 p.m.	5.20 p.m.	1		1		7	S.W.–S.	2½ / 2	high middle	10			
7	5.35 p.m.	6.15 p.m.	9	E.S.E.	1		10	S.E.	2		9			
	11.45 a.m.	6.5 p.m.												
8	7.50 p.m.	8.40 p.m.	9	E.S.E.	2		10	E.S.E.	2		10	S.S.E.	2	
	8.30 a.m.	10.00 a.m.										E.	2	
	12.10 p.m.	12.30 p.m.												
	1.15 p.m.	1.30 p.m.												
	2.30 p.m.	3.00 p.m.												
9	5 p.m.	8.30 p.m.	10	E.S.E.	1	Fog	10	E.	2	Fog	0			
	12.60 p.m.	3.30 p.m.												
10	4.35 p.m.	4.40 p.m.	10	E.	2		10	E.S.E.	2		10		2	
11			8	S.E.	2		6	E.S.E.	1		3		2	
12	3.30 p.m.	5.50 p.m.	6	E.	2½ / 1½		9	N.E.&S.E.	2 / 2		10			
	6.45 p.m.	9.00 p.m.		N.E.										
13	12.30 p.m.	2.30 p.m.	9	S.S.E.	2		10	S.E.	2		10			
	7.50 p.m.	8.5 p.m.												

the center of the belt, 6,500 feet above the level of the sea, at the *very spot* and *level* where the continuous rain should be, and the vortex discoverable, and neither are discovered. The rains are all in short showers, and 22 out of 26 in the afternoon.

Now look at a table of the *course* of the clouds during all the months that the central belt of rains was over him. The clouds, for aught that he appears to have discovered, *moved horizontally.* It could not have been so, if there had been any vortex in the belt. THAT, *you* and *I* very well know, whatever professed meteorologists may say about it. Nor can *we* "wink so hard" as not to see that the southeast trade was blowing during the seven months, as a prevalent wind, just as it does everywhere, and with no more variable winds. In those tables, and in extracts which I shall hereafter make, you have the evidence of an intelligent, acute observer "on the spot," as we say, *up in the belt,* and there *all the time while the belt was passing.* Prof. Henry may doubt his word, or think his statement unworthy of notice. He did disregard them in his Patent Office compilations, but you and I, reader, must credit and regard them, especially when we find them corroborated by *all other practical observers.* And if regarded they are conclusive.

I next refer to the observations of Lieut. Herndon, in the valley of the Amazon. Lieuts. Herndon and Gibbon were ordered to explore the Amazon and its tributaries in 1850, from Lima, Peru, to Para, at the mouth of the Amazon. At Tarma, on the Andes, Lieut. Gibbon separated from the party and turned south to Cochabamba, and from thence down the Madeira river to the Amazon, while Lieut. Herndon struck its head-waters and descended directly to the sea. There is much that is instructive in his narrative. The rainy belt passed twice over him. He appears to have had an unusual acquaintance with meteorology, and his register seems to have been kept with great care and particularity, and to be reliable. It commences at Pilcomayo, 10,500 feet above the sea, in July, 1851, and continues until his arrival at Para, at the mouth of the Amazon, in May, 1852. Occasional showers were experienced as he descended the mountains, in July

COURSE OF THE CLOUDS AT COLONIA TOVAR

Month 1857.	E.N.E. High	E.N.E. Middle	E.N.E. Low	N.E. High	N.E. Middle	N.E. Low	N.N.E. High	N.N.E. Middle	N.N.E. Low	N. High	N. Middle	N. Low	N.N.W.	N.W. High	N.W. Middle	N.W. Low	W.N.W.	W. High	W. Middle	W. Low	W.S.W.	S.W. Middle	S.W. Low	S.S.W. High	S.S.W. Middle	S.S.W. Low	S. High	S. Middle	S. Low	S.S.E. High	S.S.E. Middle	S.S.E. Low	S.E. High	S.E. Middle	S.E. Low	E.S.E. High	E.S.E. Middle	E.S.E. Low	E. High	E. Middle	E. Low
June		1		1	1						1							1						1				14	1		6		18	1	1	10	5		1	4	14
July		2		1	4	6		1		1	1	1						4				1		1	2			2 8	5		7 5	6	1	2	2	2	12		5	11	
August		1			2					2	2							1				1	2		1		1	5 8 1			1 1 7		19			3			11	1	
September		1	2		1					2 1	1 1				2	1		5				1	1					14			5		17			1	6		1	5	
October												11		1	1			1				1	2					25			2			12			1		12		
November		1 1		2 1	4	7		2				12		1				8	1				1					1	18		3			13			1		2 1	7	
December		3			1			2										8				1						10									2		11	3	
Total	3	9		8	20	2	1	2		6	30	2		2	1	1		28	1		2	9	1	2	2	3	4	97	8	2	36	6	1	98	9	1	30	22	1	34	66
Sum		12			30			3			38				4				29		2	10			5			109			44			108			53			101	

N. to E. 83

South to N.N.W. 50

E. to S. inclusive 415

and August, while the rain belt was at the north, and he was upon the river Hualaga, and the upper waters of the Amazon.

In the vicinity of Nauta, on the Amazon, in September, 1851, the belt reached him in its descent to the south. On the 9th of that month, he writes respecting the rainy season as follows:

" We have had a great deal of cloudy weather and rain since we have been upon the Amazon, and it is now near the commencement of the rainy season at this place. No one suffers from heat, though this is probably the hottest season of the year; the air is loaded with moisture, and heavy squalls of wind and rain sweep over the country almost every day. In the dry months, from the last of February to the first of September, a constant and heavy breeze blows, nearly all day, against the stream of the river; the wind, at all seasons, is generally easterly, but is, at this time, more fitful and liable to interruption; so that sail-boats bound up, make, at this season, the longest passages."

From Nauta, he diverged to the south, up the Ucayuli to Sarayacu, which he reached October 18th. Of that he says: "We had rain nearly every day that we were there, but it was in passing showers alternating with the hot sun."

General expressions of travelers are not so reliable as a record carefully and intelligibly kept from day to day. The meteorological record of Lieut. Herndon is of that character, and the only reliable record of the kind ever kept in Central South America and the valley of the Amazon, and published, to my knowledge.

The record furnishes conclusive evidence that upon Central South America, and in the broad valley of the Amazon, while the Central belt is passing over it to the south, nothing like continuous rain for days is experienced. Rain even through a single day, is recorded but once. Nearly all the rain which falls from it, falls in showers during the afternoon or night. Except in a very few instances, cirrus and cirro-stratus clouds were observed in the morning, cumulus and cumulo-stratus clouds in the afternoon and night. Sometimes, but very rarely, the rain of the night continued into the forenoon of the next day.

But to be more particular; from Sept. 10th, when he arrived at Nauta, until January 1st, 1852, making 112 days, there were recorded nine mornings as clear, and of these, one only occurred in October, and none between the middle of October and the middle of November, when the center of the belt was over him. During the same period, there were thirteen mornings when there was rain, and in seven of those, the rain was a continuation of that which commenced in the night and ceased early in the forenoon. During the whole period of 112 days, there were but six days in which rain *commenced* in the forenoon.

From this it clearly appears that as a rule, under the central belt of rains upon the Amazon as at Colonia Tovar, the rain from that belt falls in heavy showers of short continuance, occurring in the afternoon or night.

I refer in the next place to the record of Lieut. Gibbon. Passing southeast from Tarma, lat. 11° 25′ S., across the ridges of the Andes, he descended to Lake Titicaca, and from thence to Cochabamba, in lat. 18° S., where he arrived in December. At the commencement of his journey he thus speaks of the weather:

" Our course is to the eastward. The snow-capped mountains are in sight to the west. Temperature of a spring, 48°; air, 44°. Lightning flashes all around us; as the wind whirls from N.E. to S.W., rain and snow-flakes become hail, half the size of peas. Thunder roars and echoes through the mountains; the mules hang their heads and travel slowly; the thinly clad aboriginal walks shivering, as he drives the train ahead; the dark cumulus cloud seems to wrap itself around us." (July 18th, 1851.)

As he passed down the mountains in a S.E. course, he struck the S.E. trades, and thus speaks of them:

" The southeast winds that we meet here, come across the South Atlantic ocean; passing over the lowlands, they strike against these mountains. Rising from the vapors of the sea, they are wet, but after traveling over dry lands, their dampness is distributed on the soil, and there springs up a growth of forest trees and wild flowers, which otherwise would be burnt down by the fiery rays of the sun. By the time the winds reach these lofty mountains, they are comparatively dry. The little dampness remaining in them, meeting with the cold atmosphere of the mountain peaks, freezes and falls in the shape of snow or hail."

Lieut. Gibbon arrived at Cochabamba, on the 13th of December, and remained until May. The central belt of rains did not reach that place until after he did, and then the rain was not continuous, but mainly in thunder-showers. His register is imperfect in not giving the time of commencement, continuance, or ending of the showers, but there were many days without them, or any rain. He remained at Cochabamba until May, when he traveled northward down the Madeira river to the Amazon. Of his journey to the North he thus speaks:

"Our route from Tarma to Oruro was south. We traveled ahead of the sun. In December when we arrived in Cochabamba, the sun had just passed us. As soon as he did so, the rains descended heavily on this side of the ridge; it was impossible to proceed. The roads were flooded, the ravines impassable, and the arrieras put off their journey until the dry season had commenced. After the sun passed the zenith of Cochabamba, and had fairly moved the rain-belt after him toward the north, then we came out from under shelter, and are now walking behind the rain-belt in dry weather, while the inhabitants are actively employed in tending their crops."

I turn now to the continent of Africa, and the observations of the traveler, Livingstone. I alluded (page 141) to the fact that when he was about to leave Linyanti, lat. 18° S., he was urged by the chief of the Makololo, to wait for the arrival of the rainy season, before commencing his journey across the continent eastward to the Indian Ocean. But he had previously twice passed through the central belt of rains, during his journey from Linyanti to Londa, on the west coast in 1853. The meteorological record which he kept during that journey was lost, but a few extracts from his narrative, will sufficiently illustrate the point we are considering. On his way up the Zambesi in November, before the arrival of the central belt of rains from the north, and when at Gonye Falls, in lat. 16° 40' he thus speaks of the dry season:

. "November 30th, 1853, at Gonye Falls. No rain has fallen here, so it is excessively hot. The trees have put on their gayest dress, and many flowers adorn the landscape, yet the heat makes all the leaves droop at midday, and look languid for want of rain. If the country increases as much in beauty in front as it has done within the last four degrees of latitude, it will be indeed a lovely land. We all felt great lassitude in traveling. The atmosphere is oppressive both in cloud and sunshine. The evaporation from the river must be excessively great, and I feel as if the fluids of the system joined in the general motion of watery vapor upward, as enormous quantities of water must be drank to supply its place."

On the eastern sides of the Andes we have seen that showers occasionally occur during the dry season. The same is true of Africa under the Equator, where the country is elevated, as Du Chaillu experienced during his second journey. But upon the flat plains of Southern Africa, where Livingstone was then traveling, showers never occur in the dry season.

But Livingstone was traveling to the north up the Zambesi, and the belt of rains was descending to the south, and early in December they met at Naliele, in lat. 15° 24'.

"The rains began while we were at Naliele; this is much later than usual; but though the Barotse valley has been in need of rain, the people never lack abundance of food. The showers are refreshing, but the air feels hot and close, the thermometer, however, in a cool hut, stands only at 84°. The access of the external air to any spot at once raises its temperature above 90°. A new attack of fever here caused excessive languor; but as I am already getting tired of quoting my fevers, and never liked to read travels myself where much was said about the illnesses of the traveler, I shall henceforth endeavor to say little about them."

From that time he speaks of heavy rains and showers, mostly in the afternoon and night, until his arrival at Londa, but nowhere of continued and unintermitted rains. On the 16th of January, he arrived at Cabompo, lat. 12° 37', and remained until the 24th. As the rains had been moving to the south, and he to the north, passing each other, he was now near the northern edge of the belt,

and of course had the winds from the north, and he thus speaks
of the rain and wind.

"One cannot get away quickly from these chiefs; they like to
have the honor of strangers residing in their villages. Here we
had an additional cause of delay in frequent rains ; twenty-four
hours never elapsed without heavy showers; everything is affected by the dampness; surgical instruments become all rusty ; clothing mildewed and shoes mouldy; my little tent was now so rotten
and so full of small holes, that every smart shower caused a fine
mist to descend on my blanket, and made me fain to cover the
head with it. Heavy dews lay on everything in the morning, even
inside the tent; there is only a short time of sunshine in the afternoon, and even that is so interrupted by thunder-showers that
we cannot dry our bedding. The winds coming from the north
always bring heavy clouds and rain; in the south the only heavy
rains noticed are those which come from the N.E. or E. The
thermometer falls as low as 72°, when there is no sunshine, though
when the weather is fair, the protected thermometer generally
rises as high as 82°, even in the mornings and evenings."

Livingstone arrived upon the western coast at about 9° S. lat.,
in May. The central belt of rains had passed him again, and
gone to the north. In December following, he again left for Linyanti, and again the belt of rains passed over him in its transit to
the south. On page 80 he thus speaks of it:

"We had now rain every day, and the sky seldom presented
that cloudless aspect and clear blue so common in the dry lands
of the south. The heavens are often overcast by large, white,
motionless masses, which stand for hours in the same position and
the intervening spaces are filled with a milk-and-water-looking
haze."

He does not seem to have had any distinct conception of the
nomenclature of clouds or the strata of the atmosphere, but in this
paragraph he has represented the cumuli, and the milk-and-water
looking cirrus seen through their spaces and above them. He
was at that time in the very center of the rainy belt.

That belt moved to the north again in spring, and the latter part of his journey to Linyanti was concluded during the dry season. There he waited for the rainy season again, as we have already stated, and then left for the eastern coast of the continent, passing the fourth time across the country while the belt of rains was over it. Here again, he says of the rains:

"These rains were from the east, and the clouds might be seen on the hills exactly as the table cloth on Table Mountain."

There is in the Wilkes Exploring expedition, a view of Cape-town and Table Mountain, with the cloth on. If that is a correct representation, the cloth does not differ in any material respect from the foggy patches of scud which are often seen on our high hills, in all parts of the country, forming as the moist current passes over them, and dissolving again after the particular portion passes away from the influence of the hill or mountain top.

On this journey he nowhere speaks of continuous rain for a longer period than several hours, and no material difference in the character of the belt is described. It moved to the north and left him in dry weather again, before he reached the eastern coast at Quillimaine.

In the works of Du Chaillu, describing his discoveries in equa-torial Africa, we have like descriptions of these rains. During their absence at the north, they have the same dry season in the vicinity of the equator, as elsewhere. Their dry season begins in May, and ends when the rains return, in September. During his last journey into the interior, he reached an elevated section of the country, and there, as under like circumstances in South America, occasional showers are experienced in the dry seasons of the year. There, as elsewhere upon the continent, the showers passed from the east to the west, and the rain was all in showers.

The natives called the elevated country to the east of them "*the mother of rains*," conveying the same idea that the natives of Brazil did to Prof. Agassiz, when they said "*the path of the sun is the path of the storm.*"

We may now get an idea of the rapidity with which the belt of rains moves over the continent in its transit. It reaches the equator, according to Du Chaillu, early in September. It reaches the Congo, five degrees further south, about the latter part of September, and Livingstone met it when he waited for it at Linyanti, in lat. 18° 17′ 20″, the 27th of October. But as we shall hereafter see, it varies in different years.

I refer next and again to the record of Dr. Barth, in North Africa. The Dr. left Tripoli in March, 1850, and journeying to the south, crossed the Desert of Sahara, arriving at Kukawa, near lake Tsad, in April following. This was the base of his movements and observations during that year. From this base he made various short journeys during that time, but they were all between 9° 30′ and 12° 30′ N. He kept a Meteorological Register, and we have nothing of the kind more perfect for that part of the continent. An occasional brief shower was experienced during the journey across the Desert, especially near its southern border, and in September. At Kukawa, where his most important observations were made, the rainy season commences in May, and ends in September. His description of the belt of rains is substantially like that of all the others. The substance of it, so far as it relates to our present inquiry, is contained in the following paragraphs :

" It is generally supposed that storms in the tropical climes break forth in the afternoon or in the course of the night, and this certainly is the general rule; but if there has been a storm the day before, or during the night, and the weather has not cleared up, there can be no certainty that it will not come on again in the course of the morning. It is rather a rare phenomenon in these regions for a storm to gather in the morning on a clear sky, but nevertheless the reader will find several examples even of this, in my meteorological tables."

This is in accordance with the observations of Herndon, on the Amazon—where the rain continued from the night into the morning, on seven days, and commenced in the morning, on six days out of 112, during which the belt was passing over him. The exceptions are few in both cases, and prove the rule.

" Rain was very plentiful this year, 1851, and I am sure would, if measured, have far exceeded the quantity found by Mr. Vogel

in 18 4. Indeed there were twelve very considerable falls of
rain, during the month of August alone, whica together probably
exceeded thirty inches. It must be borne in mind, moreover, that
the fall of rain in Kukawa does not constitute the rule for the
region, but is quite exceptional, owing to the entire absence of
trees and of heights in the neighborhood. Hence the statement
of Mr. Vogel in one of his letters that the line of tropical rains
only begins south of Kukawa, must be understood with some re-
serve, for if he had measured the rain in the woody country north
of that capital, between Dawerghu, and Kaliluwa, he would, in
my opinion, have attained a very different result. It is evident
that all depends upon the meaning of tropical rain. If we imply
a very copious fall of rain, Kukawa certainly does not lie within
the limits of tropical rain, but if we are to understand by it the
regularly returning annual fall of rain, produced by the ascending currents of heated air, it certainly does. There was a very
heavy fall of rain on the night of the 3d of August, which not
only swamped our courtyard, but changed my room, which lay
half a foot lower, and was protected only by a low threshold,
into a little lake, aggravating my feverish state very considerably,
and spoiling most of my things."

In the following year Dr. Barth went to Timbuctoo, which
lies to the north and west in about lat. 18°. He encountered the
rainy season again in the summer, and described it by saying :
" We had a thunder-storm almost every day, followed now and
then by a tolerable quantity of rain." Timbuctoo is upon the
Niger, and there is a curious anomaly connected with the rise of
that river. Dr. Barth thus alludes to it :

" Towards the end of January, the waters of the river reached
their highest level, exhibiting that marvelous anomaly, in comparison with the period of the rising of other African rivers north of
the equator, which is calculated to awaken astonishment in any man
acquainted with the subject. For when he knows that the rising
of these rivers is due to the fall of the tropical rains, he will naturally expect that the Niger, like its eastern branch, the Tsad,
or Benuwe, or the Nile, should reach its highest level in August
or September."

The explanation of this peculiarity is found in several facts: First, the Niger heads on the northern and eastern slopes of the mountains of Senegambia, Liberia, and Ashantee, in lat. 10° N., and from thence runs Northeasterly to Timbuctoo, through 7° or 8° of lat., and from thence easterly and southeasterly, to the Gulf of Guinea. The rains which raise the river, therefore, at Timbuctoo in January, fall several degrees to the south of it. Second, the belt of rains upon the Atlantic, though it moves up in summer as high as 15° N. lat., does not move more than 3° or 4° south of the equator in winter. Its transit upon the eastern part of the Atlantic, therefore, is scarcely 15°, and it lingers in December and even into January, upon the mountains of Senegambia and Ashantee and Western Soudan, while its transit upon the continents of Africa and America exceeds 30°. Third, when the belt of rains is moving to the north in summer it can derive little moisture from the N.E. trades, as the country north of it is an extensive, sandy desert, and the moisture supplied by the S E. trades is intercepted by the Kong mountains of Western Soudan, Guinea, and Yaraba. When, however, the belt descends to the south in the fall, and the N.E. trades blow into it from the then uncovered and *drenched surface* of Soudan, they deposit their moisture upon the northern and eastern slopes of those mountains, where the Niger has its rise, and fill its banks so as to occasion the peculiar rise of the river at Timbuctoo, in the middle of the dry season.

A like anomaly is found in India. A very moderate supply of rain falls on the Deccan, east of the Ghauts, during the transit of the belt of rains to the north, for then the countries about the head of the Bay of Bengal, on which its N.E. trades then originate, are dry, but after the belt has been up over them in summer, and descending has uncovered them in the fall, the N.E. trades blowing from them are charged with the evaporating moisture of their summer rains—and abundantly supply the Deccan with rain.

From all this evidence I think it entirely certain that nothing like that *continuous rain*, under the central belt—which a *vortex in constant action* night and day *must* produce if the theory was true—is to be found, and this fact would be amply sufficient of itself, to disprove the theory. But in addition we have the concurring evidence of all travelers, that the rains of that belt are produced by showers, "gathering," to use the expressive phrase of Dr. Barth, generally in the afternoon or night, and "sweeping," to use the equally expressive phrase of Prof. Agassiz, along horizontally near the surface of the earth, precisely as they do with us, and with a like *unperforated layer* of cirrus or cirro-stratus above them. And in addition to all this we have the observations of Fendler, Gibbon and others, up in the upper horizontal trades, even in the rainy season, and *not an observed fact, anywhere, indicating the existence of a vortex.*

Having thus demonstrated, as I think, that the theory of Halley, in respect to the existence of a vortex in the central belt, is a delu-ion, it is obvious that we must examine critically the great belt for the purpose of ascertaining its elements. In respect to them there is a mass of evidence, which to my mind is entirely sufficient to show that it is a stratified and organized body, consisting of a stratum of cirrus and sometimes of cirro-stratus above, a stratum of cumulus and cumulo-stratus in the middle, and of scud, fog and other incidental condensations, beneath and at the surface.

I have already cited the graphic description given by Humbolt of the appearance of the central belt when first discovered approaching from the south, when he was on the banks of the Orinoco, but I wish to call your attention to it again, in this connection.

"When, after a long drought, the genial season of rain arrives, the scene suddenly changes. The deep azure of the hitherto cloudless sky assumes a lighter hue. Scarcely can the dark space in the constellation of the Souther Cross be distinguished at night. The mild phosphorescence of the Magellanic clouds fades away. Even the vertical stars of the constellations Aquila, and Ophiuchus shine with a flickering and less planetary light. Like some distant mountain, a single cloud is seen rising perpendicularly on the southern horizon. Misty vapors collect and gradually overspread the heavens, while distant thunder proclaims the approach of the vivifying rain."

Here we have described the elevated stratum of misty, cirrus condensation which overlies this belt of rains, extending far to the north and south of that portion where the rains are falling, and becoming cirro-stratus nearer the center. Humboldt discovered it first, not only because of its *elevation*, but because of that *extension* which brought it first into view. It moved up and over him, dimming his vision of the stars and constellations, before the rain-cloud was seen on the southern horizon. And before the rain-clouds of the belt had approached sufficiently near for their thunder *to be heard*, that stratum of condensation had " *overspread the heavens.*" Here then we have from Humboldt a description of two strata, and two kinds of condensation as characteristic of this belt—the upper stratum of cirrus or cirro-stratus, and the cumulo-stratus, or thunder-cloud below it.

To the same effect are the observations of Fendler at Colonia Tovar, modified of course by elevation. During the month of April before the rainy season set in, when there was light rain on five days only, the sky was often entirely overcast at 7 A.M., and not a perfectly clear morning is recorded. The mean amount of cloudiness at that hour, for the month, was $6\frac{7}{10}$, while in May when the rainy season had set in, and there was rain on 25 days, the mean amount of cloudiness at 7 A.M. was $8\frac{3}{10}$, a difference of only $1\frac{6}{10}$.

I turn now to the tables of Herndon, to which I have referred, and see how perfectly the same state of things appears from his unstudied but clear and intelligent description. He entered upon the Amazon, on the 4th of September, when the belt was approaching from the north. His history of the mornings from that time forth until the belt had passed him on its way to the south, is a description of that upper layer, sometimes spoken of as cirrus, and at others as cirro-stratus, as it assumed one or the other form. On the 15th of September he has this entry. " The first perfectly clear morning we have seen since entering the Amazon. At 1 P.M. cumulus clouds; showery at $2\frac{1}{2}$." So his description of the clouds at midday and in the afternoon, is of the formation of cumuli and cumulo-strati which furnish the showers of the afternoon and night. None of these cumuli were ever seen to ascend into a vortex. They are frequently described as passing by to the southward or to the northward of him, horizontally and at the same elevation. In this connection I cite a brief but interesting description by Prof. and Mrs. Agassiz of a shower which they witnessed upon the Amazon. It may be found on page 248 of "A journey in Brazil."

"October 23d. We left Teffè on Saturday evening on board the Icamiaba, which now seems quite like a home to us ; we have passed so many pleasant hours in her comfortable quarters since we left Para. We are just on the verge of the rainy season here, and *almost every evening* during the past week has brought a *thunder-storm.* The evening before leaving Teffè, we had one of the most beautiful storms we have seen on the Amazon. It *came sweeping up from the east ;* these squalls *always come from the east,* and therefore the Indians say, "*the path* of the *sun* is the *path* of *the storm.*" The upper, lighter layer of cloud, traveling faster than the dark, lurid mass below, hung over it with its white, fleecy edge, like an avalanche of snow, just about to fall. We were all sitting at the doorstep, watching its swift approach, and Mr. Agassiz sa'd that this tropical storm was the most accurate representation of an avalanche on the upper Alps, he had ever seen."

I have taken the liberty to italicise some very suggestive portions of that description. The thunder-storm occurred in *the evening,* during the *rainy season,* as did others, previously experienced. It came from the east, *sweeping,* that is horizontally, and near the earth, as all such storms do in that country. It was in October when the belt of rains had arrived from the north, and was passing over them to the south. They remained upon the Amazon and its tributaries until March, and during that period the entire belt passed over them, and they had full opportunity to observe its phenomena. To their description of this storm they append a statement of the natives, that "the path of the sun" and the "path of the storm" there are the same, and record that statement for our instruction, uncontradicted and unqualified ; and it *imports*, that *all* the thunder-storms, *under the belt*, during its transit north and south, *sweep* in like manner, *horizontally*, from the eastward to the westward.

And they also describe two layers of cloud, the upper and lighter (cirrus or cirro stratus,) moving more rapidly than the dark cumulo-stratus below it, precisely such as are often seen as our belts of showers approach from the northwest.

I have not the honor to know these distinguished travelers, except as the world knows them, by reputation. But I believe they have not unfitted themselves by a too ready and unwise committal to the Halley theory, (as too many have done,) for the discovery and reception of truth on this subject, if inconsistent with that theory ; and if I did know them, I should ask whether they discovered *anything*, during the transit of that belt over them, which indicated that it was *constituted* by a *great ascending vortex*. Of the character of their answer I cannot entertain a doubt. *That brief description speaks with the fulness of a volume.*

I have alluded to the record of Lieut. Gibbon. From the middle of December to the middle of February, as appears from that record, there were but five clear mornings, and during the 30 days preceding the arrival of the belt of rains, not a single day when it was not cloudy at 3 P.M. The overlying stratum of cirrus and cirro-stratus must have extended nearly or quite 10° S. of the southern edge of the belt of rains. After that time, not a single clear morning is recorded. There is a single instance on the 31st of January, when the record reads, "blue sky, with hazy atmosphere," but that haze was undoubtedly the cirrus or cirro-stratus of the overlying stratum, somewhat thinner and more broken than usual.

On the west of the Andes and over the Pacific Ocean, that overlying stratum extends still farther to the southward. From the statement and record of Dr. Lynes it would seem to extend nearly to the southern limit of the S.E. trades.

Turning again to Southern Africa, we have the evidence upon

this point also, of Livingstone, but as he kept no record which is preserved, and had little knowledge of Meteorology, the evidence is meagre. It sufficiently appears, however, from his general statements that the same state of things exists there. I have already copied a description (page 169) of the two strata as described on his journey to Londa. In his journey to the eastward from Linyanti to Quillemaine, and on the 27th of November, in lat. 17°, he thus writes: "The temperature was pleasant, as the rains though not universal, had fallen in many places. *It was very cloudy, preventing observations.* The temperature at 6 A.M. was 70°, at midday, 90°, in the evening, 84°. This is very pleasant on the high lands, with but little moisture in the air."

It is obvious from the fact that he could get no observations at any time of the day, because of the cloudiness, before the belt of rain was felt in its force, and while the showers were infrequent, that the same overlying layer of cirrus or cirro-stratus had extended south in advance of the rains.

After he had passed the coast range of mountains, and while descending their slopes to the sea, he met with what was to him, a new feature. Everywhere else in his travels he found the thunder-storms and showers in the interior moving to the westward, but here they seemed to him to move to the eastward, but he was mistaken, as we can all understand. Thus, while upon the hills in January, he says: "The clouds rested upon the tops of the hills, as they came from the eastward, and then poured down plenteous showers upon the valleys below." But a little farther down, on the 29th of the month, he says, "A double tier of clouds floated quickly away to the west, and as soon as they began to come in an opposite direction, the rain poured down." The lower of the two tier of clouds was the rain-bearing cloud, and passed off to the west, with the layer of cirro-stratus above it, like that observed by Prof. Agassiz in Brazil. The clouds which deceived him were scud, lower down still, running as we often see them, in opposition to the path of the storm, especial'y when here as there the belt of showers has passed over highlands. The tendency to easterly surface wind and scud under a belt or storm which has passed over the Alleghanies, is well ascertained and understood.

This superior cloud stratum was also observed by Du Chaillu, and seriously interfered with his observations when the cumulus and cumulo-stratus were not present. This was especially true over the hills of the interior, and he thus speaks of it. "Up to the present time, May 18th, I have only twice seen the sky free from cloud since my arrival at Fernand Vaz, from England."

On this point the record of Dr. Barth, in north Africa, is very full and clear. On page 702, Vol. II., during April, 1852, and from that time on until the arrival of the rains, he records with few exceptions, the morning as "*a little overcast*," and "*thickly overcast*," and after the arrival of the belt of rains, he described the gathering of thunder-showers in the afternoon and evenings, passing sometimes to the south and sometimes to the north, and sometimes over them, but always to the westward horizontally, and generally with strong or violent winds. No thunder-storms are recorded by any writer, to my knowledge, of a more violent character.

I extract for a week from the 26th of April, 1852, to the 2d of May, inclusive. This was before the belt of rain had arrived, but when it was approaching from the south.

Date.	Hour of day.	Degree in scale of Fahr.	
1852 April 26,	1.45 p.m.	101.8	Sky thickly overcast, the sun breaking through the clouds at 9.30 A.M., the atmosphere remaining sultry. In the afternoon a thunder-storm accompanied by heavy squalls of wind, but no rain.
27,	....	....	Atmosphere sultry.
28,	....	....	In the afternoon a thunder-storm gathered but brought us only a few drops of rain in the evening.
29,	....	....	The sky the whole day overcast, in the afternoon a storm gathered in the south, but not accompanied by rain
30,	....	....	In the afternoon a thunder-storm arose, followed by a considerable rain the following night, lasting for about two hours.
May 1,	....	....	Sky overcast. the sun breaking through the clouds about 10 A.M. but only for a few moments At 4 P.M. thick thunder clouds, with much heat-lightning but no rain.
2,	....	....	About 5.30 P M. dark thunder clouds gathered but passed by westward.

There is much other evidence which might be adduced in support of the proposition, but I have not space for it, and it cannot
be necessary. Lieut. Maury was undoubtedly right, when from
the mass of nautical evidence before him, he assumed that there
was a tropical cloud ring encircling the earth. Seen from above,
this layer of cirrus and cirro-stratus does undoubtedly form a cloud
ring, wide over the continents and narrower over the oceans, broken or attenuated in places or at times, it may be, but still a *girdle*
of cloud. Underneath this cloud ring, and near the center of it
and in the trade-winds, the scud enlarge and the showers form,
from which the rains descend. And beneath those showers are
the incidental and varied slant winds, which constitute the squalls
and gales and hurricanes which are felt at the surface. These
together constitute the *elements* of the *great, permanent, central*
atmospheric condition.

In his lecture before the " American Association for the advancement of Science " in August, 1859, Prof. Henry endeavored to make a strong point upon the assumption that the trades
could not pass each other, and therefore that when they met, they
must ascend and consequently that there *must be a vortex.* But
there is nothing in either the assumption or the argument. Of
the manner in which they do meet and pass there is much and
satisfactory evidence.

In the first place we have the evidence of those who have
witnessed the manner in which they pass each other. Mr. Fendler thus saw them at Colonia Tovar. He says:

" Sometimes the eastern currents may be seen in their gradually ascending but nearly horizontal course, to meet the higher
southern currents at right angles, and without mixing, to be deflected by the latter in a horizontal semicircle, or downward or
upward as the case may be. I have also seen two opposite currents meet when each endeavored to force its antagonist back
with alternate success and failure, until one got the better over
the other, and at last kept undisputed sway."

In the second place, it is by no means certain that we do not
owe the alternate character of our condition to the obstruction of

the N.E. trades, blowing against the S.E., preventing temporarily their passage, causing them to accumulate until sufficiently strong to overcome the resistance, and then move forward in larger volume and an excited state, and that thus our storms and even the hurricanes originate.

But in the third place, there are few places where the trades can be said to *meet*. They blow obliquely in under the central belt, in most cases at right angles to each other, and it is an easy matter for one or the other to slide over or under as Fendler saw them do at Colonia Tovar. There is much reason to believe that this is true, as a rule, upon the oceans, where both are of considerable strength.

But in the fourth place it is to be considered that even upon the oceans they are rarely of equal strength. Thus, in the Atlantic Ocean, the N.E. trades are very strong upon the coast of Africa, where the ocean is kept cool by the Polar currents, but weak and almost entirely wanting, west of longitude 35°. This is especially true during the summer season, when the oceanic isothermal of 80°, is carried up to the latitude of Bermuda. If you turn back to the statement of Dr. Lynes, you will see that although they struck the latitude in which the N.E. trades should be, on the 7th of July, they found very little of that description of trade, until they made easting, and a day or two before they encountered the belt of rains. Their route was nearly a direct line from New York to Cape St. Roque, with a slight curve to the east. Sailing vessels, bound south of the Equator, make a much larger curve to the east, so as to strike the N.E. trades in their strength on the east side of the Atlantic, and cross the belt of rains further east.

We have seen that in the summer season the belt of rains *curves up* over the western Atlantic, and over the West India Islands, and on to Florida and the southern portions of the Gulf States. West of a line drawn from New York, to lat. 30° upon the Equator, the S.E. trades prevail in the summer, and the N.E. trades are occasional, weak, and unreliable. The surface is usurped by the southerly trades. Of this too, Mr. Fendler speaks with that intelligence that characterizes all his observations.

" As to the trade-winds, I found on my trip from Philadelphia
to Laguayra, that *within the tropics* we had no E.N.E. wind,
which is thought to be the regular trade-wind of those regions.
After crossing lat. $23\frac{1}{2}°$ in long. $68\frac{1}{2}°$, we were becalmed for one
day, and soon after got a fresh breeze from the south, which we
kept all the way to long. 63°. By tacking, we got to lat. 22°,
long. $63\frac{1}{2}.°$ From thence we had the wind all the time from S.S.E.
which we kept to lat $11\frac{1}{2}°$, the day before we reached Laguayra.
Capt. Wilkins, who has been in this southern trade for 18 years,
assured me that within the last 8 years, he never could depend
much upon the trade-winds. He finds that between lat. 23° and
18° the south wind frequently kept on blowing very brisk for
eight days in succession.

" On the way from the colony to Caracas, along the high ridge
of the principal mountain chain which stretches east and west
parallel with the coast, at an elevation of from 7,000 to 8,000
feet, we travel about six miles over a region bare of forests, where
we, nearly at all times, find a very strong breeze from the south
rushing up the declivity and over the ridge, hurries off to the
north, towards the ocean. The ocean can be plainly seen from
this elevation. That this great current of air does not sink down
along the northern slope, but on the contrary, is somewhat projected upward by the shape of the mountain, can be seen by the
course of the condensed vapors, which, in the form of fog and
mist, are driven along. May not this current of air sink gradually lower and lower, until it reaches about lat. 18°, where it
strikes the sea? I have found this south wind at sea, always
much colder than any of the other winds in these latitudes." And
he gives the following cut, exhibiting the passage of this upper
trade to the eye.

FIG. 36.

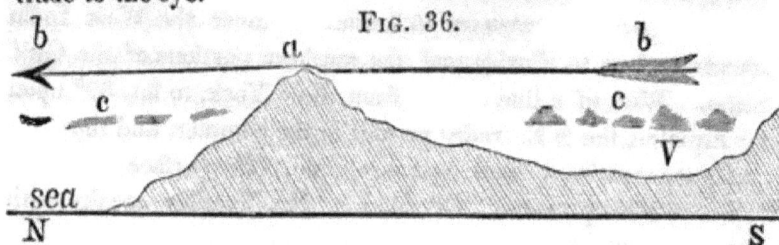

The view is a section looking east. The elevation where the upper trade is felt is at *a*, and the long arrow indicates the course of the upper trade, and the scud in the valleys the lower trade. But I will give the description in the language of Mr. Fendler himself.

"Several times I had a most excellent opportunity for observing and tracing the course of this southern current to a great distance in the direction south and north. I was then standing on the very crest of the mountains of the coast, having a view towards the north upon the sea, and towards the south over a part of the fertile valleys of Araguay. Scattered masses of clouds showed plainly by their motion, the direction of the current in a long line, whence it came and whither it went. The annexed figure may serve to give a somewhat clearer idea. It is to represent a vertical section of the territory from south to north, 'a' the place of observation, 'v' the valleys between the northern and southern ranges, 'c c' clouds moving with the eastern trade-winds towards the west, the line 'b b' the track of the high southern current, which had a velocity of about twelve miles per hour, and a somewhat sinking tendency, until it struck the northern range, where it was forced upwards for a short distance until it reached the crest, and then went on unobstructed on the other side of the mountains, in a horizontal line, apparently lowering but very little, leaving hereby the eastern trade-winds of the sea far below and undisturbed in their regular and steady course, which is nearly at right angles to that of the former. The lower clouds of the valley showed plainly a motion from east to west, as seen against the dark background of the southern mountains. The high southern current was not indicated by clouds in those places where it was vertically over the lowest parts of the valley; but when drawing nearer to the Cordilleras, on which I stood, the vapors which it contained condensed rapidly, and became visible, as drifting, incoherent clouds sweeping by, and which could still be seen on the sea side as long as they floated over the dense, primeval forest, which extends here from the mountains' tops, to the very margin of the sea."

" Here I may also remark that the great amount of cloudiness, which in some respects may be regarded as a disadvantage to observation, offers, with regard to the currents of the atmosphere, great advantages, the condensed vapors indicating the various motions and directions of these currents, and I have had, therefore, opportunities to observe them in most of their various forms. Sometimes I have seen the air ascend and descend vertically with considerable velocity, at other times pushed up the inclined planes of mountain flanks on one side until reaching the crest, and then gliding or flowing down on the other side, somewhat like a liquid, following in its course the most depressed localities and ravines in all their windings. Sometimes, the eastern currents may be seen in their gradually ascending but nearly horizontal course, to meet the higher southern current at right angles, and, without mixing, to be deflected by the latter in a horizontal semicircle, or downward or upward, as the case may be. I have also seen two opposite currents meet, when each endeavored to force its antagonist back, with alternate success and failure, until one got the better over the other, and at last kept undisputed sway."

In addition to this there is very much evidence to establish the fact that along the N.E. coast of South America, and over the eastern part of the Caribbean Sea, and the West Indies, the S.E. trades continue up in summer as a surface current 30° or more in breadth, the source of our bountiful supply of moisture, the fountain of our Atlantic system of conditions, and the field where the hurricanes of the Western Atlantic originate and operate with their greatest violence, (see charts of Lieut. Maury, and Mr. Redfield, herein before given,) and Prof. Coffin's tables of the wind of the Gulf States.

The evidence to show that the N.E. trades of the eastern Atlantic find their way to South America, covering the valley of the Amazon and pressing up the slope of the Andes, derived from the records of Herndon and Gibbon, and from other sources for which I have not space, is entirely satisfactory. There is no intrinsic difficulty in the idea that when they meet the stronger S.E. trades of the west side of the ocean, they rise and pass over them, descending again or meeting the elevated land and depositing their moisture upon the continent of South America.

If now you turn to the diagram of Prof. Coffin, on page 134, or the diagram which I have introduced on page 119, you will observe that upon the middle and eastern part of the Atlantic Ocean, south of the belt of rains, the trades become more southerly and southwesterly near the coast of Africa. The S.E. trades there are light compared with the N.E. and undoubtedly pass over them, for *they are found as an upper current from the S. W. on the Peak of Teneriffe.* Of the deflected trades, which becoming still more easterly, blow in upon the continent of Africa, upon Senegambia and the countries lying north of the Gulf of Guinea, represented upon the diagrams, I cannot yet speak without anticipating.

We see then that upon the Atlantic Ocean, there is nothing to prevent the trades from passing each other, and there is satisfactory evidence that they do indeed pass.

And so there is everywhere; Fendler describes the S.E. wind blowing over the surface trade at an elevation of from 8,000 to 10,000 feet, when the belt is south, and blowing as a surface wind when the belt was over him or at the north. Gibbon found the N.E. upper trade on the Andes, south of the belt, and struck the S.E. surface trade when he descended into the valleys at their base. Goodrich found the counter trade or return trade, as Dové calls it, on the mountains of Hawaii as a horizontal, not a descending wind, and Von Busch and others found it on the peak of Teneriffe the same.

All the phenomena of the system point to the same result. All the storms in both hemispheres, which originate in the belt of rains, and are sufficiently intense to find their way to the polar zones, move to the northwestward, until near the outer limits of the trades, when they curve to the northeastward. Instances where storms have thus issued simultaneously and moved divergently to each polar zone, are on record. (See work of Col. Reid, page 43.) *Extensive arid areas lie northwest of arid or mountainous southern continents, and wet areas northwesterly of southern oceans, and vice versa.* The Andes of South America cause the aridity of Northern Mexico, Southeastern California, and Utah, and the aridity of New Holland, causes the arid areas of Western Asia.

It has been assumed that the counter or return trade leaves the central belt as a S.W. wind, but all the reliable evidence alluded to and much more for which I have not space, proves the contrary. If it were true, the western coast of North America would be wet, and the eastern dry. So of the whole western coast of South America for the counter or upper trade of the southern hemisphere, would undoubtedly follow the same law, and move in upon that coast from the N.W. Peru, which is now dry for the want of an upper N.E. trade, because it is intercepted by the Andes, would be as wet as Chili or Patagonia. And the same would be true of Europe, of Africa, or arid Southwestern Asia, and still more arid New Holland. The theory cannot be true.

But it is in all our text-books, and I will add a few words in relation to the evidence on which it is founded. And first in relation to observations on the Peak of Tenneriffe. The existence of a S.W. wind there is explained by two facts: 1st, that the southern trades of the eastern Atlantic, are deflected, and enter the belt of rains, as S.W. trades, as we have already stated, and the S.W. wind on the Peak of Teneriffe may be a continuation of them. 2d, the upper trade commences curving to the east before it reaches the outer limit of the lower trade, and this fact invalidates the evidence derived from observations on the peak of Teneriffe and the mountains of Hawaii.

The next item of evidence on which the theory is rested, is the falling of ashes in Barbadoes, from the volcano in St. Vincent, but this evidence is not reliable. Barbadoes is 100 miles *directly east* of St. Vincent, and a S.W. wind would have carried the ashes to the N.E., and clear of it; and in the second place, "slant winds" from every point of the compass are experienced in that vicinity, produced by passing storms, sometimes too far off to be observed, yet near enough to disturb the currents of the atmosphere. These occasional slants are felt by all mariners, and one is described in the diary of Dr. Lynes from the S.W. in that vicinity, when no storm was experienced. Fendler describes these occasional slants as occurring at Colonia Tovar and in the West India Seas, and Maury's Sailing Directions are full of them. Moreover it does not appear that the ashes fell upon St. Lucia, which lies much nearer St. Vincent and N.N.E. from it. This evidence is entitled to no consideration.

A stronger point is made in relation to the ashes which were thrown out from Cosaguina and fell about Jamaica. In relation to this it may be observed, first, that they too, may have been transported by an occasional slant wind, and second, that an exceptional upper current from the southwest must, to some extent, exist in that vicinity, for the S.E. trades off the northwest coast of South America are deflected, as upon the coasts of Africa and Hindostan, and constitute there a S.W. monsoon, as represented upon the diagram. And here again, to illustrate this, I copy from the record of Dr. Lynes:

"Sunday 25. 7 A.M., ther. 64°, cloudy, cool and pleasant, as the sun shines occasionally. *Strong S.E. trades* white caps and large swell. Cape Blanco in sight at 8 A.M.; saw a brig and numerous whales to-day. *Wind in afternoon, S.S.W.* Ther. higest 70°, lat. 3° 38', long. 81° 38'. Miles 188.

Monday 26. 7 A.M., ther. 69°, cloudy, warm but pleasant; close and muggy, nearly calm and smooth sea. Passed *La Plata* island at 6 A.M., and *Cape Lorenzo* at 8½ A.M. Crossed the *Equator* at 3.30 P.M. Quite warm and *light breeze from S.W.* Ther. highest 76°, lat. 0° 27', long. 80° 58' W. Miles 196.

Tuesday 27. 7 A.M., ther. 74°, cloudy, warm and pleasant. Sun rose clear, but soon overcast. *Light wind from S.W.*, smooth sea. At 7.15 A.M., passed the English steamer Santiago from Panama, bound south. Hot this P.M., showers. *Have run out of the S.E. trades into the S.W. monsoons.* Ther. highest 80°, lat. 2° 29' N., long. 80° 16' W. Miles 180.

Wednesday 28. 7 A.M., ther. 78, cloudy, warm and frequent heavy showers. *Light S.W. winds,* close and muggy air, sea very smooth but *long S.W. swell.* Slowed engine to 5½ turns yester-

day, showers all day.   Caught a swallow lighted on ship.   Ther.
80°, lat. 5° 3′ N., long. 79° 50′.   Miles 156."

Now this S.W. monsoon, constituted by a deflection of the
S.E. trade on the N.W. coast of South America, continues in all
probability and according to all analogy as an upper trade over
Jamaica, and carried the ashes there.   The same ashes fell on
the " *Conway*," at the same time at a distance of several hundred
miles on the Pacific, carried there doubtless by the surface E.N.E.
trades.   That piece of evidence is also unreliable.

Lieut. Maury in his " Sailing Directions " and in his " Geog-
raphy of the Sea," inferred that the upper trade was a S.W. wind,
because of the deposition of dust on southwestern Europe, which
contained organic forms known to exist in South America, and
which he assumed were taken up into the upper trade by whirl-
pillars and tornadoes, and carried thence to Europe.   But dry
seasons, arid plains, tornadoes and whirl-pillars, occur in Equatorial and South Africa also.   To this effect is the evidence of
Livingstone and Du Chaillu, and for aught that appears, the same
organic forms exist there.   Moreover, the dust falls on a *curve*
which clearly indicates an African origin.   And the observations
of Fendler and others, show conclusively that the S.E. trades of
the Atlantic, except where deflected near the coast of Africa, find
their way on to our continent.

From this review I think it appears that the upper trade conforms to the direction which it had as a surface trade, until near
the outer line of the surface trade, over which it is flowing, when
it curves to the eastward, obedient to a law which is hereafter to
be considered.

There is still another important fact to which I wish to call particular attention, before closing this analysis of the great central
condition, for it has a special relation to irregularities in the conditions of our own climate.   I refer to variations in the commencement, progress, and extent of the transits of the central belt in
different years.   This matter was briefly alluded to on page 171
with the expression of an intention to resume it.   Fendler thus
speaks of it :

"The dry season commences here, generally, soon after New Years day, and lasts till the end of April. The remainder of the year is taken up by the rainy season. This is *generally* so, for there are many exceptions, and our notions about the great regularity and sharply defined seasons of the tropics, which we have received from books, are sometimes materially upset and corrected by experience. When I first came to the colony, in March, 1854, we had a dry season in its usual way. The rainy season then commenced on the 23d of April, but it did not end with the latter part of December, as is usually the case; it lasted till the end of January, and commenced again with the first of March, and then kept uniformly on till the end of December, 1855. The dry season was therefore only of one months duration instead of four. The last dry season, has been, on the contrary, unusually long, and lasted till the latter part of May."

And Du Chaillu, who travelled in Central Africa, near the Equator, says:

"The dry season this year, (1864) was an unusual one for the long absence of rain and lowness of the rivers. The negroes have a special name for a season of this sort, calling it enomo onguéro; it lasts five months, and they assure me that it always comes after a long series of dry seasons of the usual length; we have had a few showers, but they have produced no impression.

And so Livingstone when he met the descending rainy belt in December, 1853, at Naliele, on his way to Louda, in lat. 15° 24', says of it: "The rains began while we were at Naliele; this is much later than usual."

And so, when in South Africa, speaking of occasional rains in the Kalahari Desert, he says: "But the most surprising plant of the Desert is the ' Kengwe or Keme,' (Cucumis caffer,) the watermelon. In years when more than the usual quantity of rain falls, vast tracts of the country are literally covered with these melons; this was the case annually when the fall of rain was greater than it is now, and the Bakwains sent trading parties every year to the lake. It happens commonly once every ten or eleven years, and for the last three times its occurrence has coincided with an extraordinarily wet season. Then animals of every sort and name, including man, rejoice in the rich supply."

And again on page 135 he says, " Having parted with Sechele, we skirted along the Kalahari Desert, and sometimes within its borders, giving the Boers a wide berth. A larger fall of rain than usual had occurred in 1852, and that was the completion of a cycle of eleven or twelve years, at which the same phenomenon is reported to have happened on three occasions. An unusually large crop of melons had appeared in consequence."

On this point there is much other evidence, but these extracts will suffice. I enter the facts here as pertaining to the condition. Corresponding irregularities occur in the polar zones and both will be explained hereafter.

In relation to the movement of the upper stratum of cirrus and cirro-stratus, the evidence is limited. So far as the observations of Gibbon and Herndon upon the Andes, and of Livingstone and Barth in Africa, bear upon the point, they indicate that the movement is westerly, and as its movement elsewhere in the northern and southern hemispheres conforms to the path of the conditions beneath it, its movement in that direction may well be analogically assumed.

And now with a brief resumé of the principal facts, thus at length collected and arranged, I will close this analysis of the great central condition.

I. This condition consists of a central body known as the central belt of rain, and two areas or wings of wind, known as the trade winds,—the whole together, having a westerly movement.

II The trade-wind south of the body, moves from some point between east and south, towards a corresponding point between west and north, and the trade-wind north of the body, moves from some point between north and east, towards a corresponding point between south and west.

III. The central condition has an average breadth of about 50°, and a transit north and south averaging more than 25°. Nearly one-half of the central portion of the globe is therefore covered by it, at some seasons of the year.

IV. The central portion or body averaging more than five hundred miles in width, precipitates rain wherever it may be, and gives to the tropical portions of the earth, as it passes over them in its transits, their rainy seasons.

V. The areas covered by the trade-winds, while so covered, are as a rule dry, except as occasional storms or showers, issuing from the central belt, in the trade, precipitate upon them.

VI. The polar zones of rains recede before the advancing areas of trade-winds, and return after them as they retire.

VII. Surfaces which are not covered by the transits of the central belt of rains, nor the tropical extension of the polar zones of rains, continue dry through the year, and constitute the principal rainless deserts of the world.

VIII. The trade-winds blow with substantial constancy night and day, when not interrupted by passing storms, and contain scud, and both winds and scud resemble the wind and scud of approaching or passing conditions in the polar zones.

IX. On the west side of the Atlantic, and also on the west side of the Pacific Oceans, the S.E. trades exist in great strength and volume, and blow as surface winds over latitudes which would otherwise be covered by the N.E. trades, up to and connecting with the polar zones of rains. These two exceptional and remarkable volumes of trade give a large and exceptional supply of moisture and fertility, and a peculiar climatology to S.E. North America, and S.E. Asia. And in these remarkable volumes of trade, originate the intense hurricanes of the West Indies, and the typhoons of the China Sea and Bay of Bengal.

X. Where the S.E. trades originate on arid areas, like that of New Holland, corresponding areas under the north polar zone of rains, like those of southwestern Asia, are dry. And where they originate on continents that are well watered, areas under the same polar zone are less perfectly supplied with moisture than those supplied by trades which originate upon oceans. And where they originate upon continents like that of South America, or upon oceans and are met in their path by lofty mountains, corresponding areas like those of southwestern North America, and the Desert of Gobi in Asia, and Peru in South America, are found.

XI. The body of the central condition is composed, 1st of an overlying stratum of cirrus and cirro-stratus, extending in a more misty form, to a greater or less extent, out over the trades. This stratum is usually misty or fibro-cirrus in the morning, becoming dense and assuming a cirro-stratus form, as the day advances. Under this stratum the trades pass and afterwards overlie each other, and in the trades occasionally, though rarely, in the morning, but generally in the afternoon or night, thunder-showers form and furnish the rains peculiar to the belt. These showers have a westerly progression, corresponding to that of the trade in which they are formed. Beneath the trades where they meet and pass each other, it is either calm or there are squalls or gusts or slant winds, incidental to showers or storms as everywhere.

XII. The northern transit of the condition is usually completed about the first of August, and the southern transit commenced before the middle of that month. So the southern transit is usually completed about the first of February, and the northern transit commenced before the 15th. But in respect to these, and also in respect to the rapidity and extent of the transits, there are some irregularities, occasioned by a cause which could not be considered without anticipating.

XIII. There is no vortex in the central belt. The theory of Halley was originally but a mere assumption, and is not supported by any facts since discovered. Observation, analogy, and every known fact, when properly understood, are inconsistent with and adverse to it. No man, in the light of facts now known, can *believe* it, and be *honest with himself.* No man acquainted with the facts, can *teach* it, and be *honest with others.*

Having thus examined, and I hope comprehended, the *nature, extent,* and *essential features* and *elements,* of the great *fundamental, permanent,* and *controlling* CENTRAL CONDITION, we are prepared to return to the polar zones, and a further examination of them.

CHAPTER VI.

Re-statement of the fact that the normal or natural state of the polar zones, is dry, fair weather, and that they have no independent arrangement for a supply of moisture—Heat, as a mechanical agent, does not do it—It is effected by passing conditions which originate in, or result from, the fundamental, tropical condition—Some conditions, which originate in the central belt, and others which originate in the polar zones, reach the Arctic Circles, retaining all their essential elements—Copy from Dr. Kane's record during the absence of the sun, to prove it—Analysis of two recorded storms—Halley's theory requires the belief that the sun creates storms where there is no sunshine, and American Meteorologists hold and teach the absurdity—practical men, whose profession or business trains them in the logic of cause and effect, know better—Storms met with by Dr. Kane probably originated upon the Pacific—Storms originate in the polar zones, where there is a sufficient volume of upper trade—Where there is not sufficient volume, stormy conditions do not occur—Recapitulation of the proof furnished by the changes which occur in our own country—The irregularities of the central condition produce irregularities in our zone—One of the principal irregularities is drouth—Four classes of drouth attributable to such irregular action—The first class occur in spring, north of the focal path—occasioned by the retarded progress of that path to the north—most common east of the Alleghany Mountains and upon the northeastern states—Second class produced by the unusually extended transit and concentration of the focal path—This class are south of that path—such was the drouth of 1854—The third class are meridional—occur in the interior of the country—produced by the concentration of the upper trade and a succession of conditions upon the Atlantic coast—such was the drouth of 1867—The fourth class are found on the Atlantic coast only, occasioned by an extension of the outer limits of the N.E. trades—occur in July and August only—Limited and local drouths—tendency to them in certain localities—northern part of the Gulf States—south coast of New England—Configuration sometimes produces them—also the passage of successive belts of showers in the morning—Such belts if of weak intensity, do not precipitate where they are vertical in the morning—Instances of such irregular precipitation—Want of data for a thorough elucidation of drouths—a history of them by Mr. Flint, Secretary of Board of Agri-

culture, Massachusetts—Examination of that history—Are there any facts
which will guide us to a knowledge of these irregularities ?—Some of them
connected with spots upon the sun—Description of those spots and of
Schwabe's investigation—Tables and observations given by him—not perfect or satisfactory—nevertheless the connection with the irregularities of
the system apparent—Comparison of the table of spots with the transits of
the focal paths of the conditions—Effect upon the mean fall of rain—Effect
upon temperature—Extended examination of the subject, in various localities and under various circumstances—Examination of facts which seem
to militate against the connection—Properly understood, the facts are not
adverse, because consequent upon irregular transits—Extended examination of the whole subject—Connection between the spots and the irregular
action of the whole system—Inquiry whether there are any other forces
or causes which affect the system—Volcanic action—its influence probable
—Extended examination of the question—Influence of the moon upon
the weather—Popular opinion upon the subject—Such an opinion requiring careful examination—No such influence can be traced, upon such examination—No facts from which it can be analogically or philosophically
presumed.

We have in what has already appeared ample evidence of the
truth of our first proposition, that the normal or natural state of
the polar zones is dry, fair weather, and that they have no independent arrangement for producing storms, and a supply of moisture. Heat, as we have seen, as a mechanical agent, produces no
vortices or storms *on deserts*, or on fertile surfaces during their
dry seasons however hot. The distribution of moisture in the polar
zones, is effected by passing, irregular, and limited conditions, which
occasion all the incidental changes of the weather, and those conditions *originate in*, or *result from* the action of the fundamental,
central condition. That condition sends off its storms and they
traverse for days the polar zones up to and into the Arctic Circle.
Although no one has traced a storm from the central belt of rains
to the Arctic Circle, they have been traced by several American
minds, into high northern latitudes, as far as settlements or commerce have enabled them to obtain evidence, moving with diminished but still sufficient strength to reach the circle. And the
conditions appear there with every element and feature which
they possess in lower latitudes, and appear when the oceans are
locked up with ice, the surface covered with snow, and the sun

has not shone for weeks, and the temperature is between 25° and
30° below zero.   Look now upon the following record from Dr.
Kane's Register, kept during the Grinnell expedition.

JANUARY 1851, (LATITUDE ABOUT 74°, LONGITUDE ABOUT 70°).

Date.	Wind.	Force.	Ther.	Bar.	Sky and Weather.
Jan. 3	....	calm.	—26.1	29.62	blue sky, m.
" 4	W.	gent. breeze.	—21.3	29.53	blue sky, detached clouds, m.
" 5	W. by N.	gent. breeze.	— 3.9	29.59	blue sky, m. clouded over.
" 6	W. by S.	light breeze.	— 0.8	29.67	clouded over, m., snow.
" 7	W.	gent. breeze.	—14.4	29.96	blue sky, detached clouds, m.
" 8	W.S.W.	light air.	—21.2	30.14	blue sky, m.
" 29	W.N.W.	light air.	—18.9	30.19	blue sky.
" 30	N.W. by W.	light air.	—13.5	30.17	clouded over, m.
" 31	N.W. by W.	gent. breeze.	— 4.4	29.35	clouded over, snow.
Feb. 1	W.	light breeze.	—11.7	29.27	cloudy, blue sky, m.
" 2	W.	light air.	—25.1	29.62	blue sky, detached clouds, m.

Observe that on the third of January, when the sun had not
been above the horizon for 40 days, the air was calm, the sky
cloudless but misty—(that state being indicated by the letter m,)
the thermometer at 26° below zero, the barometer at 29.62—a
pleasant arctic day.   The next day a breeze sprang up from the
west, the thermometer began to rise, the barometer to fall, detached clouds to appear, whether cirrus or cirro-stratus, the record
does not state, probably the former.   The next day the breeze
continued, the thermometer rose rapidly and the sky clouded over.
During the following 24 hours, the thermometer continued to rise,
and snow fell.   During the next 24 hours, the thermometer fell,
the barometer rose rapidly, and the blue sky, with its detached
clouds (scud doubtless) appeared again.   On the next day, the
thermometer was down, the barometer up, the air nearly calm
and the sky cloudless.

Now here we have the record of a distinct, passing condition,
exhibiting every essential feature.   We have the precise condition which often occurs in our own latitude, with the same changes
of state.   The rise and fall of the thermometer from the fair day
before, to the fair day after the storm, was substantially the same
as is common with us, and all the phenomena are substantially the

same. It is rare with us, in midwinter, that the thermometer rises more than 25°, prior to, and during a midwinter snow-storm. Another storm of similar character, commenced on the 29th of the same month, differing in no respect except that the rise of the thermometer was somewhat less, and the fall of the barometer greater. It would seem too much for the intellect of the most determined and exclusive disciple of Halley to assume that such a process could be originated or continued under such circumstances, by the mere mechanical effect of heat from the rays of an *absent sun*. But it is not. The doctrine which Professors Henry and Loomis, and other American Meteorologists persistently hold and teach, includes *all* storms *everywhere*, and compels them to attribute *such effects* to *sunshine* where there *is no sunshine*. But practical reader, our professions and business have trained us in the *logic of cause and effect*, and *we know* that these storms could not have originated there from such a cause. And we also *know* that if, as is probable, they originated in lower latitudes, and travelled there, they must have passed through a cold atmosphere, and over a snow clad surface of at least 1000 miles, and *could not* have been *kept in continued activity*, by ascending currents of air, produced by sun-heated land. Nor could the distinguishing elevation of the thermometer during the storms, have been produced by any direct action of the sun. There has not been, since the serpent whispered in the ear of Eve, a grosser delusion, or one more persistently taught and sustained, without evidence and against evidence. It is impossible for an honest mind to have patience with it, and be true to its best impulses.

Before I leave these storms let me observe that they undoubtedly originated upon the Pacific Ocean. The only winds which accompanied them were from the western quarter. They had not sufficient intensity to produce a strong wind, laterally or in opposition, and in snow-storms of such weak intensity, in high latitudes, in our own country, the wind conforms substantially to the path of the storm. Doubtless it did so in those described.

But all the conditions experienced in the polar zones, do not originate within the central belt. The largest proportion of them, and some of the most violent, commence and assume form and character, at or beyond the outer limit of the surface trade. Such was the Cincinnati hurricane described. But the great central condition sends forth *the volume of upper trade* which is the *basis* of all such incidental conditions, and therefore they *result from it.* Where no sufficient volume of that upper or counter trade issues forth, or traverses the polar zones no such incidental, stormy conditions are known. It is instructive to retrace these results, (although already traced) upon our own country. Thus, in midwinter, when the central condition moves far to the south, the S.E. trades of the Atlantic are intercepted by the mountains in the southeastern portion of Brazil, and the Andean ranges of western Venezuela and Columbia, and all the states of the union, east of the Rocky Mountains, except the Gulf States east of the Mississippi, and the Atlantic states northeast of them, are comparatively dry. The incidental conditions which pass over them are occasional and feeble. As the great central condition returns to the north, and the S.E. trades originate higher upon the Atlantic, a rainy season progresses to the westward over Texas, and curves to the N.W. over western Arkansas and Missouri, and so on to the northeastward. And gradually, as the central condition moves farther north, the path of the conditions is gradually extended westward and northwestward until August and September, when, if the transit has not been interrupted, New Mexico and all the northwestern states are reached by the curving conditions, and receive their summer rains. And when the central condition descends again to the south, the path of our polar conditions is gradually contracted down again upon the southeastern states. The same thing occurs upon the Pacific Coast. The paths of their conditions, constituting a distinct system, ascending and descending with the movements of the central condition. And so it is everywhere in like circumstances, in both hemispheres, and in both polar zones. So, from the action of the central condition, all our incidental conditions, with their attendant phenomena, constituting the weather result. And irregularities in the movements and operations of the central condition, produce corresponding irregularities in the polar zones. Let us look at those irregularities as they occur in this country.

One of the most noticeable and important irregularities to which we are subject, is that of Drouth. When these are extensive, *unseasonable* and long continued, they may prima facie be attributed to irregularities in the action of the central condition ; for as our fall of rain at any given point depends mainly upon the *volume* of counter or upper trade *passing over it*, any irregularity in the central condition, which interrupts the usual supply of *that trade* at *that point*, must necessarily produce a drouth there.

There are four classes of drouth of extensive character, which can be directly traced to irregular action of the central condition. The first class occur in the spring of the year, when the transit of the central condition to the north is delayed, and the volume of upper trade remains concentrated upon the southeastern states. Such drouths are most common in the early part of the decade, following open winters, and are greatly injurious to crops of hay and winter grain. Such a drouth occurred in the Spring of 1862, and it was very dry in all the northern states, while McClellan and his army were nearly drowned out on the Chickahominy. This class of drouths are most common in the northeastern states, because the focal path does not move up there as early as it does to the westward of the mountains, and often seems rather to contract down in March and April.

Another class of drouths are produced by an unusually extended transit and concentration of the central condition to the north and west in summer, carrying the path of the conditions farther to the west and north, and leaving the southeastern portion of the United States comparatively dry. Such was the drouth of 1854, which has been fully described. That too occurred in an early year of the decade, and was connected with an excessive transit of the central condition.

The third class of drouths are meridional, depending upon a concentration of the volume of counter trade, and a succession of storms issuing out of the central belt, and passing up on the eastern coast of the United States. Such a condition of things existed in 1867, when a severe drouth covered the interior states from Texas to Canada, and the Atlantic states were drenched by a succession of tropical storms, which passed up the coast, reaching inland a few hundred miles.

A fourth class which sometimes extends as far north as New England, are confined mainly to the Atlantic coast, are accompanied by a dry northeast wind, and are evidently produced by an unusual and temporary extension of the outer limit of the N.E. trade, as high up as 41° or 42° off the Atlantic coast. I have known such a temporary extension with its easterly wind to last 17 or 18 days. But the few others that I have known have ranged from a week to ten days. The heavens are never more brassy than during the short drouths dependent upon this cause.

There are also limited drouths depending upon local causes.

There is a known tendency to drouths in the summer season, in the Gulf States, upon the area which for the time being is left uncovered by upper trade, in consequence of its extension to the north and west in midsummer. The drouth of 1854, to which I have alluded, commenced upon that area, and extended north and west, as I have shown by diagrams. The line between this local drouth and the northern line of the central belt which extends up on to Florida and the Gulf Coast in midsummer, is sometimes very sharply defined. Thus, at New Orleans, Tampa Bay, Mobile, Fort Brook, St. Augustine, and Savannah, the rain-fall may be heavy, when upon an east and west line, one hundred miles to the north of them, a severe drouth may prevail. This drouth, like the others we have mentioned, although apparently local, depends upon a degree of the same action of the central condition, as that which produced the drouth of 1854.

There is a local tendency to drouths upon the southern coast of New England, and the fall of rain is materially less in the summer than upon the more elevated ridges to the north of it. This is so well recognized a fact, as to be represented upon all the hyetal charts of the country. It is painful to hear people in New England complain of the apparently excessive rain-fall of the rainy season in May and June, when there is a certainty that the springs and wells will need it all in July and August.

Other local drouths seem to depend upon the manner in which belts of showers distribute their rain. The eastern and middle states of the union are supplied with rain in summer, in normal seasons, mainly by passing belts of showers. These belts are very irregular in their action. Sometimes, they precipitate heavily in the afternoon and early part of the night only, and then the area over which they pass during the latter part of the night and morning, will receive little or no rain from them. Thus, the first condition which I have described of August, 1859, deposited about an inch and a half of rain at Buffalo and Rochester in the afternoon and night of the 3d, and nearly three inches at Amherst in the afternoon and night of the 4th, but it deposited very little rain at many of the intermediate places where the focus was vertical in the forenoon. Harvard received a trifle more than an inch, and Providence .75 of an inch. From the second belt, which passed during the first half of August, there fell on the 11th at Buffalo, .277 in.—on the 12th and 13th, at Waterford 1.28 in.—at Amherst 1.121 in., but at Harvard only .171. These differences are very considerable, and show that the fall is heaviest from the belts where their focus is vertical in the latter part of the afternoon and evening, and inconsiderable where it is vertical during the early part of the day. In this there is conformity to the manner in which the rain falls, under the great central belt of the tropics. Configuration also, undoubtedly has an influence. The elevations of the interior of New England, receive a larger rain-fall in summer than the depressed and more level coast. Hills and mountains increase the intensity and precipitation of the cloud belts as they pass over them, and that precipitation is still further increased, by the surface atmosphere and scud, which are drawn to them by the increased intensity, especially if that surface atmosphere is drawn from an extensive moist and evaporating surface.

This branch of the subject is one of much interest and would justify a more lengthy development, if I had space for it.

It is obvious that if the successive passing belts for a month or more, should all be of weak intensity and focal in the early part of the day, over a given area, that area would have comparatively little rain, although fifty miles to the east, and the same distance west of it, the supply might be sufficient. Configuration, or location with reference to large bodies of water, may neutralize, to some extent, but not materially change the result.

The materials and data for a comprehensive and thorough examination of the drouths of this country, do not exist. The records of the weather, prior to this century are too few and imperfect, and confined to a few localities. The same is measurably true of the first three decades of this century. The records since accumulated by the Smithsonian Institution and the War Department, if continued, will furnish the next generation ample data for their elucidation, and that may throw very much additional light upon the whole subject.

Something of interest in this direction has been done. Mr. Charles L. Flint, the Secretary of the Board of Agriculture, in Massachusetts, instigated by the extraordinary drouth of 1854, sought out all the old records which could be found, examined them, and embodied the result of his examination in his report to the Board, for the year 1854. His history of the drouths, commences with the settlement of the country, and is very interesting. I have tabulated them, but have not space for the tables. About 50 severe drouths are noticed, commencing with the year 1623, and extending to 1854. Of these, nearly two-thirds were summer drouths, occurring between the middle of June and the 1st of September, ending usually with heavy rains in the latter part of August, and apparently owing, like that of 1854, to a concentration of the conditions on the focal path, to the west and north. Nearly one-third were spring drouths, apparently due to the detention of the focal path at the south, on the Atlantic coast, and its undue extension to the northwest, to which we have also referred. A few continued through the whole season, and were probably due to a meridional diversion of the upper trade, like that which occasioned the interior drouth of 1867. The remaining ones were for shorter periods and obviously local. I would commend the history of Mr. Flint to a careful examination by those who feel an interest in the subject, and desire to investigate it, in connection with the data and principles I have given, and the later and more extensive records which are, or will be, at their command. Doubtless there is much of great practical moment to be yet discovered in that direction.

Having thus alluded to the irregularities in the operation of the system, and their influence upon the polar zones, I proceed to inquire whether there are any other facts which will guide us to their cause, and aid us in the further inquiry yet to be made, how far can we prognosticate the weather?

I do not purpose to inquire yet into the nature of the *motive force* of the system, but to confine myself, as hitherto, to *the facts* and a *development* of the *system* as matter of fact; and as matter of fact, we must enquire whether there are any changes of the sun's surface, affecting its power, and whether there is any connection between those changes and the irregular action of the system.

Whatever the nature of the motive-force or its manner of operation may be, it is certain that it emanates from the sun. The semi-annual transits of the whole system from south to north, and from north to south, following the sun in its transits from tropic to tropic, as well as the time of the diurnal changes, showing in a lesser degree, but with equal certainty, its influence, all point unerringly to that great luminary, as the controlling source of the power. If the planets have any influence, it has not been detected. Of the influence of the moon, I shall speak in another place.

Irregularities in the operation of the system, must of course be referred to irregular action of the power which controls it, as affected or modified by other influences. What knowledge have we of that irregular action, or of such other influences?

Our knowledge of the sun is yet imperfect. We feel and know its *heating* power; we know that we are mainly dependent upon it for *light*, and we can detect and trace its *magnetic* influence. We know that its surface is not uniformly the same; that it is sometimes partially obscured by dark spots, and at other times its surface is mottled by dark dots or pores. The former occur in cycles, and increase and decrease with substantial regularity, and their connection with the irregularities in the operation of the atmospheric system is clearly traceable. This part of our subject has great significance, in respect to the laws of the system, as well as to the elements of prognostication, both to be hereafter considered. Let us then inquire carefully into the nature of these irregularities upon the surface of the sun, their cycles, and their connection with the irregular action of the system.

"When the sun is examined through a telescope, its surface is found to be marked by black spots, edged with a penumbral fringe of uniform shade; they appear sometimes singly, sometimes in groups. These spots are not permanent, but undergo changes of form from day to day, or even from hour to hour, indicating a gaseous form of matter. They seldom last longer than six weeks, and often only a few hours. They are seen to break out and enlarge, or to contract and disappear, and occasionally one is observed to divide into several. When they disappear, the black centre or nucleus always vanishes before the penumbra. Mr. Dawes, an English observer, has noticed a violent whirling going on in some of the spots. Father Secchi, of the Roman College, describes one as presenting a filamentous aspect, the filaments seeming like currents, and appearing to describe spiral curves. Their size is sometimes enormous. Mayer records having seen one in 1758, whose diameter was $\frac{1}{20}$ that of the sun, and Secchi thinks some spots are deeper than the earth's radius. They are not scattered promiscuously over the surface of the sun, but are almost entirely confined to a belt of 25°, on either side of the equator, the northern being much the favorite hemisphere. The zone of 3° north, and 3° south of the equator, is nearly barren of spots. This distribution suggests the idea that they may be connected somehow with the rotation of the sun; and Sir John Herschel infers the existence of a movement in the solar atmosphere, analagous to our trade-winds. According to Langier, they seem to have a motion of approach toward the nearest pole, and these motions are strikingly correspondent on opposite sides of the equator, a fact which lends confirmation to Herschel's opinion.

"Often before disappearing, the spots undergo remarkable changes. Ridges of light are seen to dart across the chasm, splitting it into many parts, and these have been known to separate from each other and dart along in new and disturbed paths. Here again is suggested an analogy to our trade-winds. Schwabe of Dessau, by a most persevering study of the spots for upward of a third of a century, during which he has recorded the number visible on each day, for nearly 300 days in each year, arrived at a remarkable law of periodicity affecting them. They are found to gradually increase in number up to a certain period, and then to decrease to a certain period, and then to increase again, and so on. The cycle is completed, according to this investigation, in 10 years."

There is much in the foregoing extract from the article " Sun," in the New American Cyclopedia, which is suggestive. The reader will perceive that the sun presents to the view of the world, a great central condition, analogous in some respects to that great central condition in our own atmosphere, which is doubtless observable upon other peopled planets. But our business now is with the spots and their relation to our atmospheric system. The following is the table of Schwabe referred to.

Year.	Groups of spots.	Days sho'ing no spots.	Days of Observation.
1826	118	22	277
1827	161	2	273
1828	225	0	282
1829	199	0	244
1830	190	1	217
1831	149	3	239
1832	84	49	270
1833	33	139	267
1834	51	120	273
1835	173	18	244
1836	272	0	200
1837	333	0	168
1838	282	0	202
1839	162	0	205
1840	152	3	263
1841	102	15	283
1842	68	64	307
1843	34	149	312
1844	52	111	321
1845	114	29	332
1846	157	1	314
1847	257	0	276
1848	330	0	278
1849	238	0	285
1850	186	2	308

To it I append his comments upon the connection between the spots and the weather.

" I observed large spots, visible to the naked eye, in almost all the years not characterized by the minimum ; the largest appeared in 1828, 1829, 1831, 1836, 1837, 1838, 1839, 1847, 1848. I regard all spots, whose diameter exceeds 50″, as large, and it is only when of such a size that they begin to be visible to even the keenest unaided sight.

" The spots are, undoubtedly, closely connected with the formation of faculæ, for I have often observed faculæ, or narben, formed at the same points from whence the spots had disappeared, while new solar spots were also developed within the faculæ. Every spot is surrounded by a more or less bright, luminous cloud. I do not think that the spots exert any influence on the annual temperature. I register the height of the barometer and thermometer three times in the course of each day, but the annual mean numbers deduced from their observations have not hitherto indicated any appreciable connection between the temperature and the number of the spots. Nor, indeed, would any importance be due to the apparent indication of such a connection in individual cases, unless the results were found to correspond with others derived from many different parts of the earth. If the solar spots exert any slight influence on our atmosphere, my tables would, perhaps, rather tend to show that the years which exhibit *a larger number of spots* had a *smaller number of fine days* than those exhibiting few spots."

I have copied the remarks of Schwabe, in respect to the effect of the spots upon the mean annual temperature, that the reader may see the result to which he came. But his observations were confined to a single locality, and as he on the one hand was correctly of opinion that if he had discovered a connection it would have been of little importance as an individual case, so, on the other hand it is of little importance that he did not discover the connection, if the fact is clearly discernible in other and different parts of the earth. That it must exist is inferrible from the effect of obscuration by an eclipse, and that it is so discernible may, I think, be demonstrated.

In the first place, the table demonstrates that there is a *cycle* or period during which the spots increase, and another during which they decrease. This appears from the table of Schwabe to be a ten year period. But there is an observable difference in the regularity of their **increase** and dëcrease, and their number in the different decades.

In the second place, **the** influence of these spots upon the magnetic effect of the sun upon the magnetism of the earth, is clearly ascertained and generally conceded. All the magnetic elements increase and decrease in strength pari passu, with the increase and decrease of spots. Different investigators have arrived at different results, in respect to the length of the period from one magnetic maximum or minimum to another, but I suspect that this is owing mainly to the fact that they have selected different periods for their observations. Looking at the table of Schwabe, we see that there is a difference in relation to the number and size of the spots, in different decades, as for example: the table commences with the year 1826, when the number of spots was but 118. In 1836, the number was 272, and in 1846, it was but 157. Again the greatest number of spots during the decade from 1820 to 1830, was in 1828, but the greatest number for the decade from 1830 to 1840, occurred in 1837, and again, in the decade between 1840 and 1850, in 1848. The difference between the maxima was in one case, 9 years, and in the other 11. I think scientists have overlooked the fact that the difference of time between the maxima may be more or less than 10 years, while the mean period may be, as it appears to be, a decadal one.

The effect of these spots upon the mean temperature of the year, upon the seasons, and upon many of the atmospheric conditions,—although somewhat masked by the operation of other causes, and although the observations of Schwabe were in important respects, imperfect, and do not afford all the information which can be obtained in that direction, because they do not give the *amount* of daily or yearly obscuration,—is clearly traceable. I trust the time is not distant, when the *sum* of the daily and yearly *obscuration*—the gist of the matter—will be ascertained.

The spots were not as numerous during the decade from 1820 to 1830, as during the subsequent decades. That was a very warm decade, the mean temperature of the northern hemisphere ranging several degrees higher, than during the preceding or subsequent one. It was consequently a period when epidemics were prevalent, and the cholera spread from India all over the northern hemisphere, arriving in this country in 1832. During the subsequent decade, which I will call for the purpose of distinction, the solar decade, from 1833 to 1843, the spots were more numerous, and the season correspondingly cold and peculiar, in both hemispheres. I copy from a meteorological register kept at Rio Janeiro, in lat. 22° S., a locality as little likely to be affected by disturbing influences as any, the following means of temperature from 1832 to 1843, comparing them with the spots, adding a few others taken from tables published in this country:

YEARS.	1832	1833	1834	1835	1836	1837	1838	1839	1840	1841	1842
Groups of spots........	84	33	51	173	272	333	282	162	152	102	68
Ann'l Tem. at R. Janeiro	77.2	79.5	77.3	76.7	76.8	75.5	76.7	76.0	77.7	77.0	78.3
" Ft. Columbus,	51.5	51.5	51.0	49.6	49.6	49.1	50.3	51.2	51.2	51.2	53.4
" New Bedford,	47.8	48.5	48.5	47.0	45.2	46.1	47.4	48.1	47.8	47.8	47.5
" St. Louis,....	56.1	56.8	55.4	51.9	54.6	54.9	53.7	55.6	55.9	55.9	57.4
" Ft. Brady,....	41.93	41.34	41.16	40.68	37.03	36.46	37.90	41.71	41.09	39.43	38.82
" Ft. Snelling,..	45.85	47.95	47.09	43.41	42.95	43.96	41.75	47.20	44.82	42.93	43.39

A comparison of the number of spots during the solar decade, from 1843 to 1854, or so long as Schwabe observed, at a few selected points, least liable to be affected by transient causes, gives the following result:

YEARS.	1844	1845	1846	1847	1848	1849	1850	1851			
Solar spots,...........	52	114	157	257	330	238	186				
Ft. Washita,..........	63.84	63.31	64.08	61.06	61.56	61.60	62.03	62.98			
Ft. Snelling,..........	42.62	45.70	48.23	41.83	42.46	42.27	44.30	46.54			
Ft. Columbus,........	52.03	53.88	52.28	52.32	52.34	50.22	50.93	52.15			
Jefferson Barracks,....	56.07	57.20	56.76		55.00	54.00	55.48	56.01			
St. Louis Arsenal,.. ...	55.97	55.71	55.86	52.95	53.49	53.13	53.72	54.63			

I have selected the foregoing observations of the mean temperature, at points where I think the accounts are most reliable, and
least influenced by other causes. Implicit reliance cannot be
placed upon such observations, as will be apparent from comparison of the observations at St. Louis, by Dr. Engleman, and those
at St. Louis Arsenal, which are but three miles distant from each
other, and at Jefferson Barracks which is ten miles from St. Louis.

Two things are very apparent from careful observation of these
tables. In the first place, that there is a decrease of the mean
annual temperature, corresponding to an increase of the solar
spots, and a corresponding increase of temperature as these spots
decrease. And second, that the temperature rises highest during
the solar decade, when the spots are least numerous. These
movements in temperature although they do not conform precisely to the increase and decrease of spots, show clearly a connection between them. It may, and probably will be ascertained
hereafter, that these irregularities which may be attributed to
other causes, are owing to irregularities in the size of the spots,
for you may have observed that Schwabe gives the *number of
groups* which he observed during the year, but does not give
their *comparative size*, nor furnish as I have said, any reliable
comparison of the amount of obscuration.

In connection with this subject, it is important you should understand a fact which is disclosed by the tables. In every decade,
the year preceding or succeeding the minimum of spots, or that
in which the minimum occurs, is colder than the latter year.

In the "Philosophy of the Weather," I ventured to assume,
that the fact might be owing to a more extended transit to the
south, of the atmospheric system in winter, when the spots were
least numerous. The depression of the thermometer in January
and February, or February and March, at stations where other
causes were least likely to operate, renders that assumption highly
probable. The following table of temperature in February, taken
from the observations at Fort Snelling, which lies so far to the
N.W. as to be less influenced by other causes, tends to show this :

In 1822 the mean temperature for February was - 19.92
" 1823 *cold year,* - - - - - - - 5.95
" 1824 - - - - - - - - 14.20

" 1831 it was - - - - - - - - 14.04
" 1832 *cold year,* - - - - - - 6.46
" 1833 - - - - - - - - 20.93

" 1842 it was - - - - - - - 19.54
" 1843 *cold year,* - - - - - - 2.01
" 1844 - - - - - - - - 22.33

And the same fact was true to a greater or less degree of the other representative stations selected.

Since that work was written, Livingstone's travels in South Africa have been published. It appears from his description, already copied, that once in ten, eleven, or twelve years, such a transit of the atmospheric system is experienced there, and the central zone of rains is carried down over the Kalahari Desert, producing a periodical and extraordinary growth of wild melons.

The importance which may justly be attached to this part of the subject, will justify and require a more extensive and careful examination of it. Reliable tables of temperature for the last century were very few, and for this century also, in the western states, until the practice of recording observations was instituted at the Military Posts of the country, by the War Department in 1826. Mr. Blodget in his Climatology, has selected from such records as were accessible and reliable, and I copy from him. The earliest he gives for ten consecutive years is from 1750 to 1759 inclusive, at Charleston, South Carolina.

1750	-	-	-	65°.0	1755	-	-	-	63°.6
1751	-	-	-	66°.7	1756	-	-	-	67°.0
1752	-	-	-	67°.3	1757	-	-	-	65°.7
1753	-	-	-	66°.8	1758	-	-	-	64°.3
1754	-	-	-	67°.8	1759	-	-	-	65°.1

Comparing this with the subsequent observations made under

the War Department at Fort Moultrie, of which the annual mean
was 66°.58, and making due allowance for the location of Charleston upon the sea, and where it is sometimes reached by the central belt in midsummer, the effect of the spots is plainly discernible. There is an increase from 65° to 67°, and a decrease to
65° again. The depression in 1755 seems to indicate that there
was an extreme southern transit in that year, as well as in 1753.

The next early observations for ten consecutive years, are at
Philadelphia, and are as follows:

1767	-	-	-	53°.2	1773	-	-	-	54°.7
1768	-	-	-	51°.5	1774	-	-	-	52°.9
1769	-	-	-	52°.0	1775	-	-	-	54°.4
1770	-	-	-	52°.0	1776	-	-	-	53°.4
1771	-	-	-	51°.8	1777	-	-	-	51°.0
1772	-	-	-	52°.5					

Here again, the increase and decrease, though more irregular,
are clearly discernible. The decrease during a year of the warm
period, happened in 1774, the fourth instead of the second or
third year of the decade. Indeed the whole decadal period seems
to be thrown forward one year, as we shall see that it clearly was
in one or two subsequent decades.

The next most reliable observations cited by him, were those
made by Dr. Holyoke, covering the period from 1786 to 1825 inclusive. I copied from those observations in the " Philosophy of
the Weather," and arranged them as follows:

Cold Period.	Warm Period.	Cold Period.	Warm Period.
1786..........48°.53	1791..........48°.963	1796..........48°.678	1801..........50°.482
1787..........47°.88	1792..........48°.44	1797..........48°.135	1802..........50°.794
1788..........47°.676	1793..........50°.96	1798..........49°.471	1803..........50°.24
1789..........47°.68	1794..........50°.768	1799..........48°.291	1804..........48°.328
1790..........46°.53	1795..........50°.173	1800..........49°.989	1805..........50°.792
Mean of period 47°.659	Mean........49°.901	Mean.........48°010	Mean.........50°.117
1806..........47°.982	1811..........50°.76	1816..........47°.113	1821..........48°.15
1807..........48°.132	1812..........46°.28	1817..........46°.277	1822..........49°.81
1808..........49°.485	1813..........47°.702	1818..........48°.009	1823..........47°.58
1809..........47°.92	1814..........46°.279	1819..........50° 75	1824..........49°.25
1810..........49°.001	1815..........47°.607	1820..........48°.70	1825..........50° 99
Mean.........48°.605	Mean.........47°.925	Mean.........48°.169	Mean.........49°.15

The observations of Dr. Holyoke show a decided difference in three of the four decades which they embrace, thus, from '86 to '95 the differences are very striking, and so they are from '96 to 1805. But the first half of the decade, from 1810 to 1820, was peculiarly and exceptionally cold; indeed the year 1812, was much the coldest of any embraced in the tables, and both 1813 and 1815 were below the mean. The peculiar depression during that part of the decade, has been attributed to volcanic action, and there is much reason to believe, correctly. Certainly, earthquakes were, so to speak, epidemic during that period, in this country and elsewhere. Turning to the table of volcanic action and of earthquakes, found in the report of the British Association for 1854, we find that year was remarkable for earthquakes in the United States and South America. In December, 1811, earthquakes commenced in the valley of the Mississippi, Ohio, and Arkansas, and were felt also at places in Tennessee, Kentucky, Missouri, Indiana, Virginia, North and South Carolina, Georgia, and Florida, though not so severely east of the Alleghanies; and they continued to occur till 1813. About the same time they commenced in Caracas, and in March, 1812, became severe over the greater portion of the northern section of South America, and in the Atlantic. No such general and continued succession of earthquakes occurred during the other periods embraced in the tables, and the mean of the following five years was very low, embracing the memorable cold summer of 1816. But I shall consider the influence of volcanic action upon the seasons, in another place.

Schwabe did not commence his observations till 1826, and we have no reliable telescopic observations of the number and frequency of the spots during the second decade of the century, but there is much historic evidence to show that they were numerous and large, and visible to the naked eye, during the years 1815, 16-17, when the depression of temperature was very great. Few traditions are more clearly remembered.

These considerations would seem to be sufficient, to show that the exceptional depression in the decade from 1810 to 1820, does not materially affect the other evidence which tends clearly to establish a connection between the spots upon the sun, or their cause, and the mean, annual temperature. And it accords with analogy that *obscuration* should diminish power and effect.

But the evidence we have adduced—and much other of like character might be adduced—shows three facts, which seem to militate against the connection:

I. In the first place the tables show an exceptional cold year in the first half of each decade. This I have supposed to be owing to a more extended transit of the great central condition to the south, and a consequént greater winter depression. The truth of the supposition, on further investigation, seems perfectly apparent.

Such a depression is found during either the second, third or fourth years of every decade, except in the fifth decade of the 17th century at Charleston, when it was in 1755. In the first decade of this century, it was in the fourth year, or 1804. In the next it was in the second, or 1812. In the next it was in the third, or 1823. In the next it was in the second, or 1832. In the next it was in the third year, or 1843. And in the next it was in the second, or 1852. The last is shown by the following tables:

Fort Columbus, New York Harbor,	-	-	1850	50.93
" " " "	-	- -	1851	52.15
" " " "	-	- -	1852	51.40
" " " "	-	- -	1853	52.24
Mean of 33 years,	-	-		51.69
Fort Snelling,	- - - -	-	1850	43.5
" "	- - - - -	- -	1851	46.5
" "	- - - -	-	1852	43.7
" "	- - - -	- -	1853	46.7
Mean of 35 years,	-	-		44.54
Jefferson Barracks, -	- - -	-	1850	53.7
" "	- - -	-	1851	56.0
" "	- - -	-	1852	54.5
" "	- - -	-	1853	55.1
" "	- - -	-	1854	58.3
Mean of 26 years,	-	-		55.46

Fort Gibson, Indian Territory, - - -	1850	60.23
" " " " - - - -	1851	61.12
" " " " - - - -	1852	59.44
" " " " - - - -	1853	60.43
" " " " - - - -	1854	62.12
Mean of 27 years, - - -		60.81
Fort Washita, Indian Territory, - - -	1850	62.3
" " " " - - - -	1851	62.98
" " " " - - -	1852	60.32
" " " " - - -	1853	61.14
" " " " - - -	1854	63.18
Mean of 12 years, - - -		62.21
Fort Leavenworth, Kansas, - - -	1850	51.94
" " " - - - -	1851	53.9
" " " - - - -	1852	51.44
" " " - - - -	1853	53.02
" " " - - - -	1854	55.84
Mean of 24 years, - - -		52.78

I add a table of the annual mean temperature of the state of Iowa, for the years 1850 to 1856 inclusive, contained in a report of the climate of Iowa, made by Theodore S. Parvin, of Muscatine, published in 1857.

1850 - - -	46.28	1854	- - -	49.81
1851	-	46.66	1855	- - °	47.92
1852 - - -	46.65	1856	- - -	44.73
1853	- -	47.31			

It will be perceived that the depression in 1852 is scarcely perceptible in Iowa, although very obvious south, west, north and east of it. But the annual increase and decrease is very apparent.

We come now to an explanation of the depression thus found in the first half of the decade. I have already alluded, as evidence that it was occasioned by a southern transit, to the fact stated by Livingstone, that once in ten or eleven years, as he says in one place—eleven or twelve, as he says in another—the Desert of Kalahari receives a supply of rain, by reason of the excessive transit of the central belt to the south. That fall occurred in 1852, at the same time that the depression occurred in this country for that decade. The language of Livingstone, which I repeat is as follows: " Having parted with Sechele, we skirted along the Kalahari Desert, and sometimes within its borders, giving the Boers a wide berth. A larger fall of rain than usual had occurred in 1852, and that was the completion of a cycle of eleven or twelve years, at which the same phenomenon is reported to have happened on three occasions. An unusually large crop of melons had appeared in consequence." This not only corresponds with the depression in this country of that year, but corresponds in another particular, for the depressions are sometimes ten and sometimes eleven years apart.

The depressions of 1832 and 1843, which are the most distinctly marked, were eleven years apart, and were accompanied by unusual depressions in January and February both, at all the stations north of the usual winter line of a curving counter trade. Thus at Watervliet Arsenal, in New York, in 1832, both January and February were below the mean. The same was true at Fortress Monroe in Virginia, at Fort Mackinaw in Michigan, and in short at all the stations north and west of, and adjoining the focal winter line of the Atlantic conditions, while at all the stations in the S.E. portion of the United States, then covered by the focal path of the Atlantic conditions, the temperature was not below the mean, and in many of them, was above it. Thus the mean annual temperature at Fort Johnson, in North Carolina, is 65.68, the temperature in 1832 was 66.22.

Augusta Arsenal, Georgia, mean temperature,   -   64.01
"   "   "   for 1832,   -   -   -   64.64
Fort Marion, St. Augustine, mean temperature,   -   69.61
"   "   "   "   for 1832,   -   -   -   70.60
Fort Jessup, Louisiana, mean temperature,   -   -   66.34
"   "   "   for 1832,   -   -   -   66.16
Fort Gibson, Indian Territory, mean temperature,   -   60.81
"   "   "   "   for 1832,   -   •   -   61.76

In this connection there is a still more striking fact demonstrating the truth of the assumption. At all the stations in the northwestern states, beyond the cold area, which adjoins the curving focal path of the conditions for the winter, on the north and west, the temperature was also actually above the mean. Thus, at

Fort Leavenworth, temperature,	-	-	-		54.77
Mean 24 years,	-	-	-	-	52.78
Fort Snelling, temperature,	-	-	-		45.85
Mean 35 years,	-	-	-		44.54
Fort Crawford, Wisconsin, temperature,	-	-			47.66
Mean 19 years,	-	-	-		47.63
Fort Winnebago, temperature,	-	-	-		50.32
Mean 16 years,	-	-	-		44.80
Fort Brady, temperature,	-	-	-		41.93
Mean 31 years,	-	-	-		40.37
Fort Gratiot, Michigan, temperature,	-	-			47.93
Mean 17 years,	-	-	-		46.29

From this it will be seen that in all the stations high up in the N.W. the temperature was higher than the mean, owing undoubtedly to a corresponding *descent* of the path of the Pacific conditions which cross the continent in winter, covering Oregon and California, and the depression was confined to the curving area, lying between the *winter* focal paths of the two systems of conditions.

With reference to the depression of 1832, little evidence can be derived from the tables of precipitation, for the observations at that time were very limited, and there are none on record for California or the western coast. Records of rain-fall were not kept at the Posts of the United States, until 1836.

The same state of things occurred again in 1843. While the depression was distinctly marked upon the tables we have given, there was little or no depression in the southeastern states. Thus at

Fort Johnson, N. C., temperature,	-	51.40	mean,	51.69
Augusta Arsenal,	"	-	64.51	"	64.01

Oglethorpe Barracks, temperature,   -   67.19   mean, 67.44
Fort Marion, St. Augustine, temperature,   68.77   "   69.61
Fort Barrancas, Florida,   "   68.44   "   68.74

Striking beyond the curving focal paths of the winter Atlantic conditions, we find however the depression distinctly marked. Thus at

Fort Smith, Arkansas, temperature, -   56.83   mean 60.02
Fort Gibson, Indian Territory, temperature, 58.84   "   60.81
Fort Atkinson, Arkansas,   "   41.66   "   45.51
Fort Scott, Missouri,   "   52.48   "   54.30
Jefferson Barracks, Missouri,   "   52.26   "   55.46
Detroit, Michigan,   "   45.57   "   47.25
Watervliet Arsenal. N. Y.,   "   47.65   "   48.07
Fortress Monroe, Va.,   "   57.47   "   58.89

On the Pacific coast, we have no data for comparison in respect to the depression of 1832 and 1843, or any of the previous ones, but in 1852 there was an unusual descent of the focal path of the Pacific conditions, marked by an unusual supply of rain. And so in February, 1854, when the focal path of the Atlantic conditions in the United States, was carried very low by an extreme southern transit, there fell the unusual quantity of $8\frac{4}{10}$ inches of rain at San Francisco, and the evidence seems clearly to establish a connection between the two.

But it may be well here to look a little more particularly at the state of things upon the Pacific coast. Mr. Hittell, in his recent work upon California, gives us the following table of annual rain-fall at different points in that state and Oregon.

Astoria, Oregon, -   -   -   -   inches 86.35
Fort Humboldt, Humboldt Bay, California,   -   34.56
Fort Reading, Sacramento valley, -   -   -   29.02
San Francisco,   -   -   -   -   21.41
Sacramento,   -   -   -   -   -   21.73
Fort Miller, San Joachim valley, -   -   -   22.18
San Diego, Southwestern corner of state, -   -   10.43
Fort Yuma, about 160 miles east by south of San Diego,
     and in the S.E. corner of the state,   -   -   3.15

This table shows several important facts. First, that there is a gradual lessening of the annual rain-fall, from Astoria down to San Diego. This will appear more strikingly perhaps if we take the rain-fall for the winter, viz.:

Astoria,	- - inches 44.15	Latitude,	46° 10′
Fort Humboldt,	- - " 15.03	"	40° 46′
Fort Reading,	- - " 12.44	"	40° 30′
Sacramento,	- - " 12.11	"	38° 34′
San Francisco,	- - " 11.33	"	37° 48′
Fort Miller,	- - " 9.79	"	37° 00′
San Diego,	- - " 5.90	"	32° 42′
Fort Yuma,	- - " .72	"	32° 32′

It will be seen that there is a regular decrease with the latitude, the winter focus of the path of the Pacific conditions, being as high as Astoria. The table heretofore inserted, page 113, from another source, and for a different period, differing somewhat in amounts because covering a less period of time, shows the position of the focal or center path of the conditions as at Astoria in the winter, and eleven degrees further north at Sitka in summer. If now that focal path is from any cause shifted farther down the coast, there will be a corresponding increase at every point except Astoria, of the annual rain-fall. And that is precisely the effect which is produced by the extreme southern transits in the first half of each decade. In 1852 and again in 1854, the fall was excessive at San Diego, the extreme southern point. This is shown by the following table;

	Jan.	Feb.	March	April	May	June.	July.	Aug.	Sept.	Oct.	Nov.	Dec.	Year.
1850	0.00	1.13	1.00	0.09	0.00	0.68	0.00	0.00	0.00	0.19	2.82	1.93	7.84
1851	0.03	1.51	0.34	0.87	0.71	0.01	0.00	0.00	0.02	0.01	0.25	3.74	7.49
1852	0.58	1.84	1.87	0.85	0.32	0.00	0.00	0.40	0.00	0.06	1.45	4.50	11.87
1853	0.50	0.20	1.52	0.25	2.10	0.05	0.00	0.21	0.00	0.00	1.28	1.77	7.88
1854	1.46	2.56	2.14	0.75	0.21	0.02	0.07	1.35	0.13	0.01	0.02	3.34	12.06
1855	2.40	4.83	1.53	1.82	0.10	0.00	0.00	0.00	0.00	0.15	2.23	0.31	13.37
1856	0.66	2.04	1.97	2.48	0.27	0.00	0.00	0.00	0.05	0.00	1.47	1.20	10.14
1857	0.21	2.12	0.00	0.00	0.12	0.00	0.00	0.00	0.00	0.64	2.25	1.38	6.72
1858	1.87	0.45	1.60	0.27	0.00	0.15	0.00	0.03	0.11	0.49	0.32	3.65	8.94
1859	Record incomplete.												

See the marked manner and extent of the rain fall—increasing rapidly to 1855 inclusive, and then decreasing as rapidly to the close.

The record for Fort Yuma is imperfect, but I give it for the purpose of showing its peculiar position between the two systems of conditions in May and June.

Fort Yuma, California.

	Jan.	Feb.	March	April	May.	June	July.	Aug.	Sept.	Oct.	Nov.	Dec.	Year.
1851	0 00	0.01	0.00	0.27	.....	.....	.....	.....	.....	.....	.....	.....	.....
1852	.....	.....	.....	.....		0.00	0.28	0.33	1.45	0.00	0.35	0.04	.....
1853	0.00	0.00	0 01	0.00	0 90	0.00	0.25	0.89	0.13	0.00	0.18	0.52	1.78
1854	0.00	0 28	0.80	0.00	0.00	0 00	0.01	2.37	0.17	0 30	0.00	0.57	4.50
1855	0.12	0.26	0.00	0.00	0.00	0.00	0.10	0.00	.....	0.22	0.10	0.00	
1856	0.00	0.00	0.50	0.00	0.00	0.00	0.48	0.00	0.36	0.00	0.19	0 00	1.53
1857	0.00	0.30	0.00	0.00	0.00	0.00	0.00	0 00	0.00	0 00	0.00	0.00	0.30
1858	0.00	1.06	0.00	0.00	0.00	0.00	0.00	0.00	0.00	0.00	0.00	1.03	2.09
1859	0.00	2.09	0.00	0.00	0.00	0.00	0.00	0.50	0.00	0 42	1.83	0.00	4.88

Fort Laramie.

	Jan.	Feb.	March	April	May.	June	July.	Aug.	Sept.	Oct.	Nov.	Dec.	Year.
1852	0.72	1.10	1.55	1.25	7.29	4.08	1.88	1.46	2.74	1.70	6.42	1 23	31 42
1853	0.08	0.57	1.78	4.53	12 19	4.95	1.86	0.55	2.80	0 68	0.08	0 71	30.78
1854	0.18	0.40	0.80	3.98	4.46	3.67	3.26	1.27	1 60	1.86	0.73	0.05	22.26
1855	0.04	1.08	1.41	0.65	2.79	3.25	1.45	2.93	3 39	0.62	0.18	1 20	18 90
1856	0 55	0 45	1.75	0.34	3.51	0.82	4.15	2.55	0.23	0.29	0.21	0.17	15.02
1857	0.33	0.53	0.00	0.07	1.45	0 12	0.04	1.87	0.10	1.53	0.05	0.06	6.15
1858	0.00	0.02	0.02	0.09	1.12	0.80	1.14	1.81	0.70	1.43	0.32	0.45	7 90
1859	0.01	0.00	0.00	0.18	2.11	0.03	1.33	0.57	0.49	0 22	1.12	0.20	6.26

I give also the record for Fort Laramie in Nebraska, conforming substantially in their results. But May, 1853, is erroneous.

The fact of a depression of the system, cannot, as we have said, be reliably shown, by the fall of rain, for want of data for the whole country, *prior* to 1843. In respect to that year however, the records of the various posts are sufficiently full and entirely corroborative. On the southern states to which the focal path of the conditions was carried by the extreme southern transit of the central condition, the rain-fall was greater than the mean. Thus at

Oglethorpe Barracks, Ga., Feb., 1843, rain 2.03 mean, 2.18
" " " March, " " 12.51 " 7.11
Fort Brooks, Florida, Feb., " " 5.70 " 3.01
" " March, " " 7.20 " 3.37
Mt. Vernon Arsenal, Ala., Feb., " " 5.90 " 6.04
" " " March, " " 9.22 " 4.59
Fort Moultrie, S. C., Feb., " " 2.09 " 2.33
" " March, " " 10.52 " 4.06
New Orleans, La., Feb., " " 3.80 " 2.90
" " March, " " 5.31 " 3.90

West and north of and adjoining the then path of the conditions, we have the following deficiencies which show as well the unusual descent in February, as the lingering return in March.

Fort Towson, February rain, - 00.25 mean 2.97
" March " - - 3.35 " 4.38
Fort Jessup, February " - 1.41 " 2.76
" March " - - 3.25 " 5.02
Fort Smith, February " - 0.26 " 2.17
" March " - - 2.29 " 2.92
Fort Gibson, February " - 0.70 " 2.26
" March " - - 1.31 " 2.54
Fort Scott, Mo., February, " - 0.20 " 1.18
" " March " - - 1.75 " 1.79
Jefferson Barracks, February " - 1.75 " 2.04
" March " - - 2.30 " 3 32
St. Louis Arsenal, February " - 1.30 " 3.37
" " March " - - 2.29 " 3.82

And here in connection with this subject, I re-insert the diagram showing the situation of the paths of the conditions or belts of rains in February, 1854, with the intervening comparatively cold and dry area, that you may have it under your eye in connection with the foregoing tables. Doubtless they were in the same situation substantially, in 32, 43 and 52.

FIG. 37.

From this examination it appears that the first of the three facts does not militate against the proposition advanced, and attempted to be proved by the tables, but an investigation of its cause, directly confirms it.

The second fact is more apparent than real, although the years 1813 and 1814 did not show such an increase of temperature at Salem as is shown in other decades. The temperature was yet far in advance of the depression of 1812, and fully up to the mean. Observations for those years are very meagre. Those of New Bedford show the years to have been up to the mean there, and as high as that of 1844, but New Bedford feels the influence of the Gulf stream and is not a representative station. We have no record of the spots for that decade, but it is a matter of history that they were very numerous, large, and observable by the naked eye, especially in 1816, which was the coldest year in the century, except 1812. There was frost during every month of the year, and the corn crop was generally lost. Now as it appears from the record of the spots which we have, that they are variable in number and effect in different decades, and as we have it as a matter of tradition and of history, that they were more numerous during that decade than in any previous one, and much larger or they could not have been so extensively seen with the naked eye, we are fully justified in assigning the unusual depression of temperature, in that decade, to that cause, and this apparent exception also proves the rule.

So too of the unusual elevation of the temperature from 1825 to 1830. We have a full explanation of that in the *comparative infrequency of the spots* and that apparent exception also proves the rule.

Again, there seems to be an exception in the decade from 1840 to 1850. The temperature continues to rise until and including the year 1846, but when we turn to the table of spots, we find that they increased with *less rapidity* than in other decades, and there again, the exception proves the rule.

And thus upon examination I have found every fact which bears upon the subject, when understood, conspiring to prove that there is a period of increase and decrease of temperature, and peculiarities of rain fall in the middle and northern portions of the United States, dependent upon the transit of the atmospheric system north and south, caused and controlled by the spots upon the sun and their effect in diminishing its controlling power.

Our examination leads to the further inquiry, whether any other forces affect the volume of trade with which this country is supplied, or the volume of moisture which they bring to us, and the consequent frequency and intensity of the conditions which make our weather.

That volcanic action affects the weather locally, is generally believed. On this subject there is a mass of evidence for which I have not space. Even earthquakes, at a distance from volcanoes, seem to exert an influence. On this subject Mr. Hittell thus speaks in the " Resources of California," page 44.

" Earthquakes, according to the common theory of Californians, are electrical in their origin, or closely connected with electrical influences. Many of the strongest shocks have been preceded by a condition of the atmosphere very similar to that which precedes thunder-storms in other lands. When the weather is sultry and oppressive in San Francisco, people say, "Look out for an ea th-quake!" And it usually comes—perhaps so faint as to be barely perceptible, and sometimes not until several hours after a change in the weather."

A writer in the London, Edinburgh and Dublin Phil. Mag. for December, 1853, gives a collection of facts bearing upon this subject, which I copy. He collected the facts as a basis of theory, but I introduce them, only as tending to show that volcanic action may have an influence upon the weather, and as confirmatory of what I have already suggested, that the cold decade from 1810 to 1820 may have been caused in part by the extraordinary volcanic activity which existed during the early part of the decade in South America, the West Indies, and this country; but we cannot trace that influence, for we have no records of temperature or of rain-fall, except at a very few points upon the Atlantic coast.

There has been occasional volcanic activity since then, in the West Indies and in South America, but I have not attempted to trace its influence, if it exerted any, upon the weather of this country. I have not had time or strength to devote to it. The subject is of sufficient interest, and I commend it to the reader as worthy of investigation. And to aid him, I insert that paper of Mr. Dobson. His cited cases may all have been coincidences, but I do not doubt the influence. At the same time I should look for it as affecting the character and volume of the trades at their *place of origin*, rather than the atmosphere in the polar zones directly. But if it affects the former, it must, thereby, indirectly, the latter.

The following is the substance of the paper of Mr. Dobson :

" 1st. The eruption of a submarine volcano has produced water-spouts.

" During these bursts the most vivid flashes of lightning con-tinually issued from the densest part of the volcano, and the vol-umes of smoke rolled off in large masses of fleecy clouds, grad-ually expanding themselves before the wind in a direction nearly horizontal, and drawing up *a quantity of water spouts.*" —(Cap-tain Tilland's description of the upheaval of Sabrina Island in June, 1811, Phil. Trans.)

With this significant fact may be compared the following analogous ones.

" In the Aleutian Archipelago a new island was formed in 1795. It was first observed *after a storm*, at a point in the sea from which a column of smoke had been seen to rise."—(Lyell, Principles of Geology)

" Among the Aleutian Islands a new volcanic island appeared in the midst of a *storm*, attended with flames and smoke. After the sea was calm, a boat was sent from Unalaska with twenty Russian hunters, who landed on this island on June 1st, 1814."— (Journal of Science, vol. vii.)

" On July 24th, 1848, a submarine eruption broke out between the mainland of Orkney and the island of Strousa. Amid thunder and lightning, a very dense jet black cloud was seen to rise from the sea, at a distance of five or six miles, which *traveled toward the northeast*. On passing over Strousa, the wind from a slight air became *a hurricane*, and a thick, well-defined belt of large hailstones was left on the island. The barometer fell two inches."—(Transactions Royal Society, Edinburg, vol. ix.)

2d. Hurricanes, whirlwinds, and hailstones accompany the paroxysms of volcanos.

" 1730. A great volcanic eruption at Lancerote Island, and *a storm*, which was equally new and terrifying to the inhabitants, as they had never known one in the country before."—(Lyell, Principles of Geology, vol. ii.)

" 1754. In the Philippine Islands a terrible volcanic eruption destroyed the town of Taal and several villages. Darkness, hurricanes, thunder, lightning, and earthquakes, alternated in frightful succession "—(Edinburgh Philosophical Journal.)

" In 180ɔ, 1811, 1813, and 1830, during eruptions of Etna, caravans in the deserts of Africa perished by violent whirlwinds. In 1807, while Vesuvius was in eruption, a whirlwind destroyed a caravan."—(Rev. W. B Clarke in Tasw. Journal.)

" 1815, Java. A tremendous eruption of Tombow Mountain. Between nine and ten P.M., ashes began to fall, and soon after *a violent whirlwind* took up into the air the largest trees, men, horses, cattle, etc."—(Raffle's History of Java.)

"1817, Dec. Vesuvius in eruption. In the evening *a hail storm*, accompanied with red sand."—(Journal of Science, vol. v.)

"1820, Banda. A frightful volcanic eruption, and in the evening an earthquake and a violent hurricane."—(Annales de Chimie.)

"1822, Oct. Eruption of Vesuvius. Toward its close the volcanic thunder-storm produced an exceedingly violent and abundant fall of rain."—(Humboldt, Aspects of Nature.)

"1843, Jan. Etna in eruption. Violent hurricanes at Genoa, in the Bay of Biscay, and in Great Britain.

"1843, Feb. Destructive earthquake in the West Indies, a volcanic eruption at Guadaloupe, followed by hurricanes in the Atlantic."

"1846, June 26. Volcano of White Island, New Zealand, in eruption. Heavy squalls of wind and hail; it blew as hard as in a typhoon."—(Commodore Hayes, R. N., in Naut. Mag., 1847.)

"1847, March 20. Volcanic eruption and earthquake in Java; and on the 21st of March, and 3d of April, violent hurricanes." —(Java Courant)

"1851, Aug. 5. A frightful eruption of the long dormant volcano of the Pelée Mountain, Martinique. Aug. 17. Hurricane at St. Thomas, etc.; earthquake at Jamaica, etc.

"1852, April 14. Earthquake at Hawaii, and on the 15th a great volcanic eruption. On the 18th *a gale of unusual violence* lasted thirty-six hours, and did great damage."—(The Polynesian, April 22, 1852.)

3d. In volcanic regions, earthquakes and hurricanes often occur almost simultaneously, but in no certain order, and without any volcanic eruption being observed.

In 1712, 1722, 1815, and 1851, earthquakes and hurricanes occurred together at Jamaica; in 1762 at Carthagena; in 1780 at Barbadoes; in 1811 at Charleston; in 1847 at Tobago; in 1837 and 1848 at Antigua; in 1819, an awful storm at Montreal, rain of a dark inky color, and a slight earthquake. People conj*ctured that a volcano had broken out. In 1766 the great Martinique hurricane, a *waterspout* burst on Mount Pelée and overwhelmed the place. Same night an earthquake.

1843, Oct. 30. Manilla.—Twenty-four hours' rain and two heavy earthquakes. 10 P.M., a severe hurricane.

" 1852, Sept. 16. Manilla.—An earthquake destroyed a great part of the city; many vessels wrecked by a great hurricane in the adjacent seas, between the 18th and 26th of September."—(Singapore Times.)

"1737, Oct. Calcutta.—Furious hurricane and violent earthquake ; 300,000 lives lost."

"1816, May 26. Bombay.—Hurricane and earthquakes; 2,000 lives lost."—(Madras Lit. Tran., 1837.)

" 1800, Ongole, India, and in 1815, at Ceylon, a hurricane and earthquake shocks."—(Piddington.)

" 1848. Cyprus.—An earthquake and a frightful hurricane."—(Hecker.)

" 1819, Bagdad.—An earthquake and *a storm*—an event quite unprecedented.

"1820, Dec. Zante.—Great earthquake and hurricane, with manifestations of a submarine eruption."—(Edinburgh Phil. Journal.)

" 1831, Dec. Navigator'sIslands.—Hurricane and earthquakes." (Williams' Missionary Enterprise.)

"1848, Oct., Nov. New Zealand.—Succession of earthquake shocks, and several tempests.

" 1836, Oct. At Valparaiso, a destructive tempest and severe earthquakes."—(Nautical Magazine, 1848.)

When an earthquake of excessive intensity occurs, as at Lisbon, in 1755, the volcanic craters, which act as the safety-valves of the regions in which they are placed, are supposed to be sealed up; and it is a remarkable and highly-suggestive fact, that *no hurricane follows such an earthquake.* The number of instances of the concurrence of ordinary earthquakes and hurricanes might easily be increased, but the preceding suffice to show the *generality* of their coincidence, both as *to time* and place.

11

4th. The breaking of water-spouts on mountains sometimes accompanies hurricanes.

In 1766, during the great Martinique hurricane, before cited.

" 1826, Nov. At Teneriffe, enormous and most destructive water-spouts fell on the culminating tops of the mountains, and a furious cyclone raged around the island. The same occurred in 1812 and in 1837."—(Espy and Grey's Western Australia.)

" 1829, Moray.—Floods and earthquakes, preceded by water-spouts and a tremendous storm."—(Sir T. D. Lander.)

" 1826, June. Hurricanes, accompanied by water-spouts and fall of avalanches, in the White Mountains."—(Silliman's American Journal, vol. xv.)

5th. The fall of an avalanche sometimes produces a hurricane.

" 1819, Dec. A part (360,000,000 cubic feet) of the glacier fell from the Weisshorn (9,000 feet.) At the instant, when the snow and ice struck the inferior mass of the glacier, the pastor of the village of Randa, the sacristan, and some other persons, *observed a light.* A frightful hurricane immediately succeeded."— (Edinburgh Philosophical Journal, 1820.)

6th. Water-spouts occur frequently near active volcanoes.

This is well known with regard to the West Indies and the Mediterranean. The following notices refer to the Malay Archipelago and the Sandwich Islands :

" Water-spouts are often seen in the seas and straits adjacent to Singapore. In Oct. 1841, I saw *six* in action, attached to one cloud. In August, 1838, one passed over the harbor and town of Singapore, dismasting one ship, sinking another, and carrying off the corner of the roof of a house, in its passage landward."— (Journal of Indian Archipelago.)

" 1809. An immense water-spout broke over the harbor of Honolulu. A few years before, one broke on the north side of the island (Oahu,) washed away a number of houses, and drowned several inhabitants."—(Jarves' History of Sandwich Islands.)

7th. Cyclones begin in the immediate neighborhood of active volcanoes.

The Mauritius cyclones begin near Java ; the West Indian,

near the volcanic series of the Caribbean Islands; those of the Bay of Bengal, near the volcanic islands, on its eastern shores; the typhoons of the China Sea, near the Philippine Islands, etc.

8th. Within the tropics, cyclones move toward the west; and, in middle latitudes, cyclones and water-spouts move toward the N.E., in the northern hemisphere, and toward the S.E. in the southern hemisphere." Such are the views of Mr. Dobson.

Popular opinion has ever attributed to the moon a controlling effect upon the changes of the weather. If it be dry, a storm is expected *when the moon changes;* or if it be wet, dry weather. Such popular opinions are usually entitled to respect, and founded in truth. But every attempt to verify *this opinion*, by careful observation and registration, has failed. Weather-tables and lunar phases, compared for nearly one hundred years, show four hundred and ninety-one new or full moons attended by a change of the weather, and five hundred and nine without. The celebrated Olbers, after *fifty years of careful observation* and comparison, decided against it. So did the more celebrated Arago, at a more recent date—summing up the result of his observations by saying—" Whatever the progress of the sciences, never will observers who are trustworthy and careful of their reputation, venture to foretell the state of the weather." Still, the moon may influence the weather, though she may not effect changes at her syzygies or quadratures, and this subject should not be too summarily dismissed. That the moon can not effect changes at the periods named seems philosophically obvious. She changes, for the *whole earth*, within the period of twenty-four hours; yet, how varied the state of things on different portions of its surface. The equatorial belts of trades, and drouth, and rains, cover from fifty to sixty degrees of its surface, and know nothing of lunar disturbance. The extra-tropical belt of rains and variable weather moves up in its season, uncovering 10°, or more, of latitude, and admitting the trades and a six months' drouth over it, as in California, regardless of the moon. Under the zone of extra-tropical rains, even upon the eastern part of the continent of North America, "dry spells" and "wet spells" exist side by side; the focus of precipitation is now in one parallel, and now in another—*storms* exist *here* and *fair weather there*, on the same continent at the same time; and as the moon's rays in her northing pass round the northern hemisphere during the twenty-four hours, they, doubtless, pass from twenty to thirty or more storms, of all characters and intensities, moving in opposition to her orbit—and as many larger intervening areas of fair weather, not one of which are indebted to her for their existence, or "take thought of her coming."

The storm, which originates in the tropics, pursues its curving way now N.W., then N.E., to the arctic circle, over gulf, and continent, and ocean, *occupying one-third the time of a lunation, and two changes or more, in is progress*, without any perceptible or conceivable influence from her   Yet every inhabitant of mother-earth, influenced by *coincidences remembered* and uninfluenced by *exceptions forgotten*, looks up within his limited horizon, and devoutly expects from the agency of some phase of the moon, a change for the special benefit of his *dot* upon the earth's surface. Upon how many of these countless dots is the moon at a particular phase, or relative distance from the sun, to change fair weather to foul, or foul to fair?   Upon none.   The storms keep on their way;—the wet spells, and the dry spells, the cold and the hot spells alternate in their time, and though the moon turns toward them in passing, her dark face, her half face, or her full orb (the gifts of the sun, which confer no power,) they do not heed her. They are originated, and are continued, by a more potent agent. They are the work of an atmospheric mechanism, as *ceaseless* in its operation as *time*, as *regular* as the *seasons*, as *extensive as the globe*.

Indeed, it seems as if it was expressly designed by the Creator that the moon should not interfere materially with this atmospheric machinery.   She is the nearest orb, her influence would be controlling and continuous; would follow her monthly path from south to north, and with changes too violent, and intervals too long; and would interfere with the regular fundamental operation in the trade wind region, where she is re. tical.   Aside from the attraction of gravitation, therefore, she seems to have been so created as to be incapable of exerting any influence.   She

is without an atmosphere ; the rays which she reflects are polarized, and without chemical or magnetic power ; and, if it be true
that Melloni has recently detected heat in them, by the use of a
lens three feet in diameter, which could not previously be effected,
its quantity is exceedingly small, and incapable of influence.
Doubtless, the attraction of her mass is felt upon the earth, as the
tides attest ; and upon the atmosphere as well as the ocean. But
the atmosphere is comparatively *attenuated*, and exceedingly so
at its upper surface. Her attraction, therefore, although felt, is
not influential. She seemed, to Dr. Howard, to produce in her
northing and southing, a lateral tide which the barometer disclosed,
but owing to the attenuated character of the atmosphere, neither
the sun nor moon create an easterly and westerly tide, that is
observable, except with the most delicate instruments. Sabine is
believed to have detected such a tide by the barometer, at St.
Helena, of one four-thousandth of an inch. But even this *infinitesimal influence* may prove an error upon further investigation. There is a diurnal variation of the barometer, but it is not
the result of her attraction, for it is not later each day as are the
tides, exists in the deepest mines as well as upon the surface, and
is demonstrably connected with the *group* of *diurnal* changes produced by other causes.

Can the lateral tide, if there be one, affect the weather ? for in
the present state of science it seems entirely certain that the moon
can exert an influence in no other way.

If the received idea of many, perhaps most meteorologists, on
which all wheel barometers are constructed, that a *high barometer*
necessarily produces *fair weather*, and a *low one foul*, were true,
she certainly might do so. But that idea cannot be sustained, and
there is no known certain influence exerted by the moon upon the
weather, in relation to which we have any reliable practical data.

Humboldt appears to have adopted the impression of Sir W.
Herschel, that the moon aids in the dispersion of the clouds.
(Cosmos, vol. iv, p. 502.) But the tendency to such dispersion
is always rapid during the latter part of the day and evening,
when there is no storm approaching, and the full moon renders

their dissolution visible, and attracts attention to them. The Greenwich observations, also, carefully examined by Professor Loomis, fail to confirm the impression of Herschel and Humboldt, and those eminent philosophers are doubtless, in this, mistaken.

The foregoing paragraphs in relation to the influence of the moon, were written for the " Philosophy of the Weather," published in 1856. Since that, I have continued the investigation, but with no change of result. Dr. Lynes, in his diary of a voyage round Cape Horn, which I have copied, remarks that the trades seemed to freshen after the moon rose, but there is no conceivable way in which the moon could increase the trades, except by aiding the motive force of the sun which causes them, nor any known way in which she could do that, unless by the influence of her light, and such an influence cannot be conceived. Howard thought he could trace an increase in the frequency and volume of rain, after the moon had reached her southing, and was coming north. If we could conceive of any method by which the moon could increase the southeast trades, when vertical to them, or the quantity of moisture contained in them, we could easily see that the supposition of Howard might be true. for the increased volume of trade would arrive at the latitude of England, before the moon crossed the equator; and an increased volume of trade would be likely to increase the frequency and constancy of rains in that climate. But I am unable to see how it is possible for the moon to exert any influence upon the volume of the trades, or their evaporating power, and am alike unable to trace any influence upon the weather in this country. To the tables of Herschel and Adam Clark, I shall refer particularly hereafter.

There is one other thing which probably affects the volume of counter or upper trade, which comes to this country, and thereby the amount and regularity of its rain-fall, which I have not attempted to investigate. I allude to the deflection of the S.E trades of the Atlantic in upon Senegambia, under and below the central belt of rains, constituting the monsoon represented upon the diagram, page 119. Lieut. Maury has said that the influence which deflects the S.E. trades upon that part of the ocean, is sometimes felt nearly across to the continent of South America. If that is true, it is obvious that a portion of the S.E. trades, which, if undiverted, would pass over this country as an upper trade, is deflected away from us, and must have an important influence upon the amount of our rain-fall. The subject is capable of elucidation by an examination of the logs of vessels which have passed that part of the ocean, which are collected at the National Observatory, and I hope it will be investigated.

While this chapter has been in the hands of the printer, accounts have been published of a summer drouth which has prevailed in the southeastern states, of the second class described, and of like character with that of 1854, but of less extent. To illustrate its character, and the fact that like all of its class it has been produced by an unusual *concentration* of the *conditions* on the focal path to the west and northwest, I copy a letter from the New York Observer, descriptive of *the drouth* at its central point in Tennessee.

"DROUTH IN EAST TENNESSEE.

"KNOXVILLE, TENN., Aug. 30, 1869.

"*Messrs. Editors:* On arriving at this point, and resting for a day or two, I have had opportunity for gaining additional information as to the condition of this section and the country South and West; and I am now to write of a distressing state of things, resulting from an almost unparalleled drouth. For nearly three months there has been no rain of any consequence in East Tennessee. The rivers and smaller streams are very low. The Holston at this place presents quite a singular appearance, the volume of water being so small that ferry-boats can scarcely cross, while steam-boats are all aground, and for the present, entirely useless. The thermometer has, during this season, ranged from 90° to 104°. Persons have been and are still complaining of the severe oppressiveness of the heat, and, thinly clad, many, with umbrellas while out of the shade, barely manage to get about. The farmers, many of them, have become so discouraged that they have turned their stock in upon their cornfields, because they felt that it was impossible to receive as much benefit from them in any other way. In some portions of the state large fields of corn which would have yielded from thirty to fifty bushels to the acre, cannot possibly yield now exceeding from three to five."

And now I copy from the St. Paul (Minnesota,) Daily Press, the following notes of the rain-fall for the month of August at that place, descriptive of *the drench:*

"The month commenced with a very heavy fall of rain, nearly two inches and a half of water being deposited on the first day. Other rains followed at pretty regular intervals, and in copious showers, so that the amount of rain for the month reached the unusual quantity of seven and two-thirds inches. This monthly fall has been exceeded three times only in eleven years—in July, 1862, in August, 1865, and in June, 1867. Looking at it in anticipation, it would have seemed that so numerous and abundant rains must produce a disastrous effect upon the harvest, coming as they did at almost every stage of its gathering. And yet, with occasional exceptions, no serious injury appears to have resulted, and the wheat crop, now almost full secured, is represented to be both very large and of good quality.

Rain fell on twelve days, to the amount of seven inches and sixty-two hundredths of an inch of water. The fair days were equal to sixteen. The winds came principally from E. and S.E."

Here is evidence of a parching and destructive drouth over a large area in the Middle and Eastern States, and at the same time a dangerous drench under the focal path over the Western and Northwestern states, conforming precisely in character to the summer drouths, produced by the transit and concentration of the conditions, *as fully developed and explained in the Philosophy of the Weather*, and this work. It is a great public misfortune that the practical mind of the country should be kept in ignorance of those transits and concentrations, and their important climatological results and consequences, by men claiming to be meteorologists, and occupying prominent and controlling public and educational positions.

CHAPTER VII.

PROGNOSTICATION.

Consideration of the question how far and by what means, a local, isolated observer can prognosticate or forecast the weather—assuming the two first fundamental propositions to be proved, the logical inquiry will be, when will the normal state be disturbed by a condition and its changes, and what its character and intensity—Certainty or regularity of interval, not to be expected, although conditions of more or less intensity pass frequently—They sometimes pass on a given day of the week for several weeks, with or without an intervening condition—No certainty as to the character and intensity of the next condition—nor any reliance on planetary influence—only reliance is upon changes of state produced by passing conditions—Signs, proverbs, and maxims not founded on or connected with those changes of state, empirical and worthless—The inquiry then is for prognostic inferences, derivable from changes of state—Certain other elements and contingencies to be considered, viz: location of observer, season of the year, and the situation of the year in the decade—Recapitulation of the changes of state, induced by the conditions from which prognostic inferences are to be drawn—*Weight* of the atmosphere, how measured—various instruments examined—Barometer the principal one to be relied on—Its indications—its mean elevation generally upon the earth—difference of elevation in different localities—mean elevation least where conditions are most frequent and intense—examination of the fact in relation to this country—Places and circumstances under which it ranges highest and lowest—Allowance to be made for elevations above the level of the sea—Rules for determining that allowance—No fair weather standard for the barometer—each observer must fix one for himself—Standard must vary with the transits of the focal path and deduction must be made for altitude—Rules for fixing such standard—collection of rules used in England, for forecasting weather by barometer—critical examination of these rules—Prognostic influences to be drawn from *temperature*—before northeasters—before belts of showers—in relation to the probability of snow or rain—exceptional warm periods in winter without passing conditions—how produced—thermometer to be consulted in reference to the continuance of storms—sudden changes of our climate—their magnitude and how produced—importance of forecasting and regarding them—Prognostic inferences to be drawn from the *winds* and their changes—winds not felt at the surface discoverable by sounds and scud—all fresh, earnest winds created by conditions—importance of this class of prognostic inferences—Prognostic inferences drawn from *clearness* or *cloudiness*—their character and importance—examination of the maxims founded on them—Resumè of the appearances of the sky from which prognostic inferences may be drawn—Prognostic inferences from the existence of *humidity*—Evaporation and hygrometry—devices by which humidity is measured—contrast between the English climate and ours, in relation to humidity—examination of weather signs founded on humidity—Prognostic inferences from *rain, hail and snow*—examination of weather proverbs in relation to them—Rules for determining whether a storm will precipitate snow or rain—Prognostic inferences from the seventh element or *electric* state of the atmosphere—Dr. Jenner's signs of rain—examination of those signs as founded upon the electrical influence of an approaching condition, on men, animals, and plants—Examination of signs and proverbs not founded on any of the foregoing seven states—all fallacies—Review of some of the principal points—importance and probability of the use of the telegraph in prognostication—other material points reviewed—Examination of the question what will be the *character* of the next condition, and how to forecast that character—also the question what will be its *intensity*, and how that question can be answered—prognostic inferences in relation to the *continuance* of the conditions and how they " clear off."

If the reader has carefully followed me in the foregoing developments, and fully comprehended and appreciated them, he is prepared to enter with me upon the further and important inquiry: " *how far* and by *what means* can a local, isolated observer *prognosticate* or *forecast the weather?*" And here again we shall find the subject capable of intelligent, logical analysis, and practical result.

In the outset of this inquiry, I assume, as I think I may well assume—that the facts I have developed and arranged, have satisfied you of the truth of my two first fundamental propositions, viz: 1st, that the normal state of the polar zones is still, fair weather, and 2d, that the changes from that state, result from the influence of forming, approaching or passing conditions. It follows philosophically and logically, that in respect to prognostication, the inquiry is and must be, how long will that normal state continue undisturbed by a passing condition, or in other words, when will the next condition approach and disturb that state by its changes? What will be the character and intensity of that condition and its incident changes? and how long will it be in passing away and permitting the normal state to return?

Certainty or regularity in relation to the intervals between the occurrence of the conditions is not ordinarily to be expected. The Californian knows indeed when the focal path of the conditions has moved to the north in summer, that it will not descend until fall, and that a long period of drouth and fair weather is before him. He sees the thin and feeble southern edge of the conditions occasionally pass over him, while their intense and precipitating bodies are far to the north, carried by a law as unchangeable as the transits of the sun. Dr. Gibbon, in the Smithsonian Report for 1854, thus speaks of these lateral, outlying, feeble, southern extensions of the conditions, though without understanding them: "In almost every month of the year during the dry season, the clouds put on the appearance of rain, and then vanish. It is evident that the phenomena which produce rain in other climates, are present in this, but not quite in sufficient degree to accomplish the result, except during the rainy season, and then only by paroxysms, with intervening periods of drouth." And here, during a class of drouths like that of 1854, occasioned by a peculiar concentration and extension of the focal path to the west and north, we have similar appearances, and say, "all signs fail in a dry time," and are reasonably certain that our needed rains will not return until the conditions become more focal over us again, late in August or early in September. But with respect to the eastern part of the continent, that class of drouths are exceptional. So too, in the rainy season of spring and early summer, when the conditions are focal over us, and both frequent and intense, and we say, "it rains very easy now," and it does so almost every day, and the intervals are short, we can calculate with reasonable certainty on the recurrence of rain. But this also is confined to a season of a few weeks, and is exceptional. And it is true, as we have stated, that certainty and regularity of interval between the conditions, is not, as a rule, for the year, to be expected.

Nevertheless, it is sometimes true, not merely in the rainy season of spring and fall, but at other times, that the intervals are, in fact, regular for several weeks in succession. It is doubtless within the recollection and experience of every one who has lived long, that at a remembered period, some particular day in the week, for several successive weeks, was stormy ; sometimes with, and at others without, an intervening storm or showery condition on some other day in the week. Those cases too are exceptional, and do not affect the proposition as a general one.

Nor is there any absolute certainty in respect to the character or intensity of the next condition, on which a local observer can rely. There is *probability*, dependent upon the season of the year, and the location of the year in the decade, as we shall see, but it is probability merely.

Nor, in respect to either, is there any reliance on any planetary influence.

Inasmuch then as there is no certainty, or regularity of interval between the conditions, or in respect to the character or intensity of the one which will next pass over the observer, it is philosophically and logically obvious that his *only* reliance is, and must be, upon the observable *changes of state* which the forming, approaching or passing conditions induce ; and as matter of fact it will be found on careful examination that all the most important proverbs and signs which men have observed or adopted, and tradition has preserved, are founded on those changes of state ; and that all which are not directly *connected with*, or *indicative* of, *some one of those states*, are empirical and worthless. We must recur then to those states and changes, to trace the manner in which they are initiated, and the prognostic influences to be derived from them.

Before we do this, however, it is essential that we should allude to some circumstances which must enter as elements into the consideration of the subject at all times, and in all places, in order to a correct and clear prognostication.

I. The first element to be considered is the *location* of the observer. Of the importance of this element, the reader is doubtless satisfied, if he has comprehended the developments made; but it is fundamentally important, and an additional illustration or two may not be amiss.

If in February the observer is in Alabama, Eastern Arkansas, Southern Tennessee, Southern Virginia, or North Carolina, he is in the centre of the winter focal path of the Atlantic conditions, subject to the winter rainy season, even when most concentrated, and the conditions will be frequent and intense. If in the same month he is in Western Texas, Western Arkansas, Western Missouri, Northern Illinois, or Indiana, he is west or north of that focal path, and will have a comparatively dry season. The conditions will occasionally widen or spread out far enough to the west and north, to cover his place of observation, but occasionally only, and be of weak intensity.

If again his place of observation is in New Mexico, Western Kansas, or Nebraska, Minnesota, Wisconsin, or Upper Canada, the northern edge of the conditions will reach him still less frequently, and the weather will be comparatively still, cold and clear, with great uniformity.

If again his place of observation be Fort Yuma, in Southern California, or at the extreme northwestern point of New Mexico, or at the extreme northwestern point of Nebraska or Minnesota, he may be reached by the southern edge of the *Pacific conditions*, or their influence, during certain years in the early part of the decade; but in the latter years of the decade it will be otherwise, and no conditions will reach him from either system; and the weather will generally be clear and unaffected by anything but latitude and altitude.

If his place of observation is west of the Rocky Mountains, and any considerable distance north of the line drawn through Forts Yuma and Laramie to northwestern Minnesota, he will be covered by the Pacific system of conditions, which increase in intensity and frequency from San Diego to Astoria, and northeastward of them, and are comparatively regular in their frequency and intensity according to the latitude. This, I think will be sufficient to satisfy the reader of the importance which the element of place bears to the problem. But if not, let us contrast a little. Is he in central or northern Alabama, in August, or September, of a normal year, he is in the dry season of that section, and if any condition passes over him, it is an exceptional belt of showers, drifting to the eastward from the focal path, or a West India gale breaking and moderating upon the coast.

That focal path is now far to the west over Texas and Mexico, curving to the north and east, spreading out in the early years of the decade, as far as Fort Yuma, and sometimes, as in 1854, as far as San Diego, to the west, and giving New Mexico and the northwestern states their summer supply of rain. And now in August the centre of the focal path of the Pacific conditions has moved up towards Sitka, and the whole coast from Astoria down, and the interior northeastward of it, are in a normal state of drouth. The cold, moist coast wind of the ocean moves in upon the land and up the sides of the Coast Range, and over their summits daily, for months without a shower, and without reaching the San Joaquin valley which is said to create them, belying the theories which at the same instant are being promulgated from at least thirty Professors' chairs in the eastern states. The intelligent reader, if uncommitted to theory, and capable of looking honestly at the matter, cannot doubt, if he would, the importance of this first preliminary element, location.

II. The second preliminary element to be taken into consideration, is the season of the year. The importance of this element is illustrated by what has already been said. The element indeed is closely connected with the preceding one, and both are connected with the question, *where at the time of observation, is the focal path of the system of conditions under which the observer is situated?* Is that focal path *over* him, or *south*, or *north*, or *east*, or *west* of him, and how far? And what are the probabilities that a condition, pursuing the path, will spread out so as to cover his locality at that season of the year?

III. The third preliminary element is, what year it is in the decade, and what the state of the sun's surface in relation to spots. We have shown the effect which the presence or absence of spots

upon the sun have in carrying the focal paths of the conditions
to the south in winter, and extending them north in summer, or
contracting them at both extremities, and affecting the rapidity of
their transit in the different portions of the decade, and in different decades. No further evidence can be required to satisfy the
honest and intelligent reader of the necessity of this element.
Indeed, his own experience must have satisfied him that there is
great diversity in the severity or mildness of the winters, and in
the earlier or later occurrence of spring, and the continuance of
the rainy or dry seasons. The facts I have developed show that
these diversities are dependent, in a considerable and appreciable
degree, upon the extent to which the power of the sun is affected
by the presence or absence of the spots, and that there may be
and probably are other causes aiding or neutralizing their effect,
which cannot now be regarded, but are sure to be hereafter developed. The Spectroscope seems to be opening the way for
further discoveries, and some of interest have already been made,
but I have not deemed them sufficiently proved, to be considered
in this branch of our inquiry.

There is another class of facts to which we have more than
once alluded, which are rather contingencies than elements, but
which bear on this subject and should always be borne in mind,
and those are—First, that changes of state may be induced by a
condition, which may *pass by* the particular locality, on one hand
or the other, without precipitating upon that locality ; and second,
that a portion of the conditions—one-fifth at least—which actually pass over any given locality, are of too feeble intensity to precipitate. When therefore, the question is whether there will be
rain or snow, and the changes of state indicate the influence of a
passing condition, these two contingencies should be borne in
mind, and the *intensity* of the induced states be regarded in coming to a conclusion.

With these preliminary elements and contingencies well understood and appreciated, we may come to the proposition which is
fundamental in this branch of our inquiry, viz : that all the proverbs and signs which are of any value in the prognostication of

the weather, are founded upon, or connected with the *states* induced by the forming, approaching, or passing conditions, and all others are empirical and worthless, and should be discarded. And the further proposition that the local observer may and must look to those changes of state, for the evidence on which to base an intelligent and reliable forecast. The time will come when the telegraph will extend his horizon and give more certain results. But it is not yet.

And now what are the states and changes to which we are to look to determine the immediate future? There are seven, and we have already enumerated them, and traced them in the order in which they occurred in two passing conditions. But we must here repeat and examine them.

First. The *weight* of the atmosphere.

Second. The *temperature*.

Third. The movements of the atmosphere, or *winds*.

Fourth. Its *clearness* or *cloudiness*.

Fifth. Its *humidity*.

Sixth. In respect to *precipitation*, rain, hail or snow.

Seventh. Its *electrical* state.

Strictly speaking, perhaps, the sixth should not be included if we were looking to a forecast merely, for the principal object in consulting the others is to determine the probability of the occurrence of that particular state; but our inquiry embraces the question of character and continuance, as well as of coming, and therefore I include it.

Let us then take up these states severally and successively, and see how far our propositions are sustained, and how we must answer for ourselves the inquiry involved.

The first in the list is the weight of the atmosphere. That it has weight is universally conceded. Like every other material thing connected with the earth, though exceedingly attenuated and easily moved, it is controlled by the attraction of gravitation, and unaffected by the revolutions of the earth.

Some philosophers have supposed that they have detected a diurnal tide produced by the attraction of the moon, but the fact cannot be said to be established, and if it was, the tide would be too infinitesimal to be of importance in this inquiry.

The weight of the atmosphere is measured in two ways—by the barometer, and the temperature at which water will boil at the time and place. The boiling point of water at the mean pressure of the atmosphere is 212°. It boils at a proportionately less temperature, when the weight is decreased by altitude or other causes. The boiling point of water is rarely used to ascertain the weight of the atmosphere with a view to prognostication, and I pass it by.

Some minor instruments are occasionally used with a view to prognostication; thus, solutions of camphor and other substances in alcohol, in long and narrow glass vessels, hermetically sealed or otherwise, are sometimes used, the greater or less solution or deposit of the camphor before or after a storm, being relied upon as indications. But upon this instrument three of the states operate, and to some extent neutralize each other. Pressure and humidity operate from without by endosmosis, but the principal agent is the increase or diminution of pressure upon the surface of the fluid, by the expansion or contraction of the vapor and air contained above the fluid in the vessel, if hermetically sealed. Another instrument similar in principle, is made by inverting a Florence flask and inserting its neck in another glass vessel partly filled with colored water. An increase of pressure in the atmosphere will drive the water up the neck of the flask, overcoming proportionately the resistance of the air by compressing it; but here again another and frequently counteracting state of the atmosphere is influential, and that is temperature. If the pressure of the air increases, and the temperature of the air increases also, the latter will counteract the effect of the former by expanding the air in the flask. When pressure increases and temperature decreases, the effect upon the instrument is very observable. So on the other hand, when temperature increases and pressure decreases, the combined effect is quite obvious, but the instrument is very sensitive to changes of temperature in the house and is not adapted to out-door exposure. Neither of these instruments are in general use or of material practical importance, though both are philosophical and not empirical.

The barometer is the principal instrument in use for the purpose of measuring atmospheric pressure. It is made, as you know, by filling a glass tube 32 or more inches long with dry mercury, and inverting it into a vessel or bag also containing mercury. The top of the mercury settles in the tube to about 30 inches at the level of the sea, during a fair-weather state of the atmosphere, in the polar zones. There is another kind of barometer termed Aneroid, which is made without mercury, motion being communicated to an index by the pressure of the atmosphere on the movable covering of a vacuum. The mercurial barometer, however, is the one principally in use, and the most certain and reliable.

The barometer indicates a slight diurnal increase and diminution of the pressure of the atmosphere in set fair weather, but that is of no importance to our present inquiry.

The barometer also indicates changes in the pressure of the atmosphere occasioned by the inducing influence of passing conditions. This indication is often the first change to be perceived. Until quite recently, and perhaps even now in England, they look to a *falling* barometer only to indicate approaching rain. For a long time the barometers were constructed with movable indices or pointers, and the word " *rain* " was placed upon the dial where the position of the pointer indicated a diminution of pressure below the mean, and the words " *set fair weather* " where the pointers indicated an increase of pressure above the mean.

It is unquestionably true, that all the passing conditions affect the barometer, and true, as a rule, that that influence is perceptible before the condition reaches the place of observation. It is further true that the effect of that influence is, sooner or later, to depress the mercury, and the depression is in proportion to the intensity of the influence. If these facts constituted all the elements of the proposition, the barometer would be a perfectly reliable instrument, but there are other facts which must be taken into the account. In the first place, the barometer at one season of the year, and in one class of conditions, is elevated by the first effect of that influence, and in respect to that class, elevation and not depression is the indication of its approach, and the measure of its intensity ; and in the second place, it has no invariable fair-weather standard, a departure from which, by elevation or depression, will indicate with certainty the character and intensity of the approaching condition.   Barometers were formerly constructed in England, as I have said, whereon the words "set fair weather" were placed above the ordinary mean, and "rain" at depressions below the mean, but they were found deceptive, and are not now so constructed.

The mean height of the barometer is generally stated by meteorologists, and in our text-books, as 30 inches at the level of the sea.   This is substantially accurate when taken for long periods and for all latitudes, but there is very considerable diversity in different localities.   A few of the mean heights are as follows :

The mean height of the barometer in England, as kept by Howard for a great many years, was   -   -   29.823
The mean of London, as kept by Daniell, was   -   29 92
The mean under the central belt is   -   -   29 92
   "   in the N.E. trades of the Atlantic it is   -   29 97
   "   "   "   "   "   Pacific   -   29.99
   "   "   S.E.   "   "   Atlantic   -   29.95
   "   "   "   "   "   Pacific   -   30.01
   "   off Cape Horn it is   -   -   -   -   29.20
At the Cape of Good Hope   -   -   -   30.22

The mean from the northern limits of the trades in the northern hemisphere is higher than in the trades, differing greatly in different places, from several causes, the principal of which is the difference in the volume of the equatorial current, or upper trade, which passes over the locality.   The annual mean of the United States does not differ materially from 30 inches.

But the mean of 30 inches, thus spoken of, is the average of *all fair weather elevations*, and *foul weather depressions*, and as the foul weather depressions are greater in extent than the fair

weather elevations, and the latter differ very greatly in different
climates in the same latitude, according to their intensity, it is
obvious that 30 inches is not a reliable fair-weather standard
for any particular place. The greatest known fluctuations of the
barometer were $3\frac{1}{2}$ inches, and of that range, at least two-thirds
was below the mean of 30. The barometer very rarely rises
above 31 in this country, and sometimes falls as low as 28, making
a range of 3 inches, and it would be safe to say that the ranges,
as a rule, are twice as great below 30 as above it.

   And there are other difficulties. The ranges and mean elevation of the barometer differ in different years, in different months
of the same year, and in different localities. The reader must
endeavor to get the true fair-weather standard of his locality.
To aid him in doing this, regard must be had to the following
rules and considerations.

   In all latitudes the mean height is low when the passing conditions are frequent and intense, and that is the main reason why
it is so low at Cape Horn. There is at that point an almost constant passage of intense conditions, with consequent alternations
of N.W. storm winds, and S.W. clearing off winds. Eastwardly,
at the Cape of Good Hope, the barometer ranges much as in the
northern hemisphere under similar circumstances, and so it does on
the west coast of South America, above the Horn. For the same
reason mainly, the mean height of the barometer is lower in
Europe than here. They have more frequent, though less intense
conditions, and the changes in the barometer are more frequent
and the ranges less. At the same time their volume of upper trade
is also less. In this country the range of the barometer is affected
by the same causes. It ranges lower, therefore, under the focal
path of the conditions, and during the rainy reason, than upon
either side of it. And its range differs in different years also,
because the frequency and intensity of the conditions differ in
different years. The first is shown by the following table:

MONTHLY VARIATIONS OF PRESSURE.

	January.	February.	March.	April.	May.	June.	July.	August.	September.	October.	November.	December.	Year.
Toronto, Canada,····	29.618	.614	.622	.677	.565	.577	.589	.655	.647	.663	.626	.643	29.621
Cambridge, Mass,···	30.024	29.998	.992	.991	.969	.915	.967	30.035	30.044	30.029	29.996	30.017	29.998
Philadelphia,······	29.960	29.967	29.912	.924	.886	.891	.916	.943	.971	.949	.941	.957	29.932
Washington,········	30.147	.112	.029	.029	29.995	29.963	29.999	30.031	.067	.133	.107	.060	30.051
Cincinnati,········	29.335	.308	.315	.295	.245	.271	.331	.356	.346	.374	.849	.356	29.828
Hudson, Ohio,······	29.584	.816	.759	.785	.710	.757	.829	.846	.870	.846	.844	.777	29.806
St Louis,··········	29.616	.618	.598	.540	.493	.495	.544	.567	.584	.619	.623	.630	29.578
Glenwood, Tenn.,···	29.726	.674	.604	.513	.567	.604	.607	.618	.592	.586	.607	.630	29.611
Sitka, Alaska,·····	29.745	.790	.804	.877	.922	.969	.909	.957	.876	.761	.676	.690	29.837
Para, Brazil,······	29.910	.926	.914	.943	.952	.982	.992	.988	.946	.931	.907	.897	29.941

The foregoing table is taken from Blodgett's "Climatology of the United States," and contains all the positions given in that table from the United States. It will be observed that there is at every station a decrease and increase during the months of the year, and through a substantially regular curve. Meteorologists have attributed the decrease in spring and summer, and the increase in winter, to the decrease and increase of temperature, but they mistake. The increase and decrease, *on the face of the table*, must be attributed to other causes, for at Para, Brazil, where the barometer is at its minimum in November and December, and at its maximum in July, there is no difference in the temperature of the place which will account for the elevation in July, and the depression in December. Para is near latitude 2° S., and the sun is not farther north of it when the maximum elevation is reached in July, than it is south of it when the minimum is reached in November and December, nor is there a material difference in the temperature of the two periods. Moreover the elevation *decreases* as the sun *approaches* from the north, and the *heat increases.*

Observe again, that at Cambridge, Mass., the minimum is reached in June, and there is a very rapid increase in July and August, which are the hottest months in the year, and the elevation is greater in August than in February. In those months the focal path is at the north, and the conditions are occasional. Observe again, that the minimum in Philadelphia is reached in May, and there is an increase during July and August. The same thing is true of Washington in the District of Columbia. At Toronto, Canada, there is an increase during the spring until May, when there is a sudden depression, continued through June, followed by an increase during July and August. At Hudson, Ohio, the minimum occurs in May, conforming to the more rapid extension of the focal path to the northwest of the mountains, and there is a rapid increase from thence until September. At Glenwood, Tenn., the minimum is reached in *April*, and there is a rapid increase from thence to August. At Sitka, the minimum is reached in *December*, and the *maximum* in *July* as in Para.

The facts disclosed in this table, conform perfectly with my observation. The barometer ranges the lowest when the focal path of the conditions is over us in its ascent to the north, in the spring and early summer. Thus, Glenwood, Tenn., has its minimum in April. St. Louis, Washington, and Philadelphia have their minimum in May, which is their rainy season. Cambridge, Mass., has its minimum in June, and in Toronto, Canada, the fall from April to May and June is sudden and great. At Para, the belt of rains is at the north in midsummer, when the maximum occurs, and the subsequent rapid fall is coincident with the return of the belt and rainy season, as it is here. The rain fall at Sitka is peculiar, as may be seen by referring back to the table we have already given ; and by referring also, to a table given of the winds it will be seen that their winds are in like manner peculiar. The rainy season does not set in, as will be seen from the table, until August, and their heaviest falls of rain are in September and October, the rainy season continuing until December. Thus there is a substantial conformity between the depression of the barometer and the occurrence of the rainy season, at Sitka also. It is unnecessary to pursue this subject farther. I am satisfied, and the reader should be satified, that neither the fact nor the cause I have assigned can be successfully questioned. Certainly if he will observe for himself, he will be satisfied.

In conformity with these facts, it may be stated that the barometer ranges highest on the north and south of the focal path of the conditions, and in the winter and autumn. It is frequently true in New England, that its mean range is so high that it does not fall below 30 during very considerable snow storms in winter. It is common also in the eastern states for it to rise from $\frac{4}{10}$ to $\frac{6}{10}$ above 30, when a northeast snow storm is approaching, and still higher before a thaw. It is during that period of the year, and generally in the winter months, that it attains its highest elevations. I have rarely known it to reach 31 at any other season. The reader must be careful here, not to mistake. I am speaking of the barometer standing at, or corrected for, its elevation at the level of the sea. The reader must therefore make allowance for *his eleva-*

tion, whatever it may be.   If his elevation above the level of the
sea is 917 feet, he should allow one inch and proportionately for
a less elevation.   If 1,860 feet, he should allow two inches, and
proportionately for a less elevation.   If 2,830 feet, he should deduct three inches.   If 3,830 feet, four inches.   If 4,861 feet, five
inches.   This is according to the formula of La Place, and includes all corrections.   But it is sufficiently accurate for all practical purposes in a matter of this kind, to say, that the barometer
sinks an inch for every thousand feet of ascent, and as a matter
of convenience, it is well enough to do so.   When therefore I
say that the barometer frequently rises from $\frac{4}{10}$ to $\frac{6}{10}$ above 30,
I mean a barometer standing, as mine does, substantially at the
level of the sea; and the reader who tests the accuracy of my
statement, must deduct $\frac{1}{10}$, $\frac{2}{10}$, $\frac{3}{10}$, $\frac{4}{10}$ or more of an inch, if his
altitude is 1, 2, 3, 4, or more hundred feet above the level of the
sea.

What I have said in reference to the elevation of the barometer in the winter season, on the north of the focal path, must be
understood as stating what is generally true.   There are occasionally severe and exceptional winters, when the focal path descends
low, and is *unusually contracted and concentrated,* when the barometer remains at about 30 and sometimes even less for very
considerable periods.   I have known two or three such winters,
the wind holding steady but light from W.N.W. to N.W.   The
most remarkable instance I have known, occurred during the cold
decade from 1830 to 1840.   In one instance there was no southerly wind, and no condition passed over us and very little upper
trade for six weeks.   President Dwight, in his Travels and History of New England, describes two such winters during the cold
decade from 1780 to 1790.   He says : " In 1787 the west wind
began to blow about the 20th of November, and continued its
progress with only four short interruptions until the 20th of the
following March—somewhat more than 100 days.   During the
whole time the weather for the season was very cold."   Also, " In
1780 the wind blew from the west more than six weeks without
any intermission, and during the whole of this time, the weather

was so cold that snow did not dissolve sufficiently to give drops from the southern eaves of houses." It is during the winter and spring that the N.E. storms are most common, and their approach is foretold by a rise of the barometer, and it is during this period when the focus of the condition passes to the south of us, that the wind backs from the N.E. into the N.W. as the storm passes by and it is clearing off.

Having thus considered the peculiarities attending the range of the barometer on the north of the focal path, let us look at them when the focal path is over us, and we are having our rainy season. The barometer does not then rise so high, and sinks lower. Nor does it always ascend in fair weather between the conditions. The fair weather intervals between the conditions are less—frequently not more than twenty-four hours,—and sometimes there are scarcely any intervals. The air does not become dry even with the northerly wind. Wood work swells and everything becomes damp and sticky, and frequently quite wet, and although the barometer vibrates up and down, it does not rise promptly to any considerable elevation after the passage of a condition, as at other seasons.

When the focal path has moved to the north of us, and we begin to get the southerly winds of the passing conditions and they "clear off warm," and for that reason "summer breaks upon us" with some of those hot days which surprise us with their contrasts, the peculiarities in the action of the barometer change again. It does not now indicate the approach of a condition by a sudden rise. It ascends in the interval to its ordinary summer fair weather position, and what that is in the particular locality the observer must determine. but rarely takes any of those high flights to which it is subject in winter and spring. When it feels the influence of the approaching condition of that season, a belt of showers, it commences falling steadily, and falls rapidly or slowly in proportion to the intensity of the condition, and the rapidity of its approach ; rising again, slowly and steadily, after the condition is past. This is true in respect to conditions which approach overland from the west. It does sometimes rise on the approach of an intense hurricane condition up the coast, and afterwards fall rapidly and very considerably at those places which are covered by the condition.

I do not intend to give you specific " Rules " for observation—you can and must deduce them from the facts—but in respect to this matter of a fair weather standard I will say:

1st. Your mean fair weather elevation of the barometer will range between 30 and $30\frac{2}{10}$ inches, deducting therefrom $\frac{1}{10}$ of an inch for every one hundred feet, or to be perfectly accurate, every 91 and $\frac{7}{10}$ feet of altitude above the level of the sea. Or to be sufficiently accurate, $\frac{1}{10}$ of an inch for every 91 feet.

2d. That fair weather elevation will average higher in winter when the focal path is south of the observer, *in normal years*, than in midsummer when it is at the north of him. It will average lowest when the focal path is over him in spring and early summer.

3d. A fair weather point cannot be fixed for either period, except approximately. If I should attempt to fix them I should say, $30\frac{2}{10}$ for the period when the focal path is *farthest* south, if not unusually concentrated, $30\frac{1}{10}$ when it is *farthest* north, and 30 when *centrally* focal, *deducting*, as in rule 1st, for *altitude*, and scaling gradually from one to the other, as the focal path changes its position. But it must be borne in mind that a great contraction and concentration of the upper or counter trade, down upon the southeastern states, will produce great cold and a low fair weather barometer west and north of them in very severe winters.

It is not my purpose, as I have said, to give you any general rules or maxims in relation to the indications of the barometer. Such a collection could not be of general application, in all parts of the country, in all years, and in all seasons alike. No collection of rules has been made for this country. Several have been made for Great Britain. That of Dr. Brand, which embraced but seven rules, was for a long time, the popular one. It may be found in the 2d Volume of the " New American Cyclopedia," page 657.

A much longer list has recently been published by Mr. Steinmetz in his work entitled " Sunshine and Showers." It may be instructive to take up this list and analyze it, and see whether any additional light is thrown upon the subject.

His rules are as follows :

I. " It the barometer has been about its ordinary height, say 30 inches at the sea-level, and is steady or rising, whilst the thermometer falls, and dampness becomes less, then northwesterly, northerly, or northeasterly wind, or less wind, less rain or snow may be expected."

This merely describes the clearing off change of a condit'on of very weak intensity.

II. " If a fall take place with a rising thermometer, and increased dampness, wind and rain may be expected from the southeastward, southward, southwestward."

This describes three indications of a coming southerly rain, a change in weight, temperature, and humidity.

III. " In winter, a fall with a low barometer foretels snow."

I have seen nothing of this kind in our climate.

IV. " A fall of the barometer, with unusually high temperature for the season, will be followed by a southerly wind, with rain ; and during the gale the barometer may begin to rise, and be followed by another from the northward ; but then the thermometer will fall for *change of temperature,* and show the *direction* of the coming wind."

This describes the programme, in part, of an *intense* belt of showers or of a southeaster, like those on pages 14 and 24.

V. "But northerly winds will follow a fall of the barometer at all times, if the thermometer be low (for the season) and southerly wind if the thermometer be high for the season."

The first part of this is a mere description, by way of contrast, of the order of events when the condition has nearly passed, and it is about to clear off—the second, the order of events when it is approaching.

VI. " The barometer is *lowest of all* during a thaw following a *long frost.*"

The same is true here in our *winter thaws,* which are *warm southeasters.*

VII. " In like manner the glass falls very low with south and west winds in general."

All this means is, that the barometer falls, as a rule, lowest on the southerly side and with the southerly winds of a condition.

VIII. " If the barometer falls with the wind in the north, we must prepare for weather of the worst description,—rain and storms in summer, snow and severe frosts in winter and early spring."

A general description of the comparative severity of north-easters.

IX. " A rapid rise in winter, after bad weather, is usually followed by clear skies and hard, white frosts."

This is merely descriptive of a clearing off, and is repetition.

X. " During frosty weather, if the barometer falls it denotes a thaw ; but if the wind goes again to the north, the mercury will rise and the frost set in again."

Descriptive of the passage of a weak, brief, and imperfect condition.

XI. " During broken and cold weather in the winter, with northerly winds, a sudden *rise* of the barometer foreshows a change of the wind to the southward, with rain."

It is otherwise here—such a sudden rise is followed by a north-easter.

XII. " If, during a northerly and easterly wind in winter, the barometer rises slowly, expect snow or cloudy weather."

Common in our winter northeasters, where the first rise is slow, but becomes rapid before the storm appears.

XIII. " In a continuous frost, if the mercury rise, it will certainly snow."

Same as preceding.

XIV. " Whilst the barometer stands above 30, the air must be very *dry* or very *cold*, or perhaps both, and no rain may be expected."

True enough probably. 30 is probably the fair weather point there.

XV. " Of course a great rise in summer means dry and warm weather, and if this be of long duration, the question is, how will it end? If a sudden fall, of two or three-tenths occur, we must prepare for a storm of rain, or thunder and rain. This usually follows a period of unusual heat, unless northerly and easterly winds are to succeed with drouth."

This is scarcely intelligible. If true there, it is untrue and is inapplicable here. A " *great rise* " does not precede our summer belt, nor is it common during our drouths.

XVI. " Thunder storms are not always foretold by the barometer ; indeed the barometer cannot indicate electricity, as some of the barometer maxims seem to infer. We must consult the clouds and our feelings for thunder. The barometer falls, but not always, on the approach of thunder and lightning. It is in very hot weather that the fall of the mercury indicates thunder. Thunder clouds rising from northeastward *against the wind*, do not usually cause a fall in the barometer, simply because they are borne onward by a *polar current*, which is dense. An approaching thunder storm is indicated by a rapid decrease of the daily *evaporation* during hot weather."

This is unimportant, in relation to the barometer here. Thunder clouds do not *rise from the northeast* with us.

XVII. " A rising barometer—with a southerly wind—is generally followed by fine weather; but then it will be generally observed that a change of wind occurs at the same time, or very shortly after."

A mere general description of the action of the barometer after a clear off, and before the wind has had time to change.

XVIII. " During storms the mercury will be seen to rise and fall continually, and to be in a state of general agitation. While this lasts, no hope of good weather can be entertained."

Sometimes occurs here in the rainy season.

XIX. " Although some rain may occasionally fall with a high barometer, it is usually of trifling amount, and of short duration."

Applicable only to *passing squalls* from the northwest, such as described, and conditions passing by at the south.

XX. " When the barometer *stands* very low indeed, there will never be much rain; although, on the other hand, a fine day will seldom occur at such times. The air must be very warm or very moist or both, and so there will be only short heavy showers, with sudden squalls of wind from the west."

Describes the action of the instrument in their rainy season. Here everything is more intense in that season.

XXI. " A sudden fall of the barometer, with a westerly wind, is sometimes followed by a violent storm from N.W. to N.N.E."

Unknown here.

XXII. " In summer, after a long continuance of fair weather, the barometer will fall gradually for two or three days before rain comes; but if the fall be very sudden, then a thunderstorm is at hand."

True everywhere in respect to a belt of showers; but it is repetition.

XXIII. " When the barometer is *high*, dark, dense clouds will pass over the sky, without rain; but if the glass be *low*, it will often rain without any appearance of clouds."

The dark, dense clouds are N.W. scud, from which no rain falls except in short squalls—rain without cloud I never saw.

XXIV. "If wet weather happens soon after the fall of the barometer, there w ll be little of it. In fair weather, if the glass falls much, and remains low, expect much wet in a few days, and probably wind. In wet weather, if the glass continues to fall, expect much wet."

The first part of this relates to belts of showers which probably do not precipitate much in that climate. The rest looks like the occurrence of the rainy season.

XXV. "The barometer sinks lowest of all for wind and rain together; next for wind, except it be an east or northeast wind."

That is to say it will not fall so low in a weak condition that does not precipitate, as it will in an intense one that does.

XXVI. "Instances of fine weather with a low glass occur, however, rarely; but they are always forerunners to a *duration* of wind or rain, if not both."

Not intelligible, unless it refers to an occasional clear day in the rainy season.

XXVII. "Our storms are generally announced by a fall of the barometer, and a tendency of the wind towards east and south ; the return of fine weather, by a rise and a pretty strong west wind, apparent in the motions of the *clouds* before it is felt below."

A restatement of the order of events on the occurrence of a class of conditions—applicable everywhere in the hemisphere.

XXVIII. "A great and sudden change, either from hot to cold or from cold to hot, is generally followed by rain within twenty-four hours ; because, in the change from hot to cold, the cold condenses the air and its vapor, which being condensed, falls in rain; and in the change from cold to hot, the air is quickly saturated with moisture, and as soon as night comes on, the temperature is lowered again, and some of the abundant moisture falls in rain."

Not true here. A great change from hot to cold *follows*, but *is not followed by rain* in our climate, unless it be a brief N.W. squall.

XXIX. "When heat rapidly follows cold, the evaporation, which was checked by cold, is carried on very rapidly, in consequence of the diminished pressure of the air by heat. The less the pressure of the air, the more rapid the evaporation of moisture."

The application of this to the barometer is not apparent.

XXX. "The barometer varies most in winter, because the *difference* of temperature between the torrid and temperate zones is much greater, and produces a greater disturbance in the state of the air. It varies least in summer, because the temperature of our island is so nearly equal to that of the torrid zone or hot regions of the earth, that its state is not much disturbed by interchange of currents."

This is philosophy, and not a "rule."

XXXI. "Heat and cold do not of themselves affect the barometer, but because cold weather is either generally dry or rough, with N.E. winds, the air being denser and heavier—therefore the mercury *rises* in cold weather, but in warm weather, the air is often moist and less dense, with S.W. winds and therefore lighter, and so its pressure is less, thus causing the mercury to fall."

Philosophy again and unimportant.

XXXII. "If the top of the column of mercury be convex, or higher in the middle than at the sides, it is rising; if lower, or concave, it is falling. This is caused by the attraction of the glass-tube in contact with the mercury."

True everywhere, and worthy of attention where slight changes are important.

"Such are the chief rules and maxims of the barometer with reference to agricultural pursuits, the seasons, and the crops." (Sunshine and Showers, pages 259–265.)

I have thus separately examined the rules copied from Mr. Steinmetz. It is obvious they might be very much condensed and much more clearly expressed. It is equally obvious that they might be much more certain and reliable, even for that climate, if the action they describe was connected with an intelligent description of the different conditions to which the different rules relate. No rules can be of value unless so connected with a description of the peculiar conditions on which they are founded. In the foregoing pages I have given my readers a description of the action of the barometer, *in connection with the description of the conditions,* and at the *various seasons,* and they will find little or nothing in the English rules that will be new, or independently useful to them.

The second state which is an element in prognostication is temperature. This is important in connection with the other elements, as an indication of the approach of a condition, and important also as furnishing an indication of its character and continuance, but still more important to be considered in relation to its changes, when the condition is passing away. Let us consider it in the three points of view.

First, as furnishing an indication of the approach of a condition. In winter, when the normal state of clear cold weather exists, among the other early indications of the approach of a stormy condition, is an increase of temperature. I have illustrated this so often in the course of the developments I have made, that it ought now to be clearly impressed upon your minds. Under the circumstances mentioned, it always *"moderates to storm."* No more perfect illustration or description of that truth can be given, than in the two descriptions of storms as they occurred within the Arctic circle, copied in the last chapter, from the register of Dr. Kane. It will be sufficient therefore to add here, that as an indication of the approach of a winter-storm, which is invariably present, and nearly simultaneous with the rise of the barometer and the appearance of cirrus condensation, it should always be looked for and regarded. Unless there is a moderation of the temperature, commencing about the same time or soon after, the other indications should be regarded with suspicion.

When the season advances, and the mean daily temperature is above 65°, an elevation of the thermometer is not to be expected upon the approach of a *northeast* storm. In the hotest seasons of the year, the temperature never exceeds 70°, during a northeast storm. When it ranges above that in the daytime, but on a given day fails to rise higher, or having risen falls to that point or below, it is as certain an indication of the character of the approaching condition, as the precedent, sudden and considerable rise of the barometer is.

During the rainy season, changes of temperature, though less distinctly marked, are worthy of note, as indications. The prevalent winds during that season,—if it occur in the spring or early summer,—are easterly, and their *chilliness* proverbial.

In relation to the approach of a belt of showers, temperature is an important indication. It always rises high, and *above the mean of the season*, on the southeasterly side of those belts; it is therefore a distinguishing characteristic, as well of their approach, as their intensity. The "*hot spells*," as our fathers called them,—"*heated terms*," as it is now the fashion to characterize them—are an elemental part of the condition, and *the excessive heat is created by the cause which organizes and continues the condition.* Very hot weather sometimes occurs in midsummer, during drouths, which is not connected with, or a part of an approaching condition, but is the mere effect of an unclouded sun, operating upon a dry and heated soil. Two characteristics, however, distinguish the incident heat of the condition, from the mere heat created by the sun. 1st, The latter does not rise so high, and it cools off by radiation, at night, rarely rising to 90°, in the day time, east of the Rocky Mountains, unless there has been a *long continued* drouth, and cooling off during the night to 70°, or below. In the 2d place, the incident heat of an approaching condition is both humid and electric—a state which is variously described as sultry, muggy, close, &c.,—and the temperature continues high through the night and into the morning when the condition is to arrive. The humid, electric, muggy heat of an approaching intense belt of showers, would be scarcely endurable, if it was not tempered by the accompanying incident, southerly wind. All this will address itself to common experience.

Temperature is also an important element in relation to the character which the storm will assume. In the middle latitudes of the country it is usually an interesting inquiry whether a coming winter storm will be one of snow or rain, and one of the elements by which this question is answered, is temperature.

Snow sometimes falls from the northwest scud in squalls, for a few moments or even half an hour, early in the spring and late in the fall, when the thermometer is considerably above the freezing point. Snow sometimes also falls in the early part of storms, after the thermometer has risen above the freezing point, but unless the thermometer falls again soon, the snow will turn to rain; for snow does not often fall for any great length of time with the thermometer above the freezing point, and when it does, it is usually in large flakes, which indicate that it is about to turn to rain. Snow sometimes falls with the thermometer near zero, but such instances are rare. We have seen that in the Arctic circle, in a storm described by Dr. Kane, the thermometer rose to about zero. So when a snow storm is approaching in the middle latitudes of this country, if the thermometer is near zero, the temperature generally rises about 20° before the snow falls. The usual range of temperature in snow storms, therefore, between the latitudes 35° and 55°, is from 20° to 30°, according to latitude, and the severity of the season. Between 40° and 41°, according to my observation, the thermometer ranges from 24° to 30° from the commencement to the close of such a storm, in the majority of instances.

Rain sometimes falls when the thermometer is low. I have often heard the expression, "it is too cold to snow," and also the expression, "it is going to rain, it is too cold for snow," and there is seeming truth in both. As to the first, it is to be observed that it is obviously founded on the experienced fact that it generally does moderate up to a certain point before it snows, but the moderation does not cause the snow; it is but an incident effect of the storm action. So it is an experienced fact that rain frequently falls when the thermometer ranges lower than the ordinary snow point, freezing to the trees and constituting what is commonly called an *ice-storm;* but there again the fall of rain is not occasioned by the continued low temperature, but by a warm southerly current in the upper part of the surface story, the storm being exceptional, and having its focus to the north. Neither expression therefore is philosophically correct.

Very warm southeasters sometimes occur, even in severe winters, and one or two may be expected in ordinary winters, in January, for "January thaws" are proverbial. They are usually southeasters, caused by a very large, concentrated, and exceptional irruption of the counter or upper trade, west and north of the then focal path; having a warm area on their eastern front, corresponding to the hot area which they present on the same front in the summer months. The one which I described in the first chapter, and which is represented by the diagram on page 24, is a fair representation of such a storm; except that, as that one occurred in autumn, its path at its commencement was farther to the west than is usual with winter southeast thaws. The thermometer, in the advancing portion of such a storm, will attain an elevation of from 40° to 50°, and sometimes more, whatever its previous depression may have been; and it notunfrequently happens that the thermometer is near zero in northern New England when it is at 45° in the same latitude at the Mississippi river, on the front of the storm. But in such cases, the warm area of the advancing storm raises the temperature to 45°, when it arrives over New England the following day.

Warm periods of several days continuance, without distinct storm action, sometimes occur, even in midwinter, especially in the early years of the decade. In such cases, the barometer attains gradually a very high elevation, and it is at such times that I have seen it as high as 31 inches. The thermometer does not then attain a sudden and excessive elevation—indeed it never does except from intense storm action in our latitudes—but it does attain an agreeable elevation, commonly and aptly expressed by the term "*mild.*" Those exceptional states of the weather are produced by an increased volume of counter or upper trade, *generally diffused* over the whole country, and not concentrated in a large body, or a long band, or upon the usual focal path, so as to make a storm.

The thermometer may also be watched during the existence of storms, with reference to their continuance. In a northeast storm in the spring of the year and at other seasons, when the focus of the storm is to the southeast of the observer, if the thermometer falls and the wind backs into the north, the rain is usually at an end. The wind will continue to back to the N.W., and it will soon after light up in that quarter, and fair weather return as the storm passes away to the eastward. If, however, it veers back to the N.E. it will continue. In those northeasters where the focus is over the observer, or to the north of him, the cessation of the rain is usually accompanied by a rise in the thermometer, and a temporary lull in the wind, followed by the wind afterwards coming out from the west and hauling slowly into the northwest. A similar lull sometimes occurs in southeasters with a fall of the thermometer. The rain is then over. If the wind has been very heavy from the southeast, this lull will be followed by a sudden change to the northwest. This is most common in the intense hurricane storms which come up the coast. In a majority of the southeasters, the wind hauls gradually round through the southwest and west to the northwest, the thermometer rising gradually.

A very sudden change in the thermometer is frequently experienced when a belt of showers is passing over us with or without a change of wind to the northward. The southerly wind usually lulls before the precipitating body of the cloud reaches us, and in many—though not perhaps the majority of instances— there is a heavy gust from the westward preceding the fall of rain and continuing during that fall. Such gusts are frequently accompanied by a very considerable fall in the temperature.

All the conditions which have a southerly wind, and a warm or hot area on their easterly or southeasterly sides, have northerly winds and a cool or cold area on their westerly or northerly sides. Changes in the thermometer in a few hours are sometimes very great. It is impossible to describe in adequate terms the importance of understanding and heeding this fact, for it is thus that the sudden changes from heat to cold are produced. Let it be understood that the frame which is now sweltering with all its pores open, in a humid atmosphere of 85° or perhaps 90° on the hot side of a condition, will, as a matter of course, and by the operation of perfectly intelligible and unalterable laws, in a few hours be exposed to the chilling temperature of 60° or 65° under the cold side of it, and the additional chill occasioned by a rapid evaporation in its peculiar dry air. And let it be understood that it is by such changes, so occurring and so capable of being forecasted, but which are *unregarded*, that a great majority of the diseases which bring suffering and death to our dwellings, are produced. A word to the wise is said to be sufficient, but how many *parents* are wise for their children, to protect them against these fatal changes? How many are wise for themselves, and aim to understand the inevitable change, and guard against its influence?

In the hope of exciting your attention to this matter as it deserves, let me direct it to the following diagram, copied from a work by Dr. Torrey, which represents these sudden changes in a striking and accurate manner. The Dr. took first the monthly and annual temperature at Key West, which is an island at the southern extremity of the Peninsula of Florida. He first represents the isothermal line of 76°, which is the mean annual temperature of the place, by a dotted line. He then represents by a large white line the mean monthly temperature, which is at 68° in the month of January, and rises gradually, though somewhat irregularly to 81° in July, and descends again to 68° in January. Crossing this line, are the connected perpendicular lines, showing the extreme changes from cold to hot which occur in each month. The changes are not very great at Key West. But in contrast with them, he gives a representation of the annual temperature and monthly changes at Fort Snelling, Minnesota, projected upon the diagram in like manner. These projections represent the temperature and changes at extreme points, and the temperature and changes at the place where the reader resides, if between the two, will vary in a corresponding degree. But he may see at a glance how very great these changes are in all parts of the country, and during every month, and the importance of watching them and guarding against their fatal influences. The horizontal lines represent degrees of Fahrenheit.

And in this connection let me say, that it is after one of these belts has passed, and after its northerly wind has blown for a day or two, that we have our unseasonable frosts. That is the time to look out for them in August and September. In a majority of cases the dreaded "first frost," does not occur early in September, unless the northerly wind continues two days; but it does sometimes come in the first night after the clearing off. Frost makes at the surface of the ground, as soon as my thermometer, hanging 5 feet from the ground, falls below 40°, and if the thermometer is at or below 50° at sundown, frost is very probable, if the night is still and clear. After the wind has blown from the northward through

FIG. 38.

the day the thermometer falls rapidly after night fall. Wind or cloudiness prevent frost, but both may disappear before morning and frost ensue.

The third state which is an element in prognostication is the wind. The importance of this e'ement is apparent. The conditions themselves are characterized as *northeasters, southeasters, etc.*, by their *special* and *pecu'iar* winds ; and that the southerly winds, according to their freshness and earne tness, are reliable indications of the approach and intensity of the showery conditions, we have more than once had occasion to observe. *Breezes are often local*, but there are no *fresh, earnest winds*, unless *created by the influence* of some approaching or passing *condition*, and the quarter from which they blow, and their force, are consequently among the most reliable indications we have. Ordinarily the wind and its direction and force are felt, or indicated by the wind vane ; but it is often observable before it is felt at the surface, by its scud, or by sounds. Thus, sounds from the point from which the wind is blowing above, heard with unusual distinctness, are sometimes the very first changes noticed.

The roar of the surf, or breaking of the waves on the shore, when great bodies of water are disturbed by a precedent storm-wind, often heard before the wind is perceived on the land, I have already alluded to. And thus Virgil:

> "When storms are brooding—in the *leeward gulf*
> Dash the swelled waves; the mighty mountains pour
> A harsh, dull murmur; far along the beach
> Rolls the deep rushing roar."

Those of us who live on the shores of Long Island Sound, or any of the harbors or coasts of the Atlantic, or the lakes, can realize the truth of the foregoing. I have often heard the roar of the surf before an easterly storm, several hours before the wind was felt at my house three miles from the shore.

The moaning or whistling of the wind all have noticed. It is not uncommon to hear the expression, " The wind sounds like rain." Jenner says :

> " The *hollow* winds begin to blow."

And Virgil:

> " The *whispering* grove
> Betrays the gathering elemental strife."

This whispering is the motion of the leaves; and they are often stirred by a peculiar motion which is not that of wind. Sometimes every leaf upon a tree may be seen *vibrating* with an *upward and downward* motion, when there is not wind enough to stir a twig. This interesting phenomenon is electrical. Trees, and all vegetables, confessedly discharge electricity, and such discharges move the leaves, when very active. But the wind soon follows.

With us, sounds can be heard more distinctly from the east or south, before storms, according to the character of the coming wind. Howard mentions an instance when he heard carriages five miles off. Steamboat paddles, railroad cars, and other sounds, are often heard a great distance. The distance at which the now common steam-whistle is heard, and the direction, is not an unimportant auxiliary indication of the weather. Howard attributes these peculiar phenomena to the " *sounding board*," made by the *stratum of cloud;* but sounds may be heard from the north-west, when there is no condensation, and the wind is from that quarter, and also from the east when it is not cloudy; and in a level country the village bells often tell the direction of the current of air just over our heads when we do not feel it at the surface. The wind is undoubtedly moving in a rapid, and perhaps invisible current, not far above us. If from the east or south, it betokens rain; if from the western quarter, fair weather.

I need not dwell upon the characteristics of the different winds or their prognostic indications. Those different characteristics and indications have been directly or indirectly before us in every stage of our inquiry, and should have become very familiar. And they must again come under review.

The next element in prognostication is that of clearness or cloudiness. We look to the movements of the barometer, thermometer, and winds, as well as to some other elements not yet considered, to discover the coming of a condition, before it is visible, but in the clouds of a condition, when it comes in sight, we have actual, visible evidence of its approach and character. The first cloud seen is ordinarily the cirrus, which overlies the condition and which is not only first visible because the most elevated, but because it extends in every direction farther than the other strata of a storm.

All the forms of cirrus are seen in the advance condensation of the conditions, except what might properly have been called the cumulo-cirrus, but what Howard termed the *cirro-cumulus*, and which you will find represented on page 34, on the left of the diagram, and indicated by two birds. It is a cirrus or incipient cirro-stratus, *broken up* into small, distinct, and separate heaps, resembling fleeces of wool laid apart, near, but not in contact. It is properly called the *fleecy-cloud*, and the *only one* that is, and it occurs independently, in set fair weather only.

Hence the proverb:

> " If *woolly fleeces* strew the heavenly way,
> Be sure no rain disturb the summer day."

But the early cirrus condensation of the conditions are not always visible. There may be enough of it to affect the brightness of the sun, or the moon and stars, or to occasion halos when it is not sufficiently dense to assume the appearance of clouds. It is a turbid or misty condensation rather than visible cloud. Sometimes it is of a smoky character, like that which attends midsummer drouths, or the shorter dry spells of autumn, called Indian Summer, and gives to the sun a blood-red appearance. But ordinarily, when it constitutes the advance condensation of an approaching condition, it changes the appearance of the sky from a deep azure to that " lighter hue " which Humboldt describes as preceding the arrival of the central belt of rains from the south. It was found by Gay Lussac, and has been by other aeronauts to have the form of cloud at the height of 20 to 30,000 feet, when not visible at the earth, except as obscuring mistiness.

The sun and moon have no immediate traceable effect in producing particular storms. Some valuable traditionary signs have been founded upon their appearance, but only as affected by interposing condensation.

Thus Virgil:

> "The sun, too, rising, and at that still hour,
>   When sinks his tranquil beauty in the main,
>   Will give thee tokens; certain tokens all,
>   Both those that morning brings, and balmy eve.
>
> *   *   *   *   *   *
>
> When Sol departs, his mighty day-task done,
>   How varied hues oft wander on his brow.
>
> *   *   *   *   *   *
>
>   If the ruddy blaze
> Be *dimm'd* with *spots*, then all will wildly rage
> With squalls and driving showers; on that fell night
> None shall persuade me on the deep to urge
> My perilous course, or quit the sheltering pier.
> But if, when day returns, or when retires,
> *Bright* is the orb, then fear no coming rain."

And again:

> "Mark, with attentive eye, the rapid sun—
>   The varying moon that rolls its monthly round;
> So shalt thou count, not vainly, on the morn;
> *So the bland aspect of the tranquil night*
> *Will ne'er beguile thee with insidious calm.*"

And again of the moon:

> "When Luna first her scatter'd fires recalls,
>   If with *blunt horns* she holds the *dusky* air,
>   Seamen and swains predict th' abundant shower."

And again of the stars:

> "*Brightly* the stars shine forth; Cynthia no more
>   *Glimmers* obnoxious to her brother's rays;

And so too Dr. Jenner:

> "Last night the sun went pale to bed,
>   The moon in halos hid her head."

Those are all descriptive of the presence or absence of that early, misty, interposing, and obscuring condensation in the upper story, at the height of from 12,000 to 20,000 feet, which spreads out in advance of the more dense condensation of the condition; except perhaps the allusion of Virgil to spots upon the sun near the time of his setting. which he probably meant to be descriptive of small patches of cirro-stratus which had assumed visible form. Thin cirrus cloud, whether misty, linear or fibrous, pales the light of the sun, especially at nightfall. Occasionally in the course of the day, when the cirrus is dense, various coronæs and halos described in our text books, appear in it; but the coronæ and halos of the sun are too infrequent to be of importance in prognostication. It is otherwise with the halos of the moon, alluded to in the last couplet quoted from Jenner. They are worthy of especial notice.

There is at all times more or less appearance of a circle round the moon, produced by the mistiness of the atmosphere, but during normal fair weather the circle is small and not very distinct, and shades off gradually into the azure. But when the circle is large, perfect, and the rim of it dense and well defined, it is a certain sign that the cirrus in which it is formed is the advance or lateral condensation of a considerable storm. It is best seen when the moon is vertical or nearly so. Occasionally an imperfect circle may be seen in an isolated, passing cirrus cloud, but that is of no importance as an indication, except that it m y excite suspicion that a more dense and uniform layer of cirrus of which it is an outlying, advanced portion, may be approaching.

In relation to the halo around the moon then, look first to the *elevation* of the circle, and be satisfied that it is in the cirrus condensation of the upper story; second, look to its *diameter*, which should be *large;* third, look to the *completeness of the circle*, showing that there is an uniform layer of condensation, and not a mere casual cloud; fourth, look to the *density* of the *outer rim*, for that determines the depth of the condensation; and fifth, to its *steady continuance* and *gradual increase*, for they are evidence that it is formed in an extensive, continuing, and deepening stratum.

Other signs have been regarded, founded professedly upon the appearance of the sun, but in reality upon the character and ex-

tent of its obscuration by condensation, or the visible reflection of
his rays from the cirrus or cirro stratus clouds.

Thus there is preserved a couplet by Darwin of the appearance of the sun at sunrise:

> "In fiery red the sun doth rise,
> Then wades through clouds to mount the skies."

This is descriptive of an appearance not unfrequently seen.
The thin eastern edge of the storm-clouds of the condition have
passed over us to the east at sunrise, but not to the horizon—the
sun, as it rises, shines upon the under surface of the cirro-stratus,
or stratus, and his red rays are reflected and give those clouds a
bright red appearance. But the body of the stratus clouds
passes on to the east, and the sun rising from the horizon *seems*
to enter and *wade* through them, until they are sufficiently dense
to hide him entirely from view. The counterpart of this takes
place in the west when the sun is about to set, and the condition
has nearly passed by, and it is about to clear off, and the western
edge of the melting stratus or cirro-stratus is above the horizon.
The sun then shines under them, tinging them with bright and
beautiful colors. It is thus that the beautiful sunsets are produced.
And as they are evidence that the body of the condition has nearly passed by, they are indications of approaching fair weather.
These appearances are seen everywhere, but vary somewhat in
different climates Dr. Lynes describes such a clearing off sunset in the southern hemisphere when he was off Cape Corrientes.
That the Jews understood this, we know, for on this subject we
have an allusion to the weather, by our Saviour while on earth,
which, like all such allusions found in the Bible, is of remarkable
philosophical accuracy. It is found in Matthew, chapter xvi,
verses 2 and 3: " He answered and said unto them, When it is
evening ye say, It will be fair weather, for the sky is red. And
in the morning, It will be foul weather to-day, for the sky is red
and lowering. O, ye hypocrites, ye *can discern* the face of the
sky," etc.

The reader will recollect that the Saviour alluded to their ability to read the signs of the weather correctly, in connection with their inability to read the signs of the times.

And there is another old proverb, founded on the contrast of the appearance of the atmosphere in the morning or at evening, the philosophy of which is the same. It is as follows:

> " An evening red and a morning gray
> Are sure signs of a fair day;
> Be the evening gray and the morning red,
> Put on your hat or you'll wet your head."

I will allude in this connection to another circumstance which frequently occurs, and has always attracted popular attention.

When the sun shines clearly, at the east or west, through a *small opening* in the clouds, the *condensing vapor* is shown by the diverging streaks of sunlight, just as the fine particles of dust are seen in a dark room, when a few rays of sunlight are admitted through a small aperture. This phenomenon is often observed, and it is said of it—" It is going to rain; *the sun is drawing water*."

Virgil alludes to this as seen in the east in the morning, thus:

> " But when beneath the dawn *red-fingered rays*
> Through the dense band of clouds *diverging* break,
> *   *   *   *   *   *
> Ill does the leaf defend the mellowing grape;
> Leaps on the noisy roof the plenteous hail,
> Fearfully crackling.

In describing the various conditions and the cloud elements of which they were composed, and their approach, and the manner in which they severally become visible, I have said all that it seems important to say in respect to this element of prognostication. Every one has impressions more or less clear upon the subject, derived from experience, even though they have paid little attention to it. There are few who do not sometimes look at the face of the sky and say: "It looks like snow," or "It looks like rain," or "like clearing off" and many by an attentive watching of the clouds, are enabled to form a very accurate opinion. It is not a difficult matter to become acquainted with these appearances, and there is a pleasure in understanding and observing them of which you have little conception.

Bearing in mind that all the conditions approach from the westward, I will give you a brief résumé of the appearances to be looked for. First, as to the season of the year. In winter and early spring, when the focal path is at the south, we look at the southwest for the first appearance of the clouds of the condition. They may be looked for at all hours of the day, but if they exist at all, will be seen most distinctly at nightfall. Later, in the spring and early summer, when the focal path has moved to the north, you will look to the west, and in summer and early autumn, when the focal path is north of us, you may look north of west, unless indeed you reside in the Atlantic or New England States, and have reason to look for an approaching hurricane condition, which is coming up the coast, and then you will look south-south-west. The appearance may be that of the misty cirrus which we have described, discoverable by the aid of the sun, moon, or stars; or the cirrus existing in visible, thready patches or wisps; or some of the forms of cirro-stratus; to be followed by the cumulo-stratus or thunder head, or the rain-bearing stratus, according to the season of the year and the character of the condition. Sometimes, though not often, the scud may be seen floating in the southerly or easterly wind, before other cloud condensation is visible except the misty, formless cirrus. Generally, however, the scud are not seen before the cirrus assumes form, and patches of cirro-stratus appear.

But I must not omit one other means of discovering the approach of a belt of showers, by direct observation.

When in midsummer, a belt of showers is approaching from the N.W., and the cloud condensation does not show itself above the horizon, before nightfall, you may sometimes discover it in the evening or night, as it is illumined by the flashes of lightning which play on its summits. And before the thunder heads have become visible above the horizon, the flashes of lightning will be seen perhaps, reflected from the milky stratum of cirrus, which will cover that part of the sky, and seeming to come from the atmosphere above the horizon. Some years ago, Lieut Maury issued circulars to the farmers of the western states, making inquiries relative to the direction and other circumstances of their rain-fall. Among the answers, it was stated by several of them, that they regarded the appearance of lightning in the northwest in the evening, as a certain indication of coming rain. Of the fact of a coming condition you may be assured, by the sign, but whether or not it will precipitate upon your locality you cannot be sure.

The fifth state or element in prognostication is that of humidity. I had not been accustomed to place much reliance upon this element when the "Philosophy of the Weather," was written, but I have since satisfied myself that it is an important one.

The atmosphere undoubtedly contains at all times a considerable quantity of watery vapor; whether combined or uncombined with the air or its oxygen, has always been, and is still a debatable question. Certain it is that at times there exists a considerable quantity which is uncombined and visible, and equally

FIG. 39.

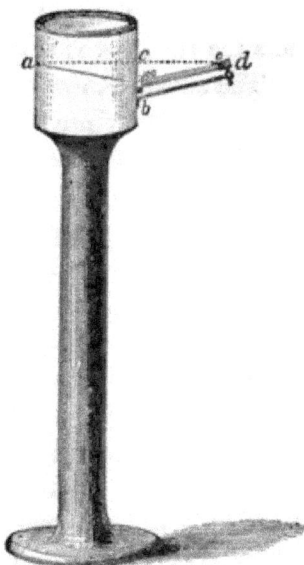

certain it is that at other times, when evaporation has been large, a considerable quantity must be contained in the atmosphere, which, if it is not combined, is at least invisible, and undiscoverable by any ordinary test. More attention has recently been given than formerly, to the subject of evaporation. Mr. Steinmetz has invented what he calls a *vaporimeter*, to measure the evaporation, which is preferable to any other instrument I have seen.

The preceding, Fig. 39, is a cut of it. It consists of a cylindrical vessel, *a*, mounted on a stand with a hollow glass tube, *d*, fixed in the side, and inclined so near the horizontal that a slight fall of the water in the vessel *e* can be measured by a scale on the g'ass tube. It is easily made with a tin cup and a glass tube. The outside end of the latter should have a cover fitted over it to prevent evaporation there. The tube is supported by the arm *b*. The diameter of the vessel should be 5 inches, and the tube long enough to measure a fall of one or more inches.

Evaporation, during a year, exceeds the rain-fall in England and doubtless here. It is exceedingly rapid, as we all know, in our N.W. winds. The excess of evaporation falls in dew.

Two very important facts are stated by Mr. Steinmetz, and corroborated by others. The first is, that "invariably the greater the evaporation, the less the rain, and vice versa, in every month and on all occasions." The second is, that evaporation decreases during the hot, sultry period which precedes a thunder storm. The importance of these two facts will be seen at a glance when considered in connection with another fact, to wit: that humidity, as measured by the hygrometer, and perceived by our senses, commences to increase with the first influence of an approaching condition, and continues to increase till the arrival of the rain. If these facts are so, and they seem to be established, and so far as I know are unquestioned, we may find in them a key wherewith to unlock the mysteries of condensation and precipitation; but farther than this, I cannot go now consistently with my avowed purpose, and without trenching upon theoretic ground.

It seems to have been early assumed, and to be still taken for granted both in England and this country, that the question whether or not rain is to fall upon any given day, depends upon the extent to which the surface atmosphere is saturated with watery vapor, and Mr. Steinmetz believes that the hygrometer is the most reliable instrument with which to prognosticate the weather. In arriving at this belief, he assumes that the condition of the atmosphere is the same from the earth upward, and ignores the existence of that rain bearing counter or upper trade, of the state of which, in relation to vaporization, we can know little or nothing, except from the character of the condensation which floats in it. Substantially all our rain, and most of that which falls in England, is precipitated from that upper trade. The hygrometer cannot tell us directly, at any given time, whether that upper trade is saturated or not, for it is flowing in a distinct and isolated stratum far above the surface of the earth. The question therefore, whether the surface story is saturated, may have a bearing upon the question whether or no an isolated cumulo-stratus will be formed out of the ordinary scud of that surface story, and drop rain ; but it can have no direct bearing upon the question of whether a condition, involving in its organization all the stories, and distributing in large quantities the rain which it has brought from the tropics, is about to pass over the observer. I believe in the hygrometer therefore, not because it indicates the state of saturation merely, or indicates a fall of rain consequent on the mere saturation of the surface story, but because it indicates the *influence* of an organized atmospheric *condition*, which influence *produces* a humid state of the atmosphere. With these suggestions borne in mind, let us examine the hygrometer and its indications.

Various devices have been adopted at different periods to ascertain the proportion or quantity of watery vapor in the atmosphere. The earliest were all founded on the principle that bodies expand or swell by the absorption of water from the atmosphere. Certain substances absorb it more readily than others, such as wood, hair, whalebone, &c., and various kinds of twisted cord. These were arranged so as to move indexes, and called hygrometers. The Dutch used catgut for this purpose, and made a toy called a weather house, in which, by the expansion and contraction of

13

catgut by the absorption of moisture from the atmosphere, small figures of a man and woman are made to swing out of and into the house alternately, as the catgut expands or contracts. The toys are still found in the market. A twisted beard of the wild oat, and a long hair deprived of its grease, were formerly used for the same purpose. Saussure's hygrometer was made of the twisted hair, and DeLuc's of a piece of whalebone. But in later years, the thermometer has been used by applying ether or water to the bulb, to ascertain the degree of cooling required to effect condensation of the vapor of the atmosphere upon the glass. The point at which the deposition takes place is called the dew-point. The following cut is a representation of the wet bulb hygrometer which is now more generally used than any other, having a dry thermometer attached for comparison:

FIG. 40.

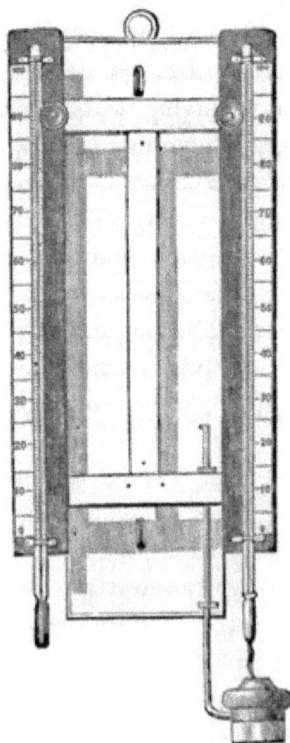

One of the bulbs of the foregoing instrument is covered with a piece of thin muslin, which is connected with the water in the cup by loose cotton threads, and they carry the water to the muslin by capillary attraction, and keep it moist. The evaporation from the surface of the muslin depresses the temperature of that thermometer, and the point at which moisture is condensed upon it above is called the *dew point*. The *difference* between the thermometers is the *complement* of the dew point, and the existing humidity is measured by *that difference*. The *greater* the humidity the *less* that difference.

Mr. Steinmetz, in "Sunshine and Showers," expresses the opinion that the hygrometer is a more reliable guide to the future of the weather than the barometer, or any or all other instruments combined. This may be true in England, but I have not supposed it to be true, nor do I think it can be true in this country. Let me, in this place, briefly contrast the climate of England with that of this country.

The area of England is comparatively small, and she is surrounded by a warm ocean. Her upper or counter trade is much less than ours, and the quantity of rain which she receives in a year is about half that which falls upon our country east of the Mississippi. According to the statement of Mr. Steinmetz, it rains there more or less, upon nearly twice the number of days in the year that it does here, and on a large proportion of the days the amount of rain-fall does not exceed $\frac{8}{100}$ of an inch. The island is situated so far north that her proper rainy season does not commence until July, and the principal proportion of her annual rain-fall occurs in the latter half of the year, as is usual in high northern latitudes. In the first half of the year, considerable storms and belts of showers like those which are common here, rarely occur; and they are not very frequent in the latter part of the year. From these facts it is obvious that a considerable proportion of their small rain-fall is derived from the surface atmosphere, the ordinary S.W., W. and N.W. scud of their feeble conditions, forming over the ocean to the west, in a nearly saturated atmosphere, assuming a cumulo-

stratus form as they pass horizontally over the island, and dropping dashes of rain. We have something like that in this country
in the spring and fall of the year, when the N.W. scud become
very dense, and furnish flurries of rain or snow, varying from a
few moments to half an hour. But not a hundreth part of our
rain-fall takes place in that manner, and nearly all of it is derived
from extensive, heavy storms of continuous rain, or from wide
belts of showers.

As so large a proportion of their rain-fall is derived from the
surface atmosphere, and in such dashes and brief showers, and in
consequence of the presence of the Gulf Stream to the west of
them, it is very possible that Mr. Steinmetz does not speak too
strongly in relation to the use of the hygrometer—for there may
be an intimate connection between the degree of saturation of the
surface atmosphere, and the change of its scud into cumulo-stratus
and nimbus. But as a different state of things exists here, and
such a saturation of the atmosphere and change of the scud rarely
takes place here, it would seem to follow philosophically, that the
reliance to be placed upon the hygrometer must be proportionally
less, and doubtless it is so.

Still it may be well for the observer to note the changes of
humidity. Doubtless that has something to do with the feeling
of oppressiveness and sultriness, felt in advance of a belt of
showers, and because it diminishes evaporation from the surface.
But this excess of humidity does not always exist in advance of
storms, and the principal changes of feeling experienced by men
and animals, are due, as we shall see, to electricity.

There is another peculiarity in the climate of England, and of
all northern Europe, which I will notice in passing, although it
does not bear precisely upon this point. In the first half of the
year, before their volume of upper trade becomes considerable,
dry northeast winds are frequent. Here, our northeast winds
generally precede storms, but we have something of the same
kind of wind, under two circumstances which probably exist alike
in both continents. First, we frequently have in the spring of
the year, before the rainy season sets in, and when the focal path

is to the south and *peculiarly concentrated*, dry N.E. winds, which are parts of the distant but contracted conditions, for many days in succession. Sometimes the lateral outlying condensation of the condition spreads up over us, and then we call it a dry northeaster. At others, that condensation does not reach us. Second, it frequently happens, when a heavy belt of showers has passed over us, followed by its northerly wind, that that wind will commence blowing in the morning from the N.N.E., and sometimes even from the N.E., gradually backing into the N.W. by nightfall. When that occurs, however, it will be observed that the southeasterly progress of the belt is slow. And the tendency to those dry N.E. winds increases with the latitude. There is enough in this occasional and peculiar occurrence of dry N.E. wind to give to Prof. Dove's theory of gyration a seeming support, but the other facts which bear upon it are conclusive of its fallacy. Those dry northeast winds occur in *high latitudes only*, and north of the then position of the focus or focal path of the conditions.

There is another observable evidence of the humidity of the atmosphere, indicative of an approaching condition, especially of a belt of showers in summer, and a southeaster in winter; such as the deposition of moisture upon tumblers and other vessels containing water, and upon flagging or other stones connected with the ground. I remember when a boy, to have seen old people, in order to ascertain what the weather would be next day, blow out their tallow candles, and immediately blow them again, in order to relight them. If the flame was restored easily, they would go to bed with the assurance of fair weather the next day. If the candle would not relight, they expected foul weather. So, I have seen them sit over the remaining coals of a large hickory fire and listen to the crackling among the coals, and in the ashes, to determine by its intensity what the weather would be. Virgil alludes to a somewhat similar sign regarded by the Romans, as follows:

> " Maidens that nightly toil the tangled fleece,
> Divine the coming tempest; In the lamp
> Crackles the oil, the gathering wick grows dim."

All these depend upon humidity.

And so do those signs in the collection of Dr. Jenner which relate to the falling of the soot, which is loosened by moisture, and the dampness of the walls and odor of the ditches. Humidity prevents the diffusion of odors and concentrates them.

The cracking of the chairs and tables before a rain is a mistake. They do not crack then, and do in very dry frosty weather.

The next element is the state of the atmosphere in respect to rain, hail, and snow. In relation to this element, there is little to be said, bearing upon prognostication. All the rules which we have been considering, or are yet to consider, have regard to the probability of the precipitation of one of the three *as a result.* There are interesting considerations in relation to the formation and occurrence of each, which have no direct connection with the subject of prognostication, and with which all are more or less familiar.

There is one circumstance, which occasionally occurs before the arrival of the rain cloud, at the place of observation, which may properly be considered in this connection. I allude to the occurrence of a rainbow at the west, in the morning.

Jenner says:

> " The boding shepherd heaves a sigh,
> *For see, a rainbow spans the sky.*"

An old almanac had the following verse:

> " A rainbow in the morning
> Is the shepherd's warning;
> A rainbow at night
> Is the shepherd's delight."

So the proverb was originally made ; but as our ancestors were not shepherds, and had a horror of ocean storms, it was commonly quoted, in this country, in the following form:

> " A rainbow in the morning,
> The sailors take warning;
> A rainbow at night
> Is the sailor's delight."

Rainbows are not reflected from *clouds*, but falling rain, and a morning rainbow at the west is, of course, evidence that it is *actually raining there*, and will, in all probability, pass over us. "Thunder in the morning, rain before night," is a common saying, and a true one. There is a belt of showers, or showery period approaching, of unusual intensity—for thunder showers in the morning are rare. The afternoon is their most common period, and they are very apt to appear then, when the morning is threatening.

It is always an interesting question. and sometimes a difficult one, when we are satisfied that a stormy condition is approaching, to determine whether it will precipitate rain or snow. Rain is the rule, except in high latitudes, and snow the exception, and very much of probability will depend upon latitude—the nearness to midwinter—and the character of the winter. Something too of reasonable certainty may be inferred from the appearance of the clouds as the condition approaches. Practical men, who are accustomed to regard the face of the sky, arrive at very correct conclusions, in most cases. It is not as easy to describe those appearances, as it is to understand them, when once familiar with them. I can give you a few facts which will aid you in observing for yourself and becoming thus familiar.

In the first place, the advance cirrus condensation is generally of the linear kind, existing in long threads or bars extending from southwest to northeast, and not in wisps or patches as in the summer season. In the second place, the layer of stratus from which the snow is to fall, is smooth and uniform, and of a lightish hue. In the third place, there is at the approach and in the commencement of the storm, usually very little wind, and that at first southwest, and afterwards northeast. In the fourth place, the barometer usually rises higher before a snow storm, and falls with less rapidity. As contrasted with these, the rain storm in winter is usually discoverable in the west or northwest. The clouds are more irregular, and of a darker color, and the precedent wind is from the southward, and there is more of it. In connection with these, the indications derived from the thermometer are very useful, but of them I have written at length, under the head of temperature, in this chapter, and need not repeat what is there said.

The seventh and last element, is the electric state of the atmosphere. Our knowledge of electricity is not yet such as to furnish alone any direct evidence of the approach of a condition, but there are many signs which are founded on the indirect effect of electricity, and are relied upon, even in less intense climates than ours. A collection of nearly all the received and credited English signs was made by Dr. Jenner, according to Howard— or as Hone says in his "Everyday Book," by Darwin—and arranged in rhymes. Some of them, I have already quoted. Nearly all of them which are of any merit, depend upon electricity. The following is the collection :

" Dr. Jenner's signs of rain—an excuse for not accepting the invitation of a friend to make a *country* excursion.

> " The *hollow* winds begin to blow,
> The clouds look *black*, the glass is *low*,
> The soot falls down, the spaniels sleep,
> And spiders from their cobwebs creep.
> Last night the sun went pale to bed,
> The moon in halos hid her head,
> The boding shepherd heaves a sigh,
> For see! a rainbow spans the sky.
> The walls are damp, the ditches smell;
> Closed is the pink-eyed pimpernel.
> Hark! how the chairs and tables crack;
> Old Betty's joints are on the rack,
> Loud quack the ducks, the peacocks cry;
> The distant hills are looking nigh.
> How restless are the snorting swine!—
> The busy flies disturb the kine.
> Low o'er the grass the swallow wings;
> The cricket, too, how loud it sings!
> Puss, on the hearth, with velvet paws,
> Sits smoothing o'er her whisker'd jaws.
> Through the clear stream the fishes rise
> And nimbly catch the incautious flies;
> The sheep were seen, at early light,
> Cropping the meads with eager bite.
> Though *June*, the air is cold and chill;
> The mellow blackbird's voice is still;
> The glow-worms, numerous and bright,

Illumed the dewy dell last night:
At dusk the squalid toad was seen,
Hopping, crawling, o'er the green.
The frog has lost his yellow vest,
And in a dingy suit is dress'd.
The leech, disturbed, is newly risen
Quite to the summit of his prison.
The whirling wind the dust obeys
And in the rapid eddy plays.
My dog, so altered in his taste,
Quits mutton bones, on grass to feast;
And see yon rooks, how odd their flight!
They imitate the gliding kite;
Or seem precipitate to fall,
As if they felt the piercing ball.
'Twill surely rain; I see, with sorrow,
Our jaunt must be put off to-morrow."

Most of the foregoing signs relate to animals, and are undoubtedly electrical, for the animal creation cannot be supposed to form any conception of an approaching storm except from their feelings. The same is true in relation to the signs founded on the feelings of mankind.

The line

" Old Betty's joints are on the rack,"

and the lines which Hone added—

" Her corns with shooting pains torment her—
And to her bed untimely send her,"

conform to common experience. It is well understood that rheumatic joints, broken bones, inflamed corns, and parts where wounds once existed, and many other ailments, feel the influence of an approaching storm many hours before it reaches us. Many persons are subject to severe headache, which comes on before the approach of a belt of thunder showers, and ceases when it is past. Howe added another couplet which deserves notice.

" The smoke from chimneys right ascends,
Then spreading back to earth it bends."

The descent of the smoke is attributed to the lightness of the atmosphere, but that is a mistake. I have observed it very carefully, and found it to descend without regard to the elevation of the barometer. I know of no cause to which it can be philosophically attributed except that it is positively electrified, and attracted by the negatively electrified earth.

The surface atmosphere, near the earth, seems to be always affected electrically by the influence of approaching conditions, and thus to affect men and animals, as stated; but it is sometimes peculiarly so, especially by the approach of belts of showers. Such a state is commonly described as " muggy," " sultry," " hot," " close," &c. The air of such a state is not necessarily *hot*, it is undoubtedly *humid*, but humidity, although it aids in producing it, will not alone account for it. Nearly all the descriptions of the Cincinnati hurricane speak of the air as being "*very hot*," and some of them as "*very close*," but the registers of temperature which I have given, show that the thermometer was not above 70°, which is not hot. No degree of humidity, with the thermometer at 70°, or at 65°,—its common spring and autumn range in northeasters—could revive the rheumatism in Aunt Betty's joints, or the pain in her corns. The renewal of the pain once felt in broken bones, or the part where wounds once existed, is felt in the dry, warm air of the house, and in bed.

Men and animals are not alone sensible to this precedent storm influence. Everything which has life feels it, and to a greater or less extent, exhibits the feeling. From an early period, certain plants have been observed to exhibit it in a marked manner. Jenner alludes to this, in the last line of the following couplet.

> " The walls are damp, the ditches smell,
> *Closed is the pink-eyed pimpernel.*"

And Mr. Steinmetz speaks of this in the following manner:

" There are five plants which have been observed from time immemorial for the signs of the weather—the dandelion, the trefoil, the pimpernel, chickweed, and the Siberian sowthistle.

" The dandelion is a very common plant, which flowers early, and remains in bloom more or less all the year. The general flowering, however, takes place about the 8th of April, and for a month it bespangles the fields, mixing agreeably with the daisy. The down of the dandelion closes for bad weather, but expands for the return of sunshine; the down of other plants may be observed for the same indications.

" The trefoil, according to the great Lord Bacon, grows more upright, with a swelling stalk, against rainy weather; and the same may be said of the stalks of most other plants, though not so conspicuously as in the trefoil. Before showers the trefoil contracts its leaves, as does the convolvulus and many other plants.

" The pimpernel is the Anagallis arvensis of Linnæus, and is found in our stubble fields, and in gardens, flowering in June, and continuing all the summer. When this plant is seen in the morning, with its little red flowers widely extended, we may generally expect a fine day; on the contrary, when the petals are closed, rain will soon follow.

" This is the plant which Lord Bacon seems to refer to under the name of Windcope, and which has also been styled the poor man's weather-glass.

" Chickweed is said to be an excellent weather-guide. When the flower expands freely, no rain will fall for many hours; if it so continue open, no rain for a long time need be feared. In showery days the flower appears half concealed, and this state may be regarded as indicative of showery weather; when it is entirely shut we may expect a rainy day.

" If the flowers of the Siberian sowthistle remain open all night, we may expect rain next day.

" We have no doubt that if the subject were systematically studied in daily observation, almost every plant would be found to indicate more or less conspicuously, all coming changes of the weather, and so it is obvious that a new charm or interest might be given to our gardens, an examination or passing inspection of which, in the morning, " before leaving for town," would lead us to infer whether " we had better take an umbrella or not "—a matter of frequent doubt in our changeable climate."

From this review of the elements of prognostication, and of

the principal traditionary signs which have come to us from our ancestors, I think the reader will be satisfied that our propositions are proved, and that all those signs which are philosophically true, are founded on one or the other of the seven states or elements which we have considered, and with which I have classed them.

There are a variety of other traditionary signs, which have no foundation whatever, in philosophy or truth. Thus, it is common to look for equinoctial storms at the equinoxes, or when, as the sailors say, the sun crosses the line. This is an absurdity. Storms of *like character* are occurring *every day* in the year, in different and numerous portions of the hemisphere. In some portions of the hemisphere, storms *never* occur. In neither respect is there any difference on that day. A storm that then visits any particular locality, especially if a high latitude, may have originated, and probably did originate, several days before. Heavy storms in the eastern states of this country, are as common in all the winter and spring months, as in March, and occur as often on other days in March, as the 20th. The same is true of September. A most violent and destructive hurricane (which I have traced,) occurred during this month, (September, 1869,) which originated in the central belt on the 5th and 6th, and passed up north of the West Indies and off the coast, mainly between 70° and 75° of longitude, and crossed southeastern New England on the 8th. Another severe one passed east of Bermuda on the 16th, which dismasted may vessels. Another originated in that belt, on the 30th, which came up the coast and extended inland, meeting another from the west, and the two *deluged the coast states* on the 3d and 4th of October. And a condition from the west which passed between the 16th and 20th, was of an *ordinary* character. *Which* was *the* equinoctial?

Moreover, the reader who appreciates the developments I have made, and remembers that the equator is a mere imaginary line, will discard the palpable though current fallacy respecting equinoctial storms.

There are numbers of other signs, more or less regarded, which have no foundation in philosophy or fact. Of this character are all almanac days or particular days of the month, or week, or other period, the weather of which is assumed to be an indication of what the weather will be at some future time. Without enumerating or alluding to the other popular signs, let me again assure the reader that all of them, without exception, are fallacies, *unless* CONNECTED WITH, *or* FOUNDED ON, ONE OF THE SEVEN STATES DESCRIBED.

I think that now my reader may form an accurate conception of the extent to which, as an isolated observer, he may prognosticate the weather, and the only means by which it can be done, viz: by a careful and intelligent observation of the changes of state, induced by the approach, presence, or passing by of the various conditions.

But it may be well to look a little more closely for a brief space, at some material points—notwithstanding it may seem like repetition—and at some things to which allusion has not been made.

Assuming first that the normal condition of fair weather exists, several questions arise. The first and most natural question is the one we have considered, viz: *when* will the next condition come, and *when* shall we perceive its influence and its changes of state? There is, as I have said, no reliable regularity of interval. The conditions, whether originating in the tropical center of the system, and coming thence to us, or whether originating over our continent, in the upper trade which is sent to us from that center, are organized by some force whose character and manner of operation are not fully understood. When we hear of an existing and distant storm, we can tell within a day, when— if it originated east of the Windward Islands, or over the Caribbean Sea, or the Gulf of Mexico, or Texas,—it will pass over any particular point in its path, for we know substantially, what its movement per hour, according to its intensity, will be, and what its course will be at that season of the year. Thus much we shall hereafter certainly know, or the next generation will know, when the telegraph is intelligently used to disclose it. But until the character and manner of operation are fully understood, we cannot tell *in advance*, when or where a condition *will be or-*

ganized. The time may come when even that degree of knowledge on the subject, will be attained, but I fear it will not be until our present generation of closet-theorists are gone, and their theories discarded. Just now their control seems absolute. But I have faith to believe that by the next generation, if not by the present, the West India Islands will all be connected by wire, and that important *representative* points of observation in this country, will be connected in like manner—that these wires will be conducted to some convenient place, where a meteorological department or bureau will be organized, and the whole be under the control of capable, practical men, and the organization and progress of the conditions be reported to the country from hour to hour. The advantage of such a system to all our industries, and its assistance in the advancement of knowledge, will be beyond present comprehension.

Such organizations have recently been formed in England and France, but an ocean, without stations for observers, lies to the west of them, and they have scarcely any opportunity to discover and report in advance. It is otherwise with us.

Before such a system of observation can be established in this country, however, *the practical mind of the country,* and especially its *Journalists* and *National Legislators,* must come *to understand the subject,* and appreciate the importance and feasibility of such a telegraph system, and its certain, useful result, if placed under the control of capable, practical observers. The men to organize and successfully conduct such a system are not now to be found among the professed Meteorologists of the country. It requires a practical knowledge of the atmospheric system they do not possess and do not seem willing to acquire.

As matters now stand, therefore, reader, you must be content to be an isolated observer, and look to the changes of state I have described, for evidence of the coming condition. You cannot tell with certainty, before its influence is felt around you, or it is seen, when it will appear.

Something of probability there is, which may be regarded. That probability depends upon the location of the observer, and the season and character of the year.

What then is your location with reference to the focal path of the conditions? Is it anywhere east of the Alleghanies, in Eastern New York, or New England? Then the probability is that in the months of January, February, and March, the conditions will come as often as once a week in normal years, oftener in some years, and less often in others, particularly in such winters as those described by President Dwight. If your position is in any of the northwestern states, the conditions will come less often in normal years, and this it may be well to understand, although it may be repetition. Look back now to the rain chart for the winter season, and see where the rain-fall is for those months. It is trifling in the northwestern states, it is heavy in the Gulf States where the conditions first strike the continent, or commence. It is considerable along the Alleghanies to New England. The conditions generally follow up those mountain ranges in the winter season, and their direction is more northerly than it would be if the country was level. The conditions would then curve more to the northeast, and would leave Pennsylvania, New Jersey, Southeastern New York, and New England, with much less precipitation than they now receive. In proof of this, observe how upon the chart the line of seven inches of rain-fall for the winter curves abruptly to the east below the great lakes and *where the Alleghany ranges lose their high elevation.* Observe again, in the tables, how, when the focal path is carried to the west and north in April, and curves over the more level land west of the Mississippi, it takes its normal, northeast direction, and crosses the mountains and extends to the coast in the same direction. Then the conditions are less frequent and less intense in Pennsylvania, New Jersey, New York, and New England. It is then to this cause that the spring drouths are due in those states. I have before alluded to the fact that the focal path appeared to extend to the north and west in the spring, over the central states much faster than upon the states east of the mountains. And now you have the promised explanation of it.

In the state of Pennsylvania and in the states northeastward of it, the conditions may continue to recur during January, February, and early March, as frequently as in autumn, but after that, for weeks they will be less frequent and less intense. It is rare that there is not a dry time for oat-seeding the latter part of March and the first part of April, and the gentleness of the rains in that section at that season is proverbial. In the western and northwestern states, the conditions are infrequent and weak in the winter months, gradually increasing in frequency and intensity as the focal path extends up over them. The middle states become very wet in the spring of the year, wherever the centre of the focal path happens to be, and where it rains almost every day, and it is much more common for seed to rot in the ground there, particularly in Illinois, Indiana, and Ohio, than in the eastern and northwestern states.

Of course wherever the path is *focal*, the conditions are *more frequent*. This is true of the spring when the path is ascending west and north. Heavy freshets are common under it, and rare above and below it.

The path is not so concentrated in its descent in autumn as during its ascent in spring. The conditions therefore are not as frequent, or so concentrated on the path and focal in character in the autumn, as they are in the spring, and for that reason, autumn is much the most pleasant season of the two.

Then occur the dry, mild, equable " spells," known as the Indian Summer, with which all are familiar. But they alternate with very long and heavy storms. The New England Indians informed the first settlers that "*winter no come till swamps full,*" and it is generally so. Winter sets in permanently, and sometimes very abruptly, when the focal path has passed south ; and the swamps are filled by those heavy rain-falls *while it is gradually descending.*

In summer where the focal path has ascended to the north, the conditions are much less frequent than in spring, and if it were not for the provision that *very heavy rains* should fall in a *short time* from the summer *belts of showers*, there would be drouth south of the focal path every year. As it is, there is generally a tendency to it, especially in the early years of the decade.

Thus much in relation to the frequency of the conditions. We cannot tell without the telegraph, when they will come, until we see them, or their influence is seen or felt in changes of state. The telegraph would not be of much assistance without *trained operators*, and we cannot have those trained operators under the present scientist method of training young men, and stuffing them with the cluster of fallacies which constitute the Halley system. The telegraph can only communicate advantageously the observations of intelligent men, each having his attention directed to the *same* and *all* the important points, and intelligently and reliably communicated to the qualified head of a Bureau. To that end practical training is necessary. These truths will ultimately be realized, and the arrival at port of a *steamer* which has been spoken and telegraphed, will not be calculated with so much certainty as the arrival of a *storm*, observed days before on its distant path, and *all its characteristics* reported by telegraph.

The next question is, what will be the character of the condition? Here we have not only probability but some elements of certainty. Thus, in midwinter, in all the northeastern states, the probability is that the condition will be a northeaster, and accompanied with more or less snow, and in most cases the N.E. wind will be felt before the storm arrives. In the northwestern states in winter, the conditions are usually snow storms. Nearer to the focal path, there is snow on the north side of them, and rain on the south. At the centre of the focal path there may be either. South of the center it usually rains. The observer therefore can infer the probability from his position, and the observed position of the approaching condensation. There is a difference in the appearance of that condensation also, which it is difficult to describe with accuracy; but, as a general rule, the *lighter* and *more smooth* and *uniform* the overcasting stratus cloud, the greater the probability of snow—the darker and more irregular, the greater the probability of rain.

Occasionally there is an irruption of the upper trade in large volume, extending or passing north of the focal path, constituting a southeaster, with warm winds upon its tropical side, blowing under and across it, breaking up for a time the uniform cold of winter and constituting what is cal'ed a "thaw," followed soon by the cold N.W. wind and winter weather. In open winters these conditions are sometimes frequent. In normal winters, there are usually one or two of them in January, and the same number in February. In some severe winters which I have known, and others which are recorded, they did not occur at all.

Under the focal path, the conditions will be irregular. East of the mountains and in Virginia, and as low down as Hatteras, probably the wind will be northeast. West of the mountains it will probably be southwest, and in South Carolina and Georgia, probably from some southerly point inland, and some easterly point on the coast, according to the character of the season, and the location of the year in the decade.

The next question is, what will be the character of the approaching condition with respect to intensity? I have said that about one-fifth were so weak that they did not precipitate at all; and there are as many degrees of intensity up to the deluging storm or shower which produces a freshet.

Something may be learned from the barometer and the extent of its fall,—something from temperature when it rises rapidly and high before a belt of showers—something from the degree of humidity—and very much from winds and scud. The freshness and strength of the wind, and the appearance of its scud, is very much in proportion to the intensity of the condition to which it belongs. This is true of all the conditions but especially of the summer belt of showers. Sometimes the S.W. wind is very light, dying away entirely at night-fall, and commencing again about 8 o'clock in the morning, and the condition correspondingly weak.

One thing should be here again noted in relation to these summer belts of showers. When they are intense, they precipitate as they pass along, persistently night and day, but in much larger quantities and with more frequent and intense thunder and lightning where they happen to be passing in the afternoon. When they are of less intensity they sometimes precipitate in the afternoon and evening only. At other times, when very weak, the winds will be light and perhaps variable, and there will be no condensation visible, except the misty cirrus in the early part of the day, but showers will form in the afternoon and evening where the condition may then happen to be. I have frequently seen them thus pass over the country, precipitating where they were vertical, in the afternoon and evening, and showing little condensation except that description of cirrus, in the early part of the day. Such was the character of one of the belts described by Mr. Basnett. As a rule, however, the freshness and persistency of the southerly wind, the temperature, and the electric state of the air must be your principal reliance in judging of the intensity of an approaching belt of showers

So you may regard the presence and appearance of the scud in the southerly wind, running towards the approaching belt of condensation. If very numerous, and large, and *ragged at the edges*, and moving rapidly, a heavy fall of rain may be anticipated. That description of scud from a southerly or easterly point, running rapidly, are conclusive indications of a heavy rain-fall *at the point to which they are running*. I have frequently predicted freshets at points north, west, and southwest, when those scud were thus running, and without failure.

So the length of time, and the freshness and strength with which a southeaster has blown, in advance of the body of a condition, and in winter its warmth and humidity, and at all seasons the number and character of its scud, are indications of the intensity of the body of the condition toward which it is moving The same thing is true of the northeaster. The strength and freshness of the wind which blows towards the focus of the storm, and the number and character of the scud, are all indicative of its strength. When the wind, however, backs into north by east, it is an indication that the body of the storm is to the southward, and will pass by without further precipitation at the point of observation. There are no signs more important and more reliable east of the Alleghanies—and so far as I know, west of them—than the character of the wind and scud. When fully developed

they indicate with very great certainty, where the observer has a
favorable horizon, the *intensity* of the *force* which is *situated in
the body of condensation*, and has *created* them, and is *drawing
them to itself*.

The next general question which may fairly be considered with
reference to prognostication is—how long will the condition, when
it has reached you, continue? This depends very much upon its
distinctive character, its intensity, and the season of the year.
Very weak belts of showers in summer, or southeasters in autumn
—which belong to the same class—are of short duration. The
condensation of a belt of showers may not be more than twenty
miles wide, and pass over in less than two hours. I have seen
the S.E. edge of one which precipitated lightly, extending from
horizon to horizon, across the northwest, at an angle of 45°, and
in two hours afterwards it had passed to the E. and S.E., and its
well-defined northwestern edge was S.E. of the zenith, extending
in like manner from S.W. to N.E., the entire N.W., half of the
sky being cloudless. That condition was narrow, and had a rapid
motion. At other times, I have known those belts of showers to be at least 150 miles wide between their southeastern
and northwestern edges, having showers in the eastern portion
and patches of cumulo-stratus in its central portions, giving
dashes of rain, with heavy drops and a darkened atmosphere as
they passed along, and having bands of cumulo-stratus in the
northwestern portion, giving what are termed clearing-off showers,
as that northwestern portion passed over to the east, and the sky
became clear in the west and northwest. You will see from this
that there is great variety in the width of the belts of showers.
We have no means of judging of that width with certainty, or
whether additional showers are contained in the other parts of it,
except by telegraph. But we may know something of it from
the character and *continuance* of the southerly wind, and its
scud—the continuance of humidity, and our feelings. The condition which passed over Springfield, in August, 1859, was a
wide one, but not a very intense one. It was more than 24 hours
from the time when its eastern edge became vertical at that place

to the time when its western edge attained the same position. In its movement from Buffalo to Springfield, it precipitated most, where its eastern portion—which contained the cumulo-stratus—was vertical in the afternoon and evening, and least, where that portion was vertical in the early part of the day. The character of the precedent southerly wind and scud, and the length of time which they continued, and the degree of humidity, in connection with the season of the year, formed a very good indication of the length of time which the condition would occupy in passing, and I had no difficulty in describing it in advance.

The average length of time during which rain falls in southeasters, is about 15 hours, but their condensation also frequently thins out to the westward, very far beyond the precipitating portion of it, and it remains cloudy for a much longer period, the wind *in such cases* hauling slowly around through the S.W. to N.W. Some idea of the length of time for which the cloudiness will remain, may be formed from the rapidity or slowness with which the wind veers in that direction. Where the southeaster is narrow, in the fall and winter months, the N.W. wind blows under the N.W. side of it, and in such cases the shift of wind is very sudden, with a temporary change of the precipitation to snow. In such cases—as in the condition passing over Springfield, Ohio,—you may know that the condition has nearly all passed. When one of this class of conditions is met at sea off our coast, and the lull which occurs between the winds is passing over, as it frequently does with us, the mariner supposes that he is in the centre of the storm, but that lull frequently takes place near or at the western edge of the condition, and then when the cloudiness has all passed by, the storm clouds of the condition are seen to the southeastward, and heavy scud are seen to the northwest, while it is clear overhead. In such cases the mariner imagines or is told by theorists that he sees a " bull's eye " in the centre of the storm. But when the storm is passing over land, we see and know that lull and " bull's eye " to be just what I have described them. And as we see the same storms pass out on to the Atlantic, we may safe'y presume that the phenomena are, sometimes at least, of precisely the same character there.

The northeaster is of longer continuance than any other of the distinct conditions. It is rarely less than 48 hours from the time when the first cirrus condensation is visible and its wind begins to blow, to the time when its wind and rain cease. It is frequently twice as long, and sometimes longer still. In watching for the vortex in early life, with as honest a belief in its existence as any disciple of Halley ever had, expecting sooner or later to see the scud go up perpendicularly or obliquely, I have seen them run continuously and at the same elevation for more than sixty hours, and until the body of the storm cloud or stratus had all passed over to the east.

The wind not unfrequently blows 48 hours towards an approaching northeaster, before its precipitating portion reaches us. It is more difficult therefore to estimate the continuance of a northeaster than any other of the distinct conditions. The cessation of the wind, its backing into the N.N.E. and N., when the focus of the storm is south, or the appearance of a light streak in the south, when the focus of the storm is at the north, are indications that the body of it has passed by. The wind does not in this condition change suddenly to the N.W. It either backs slowly through the north to that point, or veers round by the south, or ceases with a lull, coming directly out from the S.W., according as the focus of the storm is S. or N. or over the observer. From such backing, or veering, or lull, a clearing off may be inferred.

The length of time which an irregular condition will occupy, it is still more difficult to estimate. The occasional, but very rare, isolated shower which forms in the surface story, like an English shower, will last but a few minutes, and it may be safely calculated that the squall constituted by the coalescing scud of the N.W. wind will not last more than half an hour, but there are irregular conditions which may last for many days, and sometimes more than a week. There seems to be a continued supply, in large volume, and a general diffusion of counter or upper trade, more or less evenly all over the middle and eastern states, without the formation of distinct conditions or strong winds, with alternating areas of cloudiness, fog, rain, or snow, according to the season of the year, the varying volume of the trade, and the topography of the country. Such irregular conditions have been traced over Europe, as well as this country. It is impossible to estimate even approximately the continuance of such a condition, without the aid of the telegraph.

Before I close this chapter, I will give you the table of Dr. Herschel alluded to, with alterations by Dr. Adam Clark, as found in our almanacs and other popular publications, with my views of it. The following is a copy:

"HERSCHEL'S WEATHER TABLE,

" *For foretelling the Weather, through all Lunations of each year forever.*

" This table and the accompanying remarks are the result of many years' actual observation, the whole being constructed on a due consideration of the attraction of the sun and moon, in their several positions respecting the earth, and will by simple inspection show the observer what kind of weather will most probably follow the entrance of the moon into any of its quarters, and that so near the truth as to be seldom or never found to fail.

If the new moon, the first quarter, the full moon or last quarter happens.	In Summer.	In Winter.
Between midnight and 2 in the morning.	} Fair.	{ Hard frost, unless the wind is S. or W.
——2 and 4, morning	Cold, with frequent showers	Snowy or stormy.
——4 and 6 "	Rain	Rain
——6 and 8 "	Wind and rain.	Stormy
——8 and 10 "	Changeable.	{ Cold rain, if the wind be W, snow if E
——10 and 12 "	Frequent showers.	Cold, and high wind.
At 12 o'clk noon, and 2 p m.,	Very rainy.	Snow or rain.
Between 2 and 4 p.m.,	Changeable.	Fair and mild.
——4 and 6 "	Fair.	Fair.
8 "	Fair if wind N W. Rainy, if S. or S.W.	{ Fair and frosty, if the wind is N or N E. Rain or snow, if S. or S.W.
—— 8 and 10,	Ditto.	Ditto.
——10 and midnight.	Fair.	Fair and frosty.

OBSERVATIONS.—1. The nearer the time of the moon's change,

first quarter, full, and last quarter, are to *midnight*,the fairer will be the weather during the seven days following. 2. The space for this calculation occupies from ten at night till two next morning. 3. The nearer to *midday*, or *noon*, the phases of the moon happen, the more foul or wet weather may be expected during the next seven days. 4. The space for this calculation occupies from ten in the forenoon to two in the afternoon. These observations refer principally to the summer, though they affect spring and autumn nearly in the same ratio. 5. The moon's change, first quarter, full, and last quarter, happening during six of the afternoon hours, *i. e.*, from four to ten, may be followed by fair weather; but this is mostly dependent on the *wind*, as is noted in the table. 6. Though the weather, from a variety of irregular causes, is more uncertain in the latter part of autumn, the whole of winter, and the beginning of spring, yet, in the main, the above observations will apply to those periods also. 7. To prognosticate correctly, especially in those cases where the *wind* is concerned, the observer should be within sight of a good *vane*, where the four cardinal points of the heavens are correctly placed.

The above table was originally formed by Dr. Herschel, and is now published with some alterations founded on the experience of Dr. Adam Clarke."

The table is clearly empirical. 1. It professes to be founded on "a due consideration of the attraction of the sun and moon, in their several positions respecting the earth." But no other observable, or at least *material* effect produced upon the atmosphere by the mere attraction of the sun, or moon, or both, has been detected. Certainly the table makes no reference to, and admits no difference of effect from the Apogee and Perigee positions, although they influence attraction and the tides to the extent of one-seventh or more. If the mere attraction of the sun and moon, which produce the tides, have any effect on the atmosphere, why not create a tide in it, and why not a greater tide, or greater or less effect on the weather at Apogee or Perigee?

2. The table is absurd on its face. What possible difference can it make with the existing conditions, or with the organ-

ization of new ones, whether the moon arrives at a given point of its orbit a half an hour earlier or later in the day? None can be conceived of. Yet the table says a difference of half an hour, or an hour, will make the weather fair or foul for *seven days ensuing*. I cannot imagine a greater absurdity. Again, four of the twelve periods are made contingent on the wind, but the wind is produced by, and does not produce the conditions or the weather. And finally, in relation to the weather, coincidences are remembered, and exceptions are forgotten; and during the rainy or dry seasons, there are periods when such coincidences must exist and those are remembered, while the failures are forgotten, or are attributed to some other cause. The table is, I repeat, empirical, and not to be relied on.

There is very little on the subject of prognostication to be found in the writings of American Meteorologists.

Prof. Loomis in his recent treatise, has several sections respecting predictions of the weather, which I copy. As containing what a leading meteorologist teaches his classes, and publishes to be used in our colleges and schools, they require examination.

The following is an accurate copy of the whole:

SECTION V.—PREDICTIONS OF THE WEATHER.

303. "The character of the weather at any place is affected by so many circumstances which may transpire at distant parts of the world, and which can be but very imperfectly known to us, that it is impossible to predict, except very imperfectly, what may be expected at a given time and place. To a limited extent however, such predictions are possible.

304. *Predictions founded upon the constancy of climate.* Relying upon the constancy of climate, which has been established by observation, we may predict the probable general character of any month of the year.

The climate of a country remains permanently the same from age to age. Observations continued for an entire century at various places in the United States and Europe, indicate no change in the mean temperature of the year, or that of the separate

14

months ; no change in the range of the thermometer ; no change in the time of the last frost of spring or the first frost of autumn ; in the annual amount of rain or snow, or in the mean direction of the wind. It is not certain that the climate of any country, in either of these respects, has changed appreciably in 2000 years. By the destruction of forests, the earth is more directly exposed to the rays of the sun ; the moisture of the ground is more readily evaporated ; streams more frequently dry up in summer, and drouths become more frequent and severe. But these changes do not seem to affect in a sensible manner the mean temperature of any place, or the annual amount of rain.

Assuming then, the established constancy of climate, we can predict beforehand the probable character of any month of the year. Thus, at New Haven, the probable mean temperature of any future January, will be 26°. We may be tolerably sure that it will not be higher than 36°, nor lower than 17°. The thermometer in January will never rise above 64°, nor sink below 24°. The entire annual amount of rain at New Haven will not exceed 55 inches, and will not be less than 34 inches.

305. *Conclusions drawn from anomalous months.* Moreover, if several months in succession have been unusually warm, or unusually cold, instead of concluding that the climate has permanently changed and that the succeeding months will be similar in character, we should rather anticipate months of the opposite description, since the mean temperature of the year fluctuates within very narrow limits, and the longer a period of unusually warm weather continues, the greater is the probability that the succeeding months will be unusually cold. Predictions of this kind are legitimate deductions from scientific data.

306. *Predictions founded upon the established laws of storms.* Since great storms have been found to observe pretty well-defined laws, both as respects the motion of the wind and the direction of their progress, we may often recognize such a storm in its progress, and anticipate changes which may succeed during the next few hours. When it is possible to obtain telegraphic reports of the weather, from several places in the valley of the Mississippi and its tributaries, we may often predict with confi-

dence the approach of a great storm, twenty-four hours before its violence is felt at New York.

307. *Observations of the Meteorological Instruments at a single place.* When we are restricted to observations at one locality, our predictions of the weather must needs be more uncertain, and the conclusions to be derived from a motion of meteorological instruments are not the same for all parts of the world. Along the Atlantic coast of the United States, the approach of a violent N.E. storm is generally indicated by the barometer rising above its mean height; at the same time the wind veers to the N.E. and the atmosphere grows hazy. After the rain or snow commences, the barometer begins to fall; when the barometer reaches its lowest point, the wind changes to N. or N.W., after which the barometer begins to rise.

If a gale sets in from the E. or S.E., and the wind veers by the south, the barometer will continue falling until the wind becomes S.W., when a comparative lull may occur, after which the gale will be renewed, and the change of the wind toward the N.W. will be accompanied by a fall of the thermometer as well as a rise of the barometer.

A considerable and rapid depression of the barometer—for instance a fall of three-fourths of an inch in twenty-four hours—indicates an approaching storm, with rain or snow. The wind will be from the northward if the thermometer is low for the season, from the southward if the thermometer is high. If the barometer falls with a rising thermometer, and increased dampness, wind and rain may be expected from the southward.

A rapid rise of the barometer indicates unsettled weather; a slow rise indicates fair weather. The result of all rapid changes in the weather, or in any of the instrumental indications, is brief in duration, while that of a gradual change is more durable.

308. *Prognostics from the face of the sky, clouds, &c.*—When the upper clouds move in a direction different from that of the lower clouds, or that of the wind then blowing, they foretell a change of wind.

When the outlines of cumulus clouds are sharp, it indicates a dry atmosphere, and therefore presages fine weather.   Small, inky looking clouds foretell rain.   A light scud driving across hazy clouds indicate wind and rain.

Remarkable clearness of the atmosphere near the horizon, and an unusual twinkling of the stars, indicate unusual humidity in the upper regions of the atmosphere, and are therefore indications of approaching rain.

Halos, coronæ, etc., presage approaching rain or snow.   Dew and fog are indications of fine weather."

And now let us examine this with care, and seek for clear ideas. Section 303, so far as it is a confession of ignorance, is well, but the assertion that the circumstances which affect the weather at any place, " can be but imperfectly known to us " is ill.   They can and will be known to the next generation if not to this.   But they will not be understood by the present generation of meteorologists, if they persist in ignoring developments made by practical men because they conflict with their theories.

Section 304, asserts the constancy of climate, and that is true. The variations during the decade, and between the different decades, caused by variations in the power exerted by the sun, dependent in part at least on the presence or absence of spots, explain all the supposed inconsistencies of climate.   But to gravely inform a *graduating class*, that the weather and rain-fall for a given month will *probably* be within the *extreme limits* of either during any like month *for a century*, is not saying very much in the way of *prediction*.

The next Section 305, relating to " conclusions drawn from anomalous months," contains a fallacy which is said to be deduced from " scientific data."   What the *data* can be I cannot conceive.   It is not true as a rule, that " several months in succession " which are *unusually* cold or unusually hot, are followed by succeeding months of an opposite description, as the following table, showing the mean temperature at stations taken from all parts of the country, and in connection therewith the temperature of the *seasons* and year during the *hottest* and *coldest* years of the period covered by the records respectively, will de-monstrate. On examining it and comparing the temperatures with the means, you will see that during the hottest years the *elevation* of temperature above the mean, is *carried through all the seasons*, and in like manner in the cold years the *depression* is carried through them all and *the asserted contrasts or compensations do not exist.* Comparisons of any two of the seasons, in those or other years, lead to the same result.

TABLE OF MEAN TEMPERATURES.

Fort Sullivan, (Maine.)

	Spring.	Summer.	Autumn.	Winter	Year.
Mean of 25 years,	40.15	60.50	47.52	23.90	43.02
1826	41.58	64.83	48.88	24.57	44.96
1828	38.51	59.87	44.66	22.21	41.31
Range,	-	-	-	-	3.65

Fort Independence.

	Spring.	Summer.	Autumn.	Winter.	Year.
Mean of 17 years,	46.02	68.50	52.45	28.63	48.92
1825	48.74	71.87	53.12	30.60	51.08
1836	44.69	67.13	48.55	24.98	46.34
Range,	-	-	-	-	4.74

Fort Columbus.

	Spring.	Summer.	Autumn.	Winter.	Year.
Mean of 33 years,	48.74	72.10	54.55	31.38	51.69
1825	52.52	76.62	56.21	32.36	54.43
1836	44.85	67.74	49.50	28.17	47.56
Range,	-	-	-	-	6.87

Fort McHenry.

	Spring.	Summer.	Autumn.	Winter	Year.
Mean of 24 years,	52.70	74.32	56.20	34.24	54.36
1850	51.51	75.73	59.16	39.44	56.46
1836	49.25	70.50	52.78	33.81	51.58
Range,	-	-	-	-	4.88

Fort Moultrie.

	Spring.	Summer.	Autumn.	Winter.	Year.
Mean of 28 years,	65.85	80.59	68.11	51.76	66.58
1828	68.80	83.88	69.92	61.96	71.14
1841	63.18	80.26	67.07	50.64	63.16
Range, - - - - - - - - -					7.98

Fort Towson.

	Spring.	Summer.	Autumn.	Winter.	Year.
Mean of 20 years,	62.89	79.16	61.27	43.91	61.69
1839	65.79	80.70	62.12	44.39	63.25
1835	61.11	76.47	56.98	41.44	59.00
Range, - - - - - - - - -					4.25

Jefferson Barracks.

	Spring.	Summer.	Autumn.	Winter.	Year.
Mean of 26 years,	56.15	76 19	55 63	33.85	55.46
1827	59.24	78.48	60.78	38.58	59.27
1843	47.83	74.67	54.06	32.35	52.23
Range, - - - - - - - - -					7.04

Fort Snelling.

	Spring.	Summer.	Autumn.	Winter.	Year.
Mean of 35 years,	45.57	70.64	45.89	16.07	44.54
1830	48.49	75.06	52.08	17.81	48.36
1843	33.49	66.48	44.09	15.28	39.83
Range, - - - - - - - - -					8.53

Fort Brady.

	Spring.	Summer	Autumn.	Winter.	Year.
Mean of 31 years,	37.60	62.01	43 54	18.31	40.87
1830	40.86	64.23	49.60	18.84	43.48
1837	29.82	57 37	42.62	15.99	36.45
Range, - - - - - - - - -					7.03

Nor is it true, as asserted in that paragraph, that "the mean
temperature of the year fluctuates within very narrow limits."

None of the records from which the above table was compiled, embraced the cold decade from 1810 to 1820, and several did not include the cold decade from 1830 to 1840. The greatest annual range at New Bedford, which feels the influence of the Gulf Stream, and is very equable, is 6 degrees, viz:

1825, - - - - - -	51.02
1836, - - - - - -	45.02
	6.00

If we had accurate registers from the interior as well as the coast, including the cold decades from 1810 to 1820, and 1830 to 1840, the extremes of annual range for the whole eastern half of the continent would average more than seven degrees, and that is far from being a "fluctuation within narrow limits."

Contrasts occasionally occur in the English climate, between open winters and *late* springs. To such a contrast the English proverb refers :

> "If Janiveer (January) kalends be summerly gay,
> It will be wintry weather in kalends of May."

I alluded to them in the *"Philosophy of the Weather."* But a careful examination of the records since, has satisfied me that they rarely if ever occur here, and never then, or at other seasons, *as a rule* on which *prediction* can be *founded.* Indeed, the developments I have made, show that it is *impossible.* The seasons are too perfectly controlled by intelligible laws.

The next Section (306) relates to predictions founded on the established laws of storms, and is all in one sentence, and that in the most general terms. And there is very little else in the body of the work, which bears upon it. It was not possible for the Professor to say *anything* and say *less,* and it is obvious he has no definite or digested comprehension of this branch of the subject.

With reference to the observation of meteorological instruments he has one general remark, and the fact that the barometer rises along the Atlantic coast before a northeaster. If he means what his language by implication imports—that the same thing does not take place elsewhere, he is mistaken. It occurs, but less frequently, west of the mountains and in Canada, and occasionally in England. He is also wrong in stating, broadly, that when the barometer reaches its lowest point, the wind changes by the N. to N.W. It does so in the latter part of winter and in the spring, in overland storms, when the focus of the storm is south of the observer. But when the focus passes *over* the observer it usually lulls and comes out at the S.W. or W., and when in summer the focus is at the north, the wind in a majority of instances hauls round by the south to the S.W., and *does not go to the N. or N.W. at all.* A further discrimination must be made in relation to hurricane storms, which hug the coast. Their northeast winds back into the north when very distinct and violent.

The next three paragraphs of that section contain English rules, and are in substance and expression substantially identical with some of those given by Mr. Steinmetz. The first of them is sometimes though not always true, where the involved condition is true, viz: that "*the wind veers by the south.*" But it should be added that the wind often changes in such storms to the N.W. without veering through the S.W. The second paragraph is true, with this qualification, that whatever may be true in England, *storms never set in here with the wind from the northward or north of northeast.* Nor do they, as a rule, precipitate after the wind has backed from the N.E. into the N. The third paragraph is true but very generally expressed.

In the subject matter of the 308th Section, the Professor is still less at home. The first paragraph is untrue. The upper clouds in storms are cirrus clouds, and have a direction substantially invariable, which, north of 35°, is from some point between S.W. and W. The lower clouds or scud float in the existing wind and are characterized by it, and in *most instances* have a different direction. A change of wind therefore, cannot be predicted on the mere difference of direction between the upper or cirrus, or the middle stratus or storm clouds, and the lower or scud. As a rule, and as expressed by the Professor, who obviously is not a *practical observer*, it is incorrect and deceptive.

The author from whom he copied it meant to say that when (as is often the case) the wind is about to change, that change will be indicated by the *scud* in the *new* wind, running under the *scud* of the existing wind, before the new wind indicated is felt at the surface of the earth. It is therefore a contrast between two layers of scud, both in *the surface story*, and not between a layer of scud and the "upper clouds" of the storm. It is well to look for it when the wind is about to change to N.W. in a southeaster, and it is well to observe carefully the running of the scud in all storms. An occasional glance which requires little time will suffice. It is greatly to be regretted that our professed meteorologists should so generally fail to appreciate the importance and *duty* of *habitual personal* observation.

The second paragraph of that section contains three propositions. The first refers to the smooth rounded form of the cumulus clouds, which is a characteristic of the fair weather scud of the N.W. wind. They are *concurrent incidents* of the *fair weather* wind of a condition, and do not "*presage*" anything, unless it is correct to say that they *presage* an existing concurrent thing. A specimen of them may be seen on page 40, (Fig. 9,) and they are there fully described. What is meant by "small *inky* looking clouds," referred to in the second paragraph, I cannot conceive. I have seen the expression in some school book, and it appears in Fitzroy's "*Weather Guide*," but it is not intelligible. "The light scud" referred to in the third clause, "driving" or floating in an incident wind, toward an approaching storm, is an important indication always, but what he means by driving *across hazy clouds* is not so intelligible. The expression is altogether *too hazy* and indefinite for *accurate* comprehension.

The third paragraph may be true in England where it originated, but it is rarely seen here, and is of very little consequence. Reliance must be placed upon other and more important signs.

The last paragraph, so far as it relates to halos and coronæ, is deceptive. Coronæ are not always followed by rain or snow, for they are frequently seen in the *lateral* outlying cirrus of a storm which is passing by to the south. The same is often true of halos.   The rules for distinguishing those which do presage approaching conditions, I have already given you.   A discrimination of which the Professor has obviously no conception, is necessary in relation to halos.

The rest of the paragraph is not all true.   Dew is ordinarily an *accompaniment* or consequence, instead of a *presage* of fine weather.   The absence of it is suspicious, unless it is occasioned by the N.W. wind and scud of a condition which has just passed, continuing through the night.   Dew is often noticed as it rests *in large drops* upon the spiders webs, which are built by that insect upon the grass.   The popular idea is that the webs are built *in anticipation* of fine weather, but it is a mistake.   The young spiders build their web-traps in the grass to catch their food, because they are hungry, merely, and the traps may always be found unless a recent heavy rain has destroyed them.   The heavy dew-drops resting upon them, render them visible.   And so will a very wet morning fog, or the fog which accompanies southeasters, when they deposit moisture.

And fog sometimes attends storms, and is often produced by their *lateral* influence; and when the storm or belt of showers which induces it is at the N.W. rain may follow it.   There is a tendency to fog, at and after four o'clock in the morning, during set fair weather, in summer and autumn, and *in valleys*.   But fog forms in the day time even in winter, in connection with southeast thaws, which aids in melting the snow but is chilly to the senses, and hence the proverb:

> " A winter fog
> Will freeze a dog."

Now fog does not occur in winter except in connection with storms.   And at all seasons of the year fogs and mists may form before storms *on the hills ;* and *with* or *during* easterly storms at all elevations.   The expressions " *it set in thick* "—" *it came on thick* "—before a storm,—and " *the fog lifted* " when about to clear off, refer to such fogs, and are often seen in the logs of seamen.   And when it so " *sets in,* " during a cloudy afternoon, on the land, we anticipate a rainy night; and when in the forenoon following, "*the fog lifts,*" we know it is about to clear off.

In view of the developments heretofore made on this branch of the subject, by myself and others, it is to be lamented that such a *meagre, undiscriminating,* and, in many respects, *inaccurate* description of the means of prognostication is all that a leading meteorologist has to offer in his lectures to a graduating class of young men, in one of our principal colleges, and in a treatise intended for use in other colleges and schools.

There is extant in England another collection of rules, prepared for the Board of Trade by Admiral Fitzroy, but it contains nothing of material value to us, which is not embodied in the foregoing development.

I know of no other collection deserving your notice.

I have thus given you a concise, but substantially accurate development of the system as made by the Creator. In doing this I have avoided, as far as possible, the consideration of theories. I have come to regard them, and the men who persistently and against new truths as discovered, promulgate and teach them, with a degree of disgust which I have not been able to restrain. I sinned somewhat myself in that way, in the "Philosophy of the Weather," but I do not intend to so sin again. In my judgment there is no apology to be made for the persistent theorist who does not test his theories by newly discovered truth. Indeed, with our present knowledge or means of knowledge, there is no apology for theorizing at all.

And here I might with propriety consider my promise fulfilled and close the volume. But it has been my intention from the outset to invite you to accompany me in an inductive consideration of the *motive force of the system, and its mode of operation,* in a concluding chapter, but not, it must be understood, as theorists. I think the practical mind of the country, trained by *business* in the logic of cause and effect, peculiarly well qualified to judge correctly, upon the facts, in regard to the nature and character of the force. We will together then enter upon that enquiry in another chapter.

CHAPTER VIII.

The organization and motive force of the system invisible—its existence recognized in its effects—a knowledge of it to be acquired by inference from the nature of the organizations and their actions. Elements of the tornado and their mode of operation—place where it originates—manner of formation and form—its appearance and substance—its apparent manner of action in the air—its action and mode of operation in contact with the earth—estimates of the force employed—various descriptions of the power of that force—operates in two lateral lines or currents. The right hand current crosses the center in advance of the left hand, curves backwards and ascends over it. The left hand current curves behind the other and rises over it,—both together constituting the whirl in the air. These currents thus existing and operating, constitute the law of the tornado. This law of the small is also the law of the great, and is traceable through all the varied atmospheric organizations—Traced first in the belts of showers and in elliptical storms of the Northern Hemisphere—in those of the Southern Hemisphere—nearly all storms of that Hemisphere of this character. Critical examination of Col. Reid's chapter on the " Gales of High Southern Latitudes "—in every instance he purports to describe the northern half of the gale—southern half of a revolving gale not experienced—all the gales described by him, were, with a single exception, elliptical—that exception a straight line southeaster, corresponding to our straight line northeasters. " Ojo " of the Spaniards, not an indication of the center of the storm—constituted by a clear interval after the storm clouds have all passed by, and before the scud of the fair weather wind obscure the sky. The same kind of storm's eye visible with us, at least ten times a year,—never seen in the center of the storm.—All the descriptions cited by Col. Reid, belonged to elliptical storms with lateral winds—described as such by Mr. Meldrum. Extended examination of the theories of Redfield and Espy. Certain amount of truth in both of them. Some storms elliptical as claimed by Espy—others semi-revolving as claimed by Redfield—instances and illustrations of both. Hurricane of 1821. Hurricane of 1869. Examination of the views of Prof.

Henry as promulgated in the Patent Office Reports—those views crude and untruthful. Intolerance and persistency of meteorologists in relation to this matter. Return to the inquiry what is the force. Examination of the group of the diurnal changes—instructive but not conclusive. Must find a force that originates the conditions and prescribes their paths unaffected by the rotation of the earth. That rotation does not control or affect the circulation of the atmosphere—current theories on that subject all fallacies. But one such force known, and that is electricity. General view of that agent and its operations or phases. Various propositions and illustrations, showing the manner in which it operates in organizing the system and producing the varied phenomena, and in that connection an extended examination of the magnetism of the earth, and how it is constituted. Its associated currents of electricity, and their mode of operation. Particular examination of the manner in which magnetism is diffused over the earth, affecting climate, disease, and the activity and energy of its inhabitants. Appeal to the various classes of practical men to aid in reforming the science and extending the truth.

The organizing and motive-force of the Atmospheric System is invisible. We recognize its existence only in its effects. A knowledge of its nature and mode of operation can only be acquired therefore, by inference from the nature of the special and general organizations we have contemplated, their action, and the resulting phenomena and effects, aided, perhaps by an analogical view of its mode of operation and effects in other parts of the physical world.

Let us then look critically and carefully at the structure of the special and general organizations constituting the Atmospheric System, and the mode of operation and phenomena peculiar to each.

It was sensibly said by a very sensible man—Prof. Dana, of Yale College, in commenting, editorially, on the Schuyler Tornado, that "*throughout nature the small and the great have a common law*," which "*is often best read in the small*," and we will commence with the small.

The least and most distinct of the atmospheric organizations is the tornado. What are its elements and their mode of operation, or, in other words, what is its law? To determine this, we must analyze those elements, and their mode of concurrent operation and effects. I have alluded to them in a general way, in Chapter 3d, but we must now be critical and particular.

I. FIRST, THEN, AS TO THE PLACE WHERE IT ORIGINATES. It forms at the inferior surface of an extensive stratus or cumulo-stratus cloud of the second story, and extends itself downward to the earth. This is invariably true, where its path is narrow, and its form distinct, so far as observation has extended. The rain-bearing clouds of a general storm or of a summer belt of showers, have been present in more than fifty such instances, and in all which are upon record in this country since 1809. Of one which occurred during that year in Cincinnati, described by Dr. Drake, it is said, "A general rain was falling."

That which occurred at Charleston, September 10th, 1811, is described as occurring in a violent storm of West India origin. That most destructive one at Natchez, in 1840, occurred upon a day which is described as warm and cloudy. That described by Mr. Chappelsmith, at New Harmony, Indiana, April 30th, 1852, occurred upon a day which was generally cloudy and threatening. One at Brandon, Ohio, Jan. 20th, 1854, is described as the most violent of several partial tornadoes, forming in a widely-extended general storm in which the temperature was very high. The Schuyler Tornado, which occurred in June, 1857, formed beneath the cumulo-stratus of a belt of showers, and so did the New Brunswick Tornado, which occurred in 1838. Many others might be mentioned, but these are sufficient upon the point.

M. Peltier, in his elaborate work upon Spouts and Tornadoes, has a list of one hundred and thirty-seven, of which ten are said to have occurred in a sky without clouds. None such have been described in this country. And of the ten exceptional cases mentioned by Peltier, some were mere whirlpillars, and in the other cases there was a cumulo-stratus beneath which the spout formed, and which was taken by the observer to constitute a part of the spout.

As the tornado and water spout form at the inferior surface of rain-bearing stratus clouds of considerable or very great extent, more than one may form simultaneously or during the same day. Many of these described by M. Peltier, so occurred, and there have been many such in this country. Thus, the Brandon tornado, which has been alluded to, was one of several which occurred on the same day in different places, and under the same storm.

As this tornado occurred in January, during a thaw, and in a storm which extended somewhat above the focal path, I take a few extracts from Prof. I. N. Stoddard's very intelligent description of it.

"The whole breadth of the State of Ohio from S.W. to N.E., was swept on the 20th of January, 1854, by a storm of unusual violence."

"Traces of the same storm have been obtained from a point 27 miles N.E. of Little Rock, Arkansas; also from the Western part of Pennsylvania. The whole length cannot be less than 800 miles. The breadth I have not been able to determine. At Dubuque, Iowa, on the 20th of January, it was clear and very cold, with the wind from the N.W. At the point named in Arkansas, heavy rains from the S.W. occurred on the 19th, followed by a clear and cold atmosphere on the morning of the 20th. On this day, the 20th, the storm passed over Ohio."

"The temperature became mild on the 19th, and on the next day at noon, the thermometer stood at 70° in Cincinnati, and 68° in Oxford, the latter place more elevated than Cincinnati, and 30 miles from it N. by W. The barometer fell gradually during the 19th, and rapidly on the 20th; and at 45 minutes past 12 M., the time when the storm began at Oxford, it stood 28.21, lower than at any period during the last twelve months. The air was saturated with vapor, and the walls of brick buildings were dripping with moisture. Three strata of clouds were distinctly observed, the highest cirri light and fleecy, moving toward the N.E., the second, the proper storm cloud, in dark, heavy masses, moving rapidly in the same direction, the third and lowest the scud of sailors, flitting violently past a little east of north. Along the track of this wind, there were at different times during the day violent rains, vivid lightning, heavy thunder, and in some places, large hailstones fell though not in great quantity. In the North-eastern part of the state, the storm assumed the form of a tornado of great violence."

"It first struck the earth in the S.W. part of Miller Township, Knox County. N. latitude 40° 18′, Long. 5° 30′ W. of Washington."

"Its course in that County was N. $56\frac{1}{4}$ East. Traces of it are found in some of the counties further east, where its path gradually curved more towards the east, presenting its convex side to the north. The tornado in Washington County, Penn., on the same day, was not probably a continuation of that in Ohio, as its location was several miles farther south.

"It appears to have passed over one tier of counties, without touching the earth, and subsequently to have descended again with its original force."

Such is the description, and the reader is requested to notice in passing, certain facts particularly in relation to this general storm, for I shall allude to them again.

1st. That this was a winter thaw, extending somewhat above the focal path, but not as far to the N.W. as Dubuque, Iowa. Its heaviest rains ranged from Arkansas on the 19th to the N.E. and in Ohio on the 20th.

2d. The temperature where the tornado occurred was 68°. The barometer fell about an inch below the mean of the place, before the storm.

3d. The three cloud strata of the general storm under which it occurred, were distinctly observed. The cirrus in the upper story; dark, heavy masses of stratus and cumulo-stratus in the second story, and the southerly scud in the surface story moving from the S. to a little E. of N. Note carefully the strata as shown by the following diagram.

Fig. 41.

II. Second, as to the manner of formation and form. Those have been frequently and sufficiently observed. A few descriptions are all that our limits permit. Towards the end of August, 1838, about 3 P. M., a tornado passed near Providence, Rhode Island. The rain was falling, says Mr. Z. Allen, in a letter to Prof. Hare, with violence, when he saw a black cloud (cumulo-stratus) in the midst of other brilliant and fleecy clouds, assume a terrible aspect and *form itself* into a black, elongated cone, extending down to the surface of the earth."

The following is a description condensed from an account by Mr. C. F. Brooks of the Medford, Mass., tornado.

A very distinct "form" was exhibited by the cloud in this case,

usually that of an inverted cone, though often that of a double
cone or hour glass. Several "concur in saying that the conical
point *let down* from the cloud moved about at short distances, now
pushing down to the earth, and now rising from it. Its side
motions were compared to those of an elephant's trunk. This
action was like the *descending* tube in a *nearly completed* water-
spout at sea." Its width was from 50 to 70 rods, its course from
W.S.W. to E N.E., curving slightly, its rate of motion nearly
"fifty miles per hour," duration "five or six seconds."

The following is a description of the Schuyler tornado, as,
observed by Dr. Mower.

" At Schuyler Corners the phenomena was very intelligently
and carefully observed by Dr. Mower.   He as well as others saw
it first nearly north of the Corners, but a little west, then presenting the appearance of a huge cone, apex down, *suspended* by
its base *from the cloud*, its lower point apparently twenty or thirty
rods above the earth, and not more than forty rods from their
position to a point directly under it.   It passed on to the east,
lengthening as it went. Having passed out of sight from an intervening house, he changed his position for a moment to gain a new
view, and when he again saw it, it had touched the earth. Its
course was now attended with a huge cloud of dust, and he could
no more see the apex of it. At one time it swelled out in the
middle, forming as it were, two cones, base to base, the apices
above and below, and a smaller and similar cone, separated from
it towards the south ; this second phenomena was traced by its
progress upon the earth as noticed further on. He says he saw
it ten minutes and it seemed to pass slowly along. It first struck
the ground just back of the little hamlet of Schuyler Corners."

And the following of Water Spouts as seen over the Gulf
Stream extracted from a description in Espy's Philosophy of storms
page 346.

"AT SEA, AUGUST 14th, 1836."

" Captain ! Spouts over the lee bow ! cried the voice of a
sailor down the companion ladder, yesterday at 2 P. M., while
sailing along the Gulf Stream, in about latitude 25° 30'.

The cry of " Spouts over the lee bow " naturally excited some little alarm among the passengers. The Captain was on deck in a moment. I was anxious to witness the magnificent phenomena, and therefore followed him. On our arrival there, the spectacle presented by the heavens to leeward, was indeed of an imposing and awful character. A dark cloud, which every moment became blacker and blacker, was fast extending over the leeward sky. From the lower part of this ominous and stormy curtain, projected three jet black columns, which kept curving and swinging backwards and forwards as if they were endowed with life.

* * * * * * *

" Brace round the yards ! come, be quick ! Haul aft and load the gun, some hands," cried the Captain, while he himself assisted in performing these important services. Every second was of consequence, a minute or so might have sealed our doom. On, on, went the ship; and before she turned, we were frightfully near the dreadful spouts. Onward and downward these gigantic hose pipes of cloud and water uncoiled. Now they curved like a reaper's hook. Anon, they twisted like a serpent's tail ! I could imagine that two of them were at least a thousand feet in length, with a body as thick as the Washington Monument at Baltimore. Their contortions and convulsions were interesting and wonderful, and I found it impossible to withdraw my attention, even for a moment, from the grand phenomena; at length the ship was put about, and we began to increase our distance from what we had regarded as a watery death. The spouts straightened out, and the lower ends of two of them approached the surface of the deep. The sea beneath rose in a hillock of waves, as if attracted or twisted into a rising *tumulous* by the cloud, or formed by the whirlwind. And now two of the columns were perpendicular, resting upon a mount of foaming, roaring waves— a perfect

" Hell of waters."

I should say that from one hundred and fifty to two hundred feet above the sea, these columns were transparent as crystal, and the water might be seen swiftly traveling up them. This appearance lasted for six minutes and a half, *the third spout never reaching the sea at all.*"

A single extract from Peltier, descriptive of the formation and character of a tornado which occurred at Chatenay, near Paris, will be sufficient on this point. The day was cloudy and the description of weather was that of a summer belt of showers. A sort of combat is described as having taken place between two showers. It is then added :

"A great agitation then manifested itself in the intermediate parts, and the thunder rolled violently, when all at once, the clouds of the second storm *lowered themselves toward the earth,* and put themselves in communication with it. At this instant the thunder appeared to cease, and there arose a frightful whirlwind of dust and light bodies, with an extraordinary, confused rolling."

＊　　＊　　＊　　＊　　＊　　＊　　＊

"Arrived at the Croix du Frèche, the descendent cloud had great dimensions ; it was then a terrestrial spout, well formed, which, according to the account of several inhabitants of Fontenay, had the form of an inverted cone, having its base in the upper clouds, and its apex, about seven metres from the earth. The vapors which composed it had a grey tint, and rolled one on the other with great impetuosity, letting some points of their pale light be seen, and causing a confused rolling to be heard."

Many other similar citations might be made, but these are sufficient upon the point. Let its manner of formation at the inferior surface of the stratus clouds of the second story, its extension or elongation downwards to the earth, its subsequent alternating elevations and suspension above the earth, and the fact which may be added, that when it breaks up action ceases first at the earth and last at the cloud—be particularly noted and remembered.

III. Let us look now, in the third place, at THE APPEARANCE AND SUBSTANCE OF THE ORGANIZATION. These are variously described by different observers. Its form is uniformly described as that of an inverted or funnel-shaped cone, when it has been narrow (less than a hundred rods) and distinct. When, however, the breadth has been much greater, the conical shape has not been so observable.

In *color* and *substance* it differs from ordinary cloud, and this is important. Of the Charleston tornado it is said that " it exhibited the *lurid* appearance, and funnel-shape peculiar to these tornadoes, shifting its position very much."

An abridged description of that which occurred at New Harmony, Indiana, April 30th, 1852, is as follows:

" This is traced from near Paducah, Kentucky, northeast of New Harmony, and from that point 200 miles nearly east to Georgetown, Ky., apparently following the general line of the Ohio river. The day was generally cloudy and threatening, with a low barometer—the tornado occurred at 4½ P. M. At its point near New Harmony, the track of fallen trees was one-half a mile to a mile in breadth, and the rate of progress calculated at nearly 60 miles per hour. The destruction was the work of a moment, and intense electrical energy was apparent, an observer says: "*the cloud appeared on fire at the bottom*, like a large pile of burning brush," others describe it as "a cloud with *green and red flame*," others "*green and blue.*"

Mr. Stoddard describes that which occurred at Brandon, as follows:

" Persons just outside the path describe the storm as a column of vapor or *smoke* whirling in indescribable confusion, accompanied with a deafening roar, so that the thunder, if any, was undistinguishable amid the general din and confusion."

He also describes the one which occurred at Harrison, Ohio, as follows:

" The storm approached the house nearly from the west, and yet it was struck on the eastern side, by the current from the southeast. The proof of this is explicit. The *whirling mass of vapor*, attended with a thundering roar, was seen by Mr. Graham, approaching from the west; he shut the door, a few moments of awful suspense followed, then a window over the east door was driven in with a loud report, then the door followed, and the next instant the house was torn from its foundations and shivered to fragments."

Prof. Olmsted thus speaks of the New Haven tornado:

"The appearance of the storm as it approached was deliberately contemplated by numerous observers, who saw it coming over the plain. All describe it as a *strange cloud* of *terrific aspect*, *white* like a driving snow-storm, or light fog, and agitated by the most violent intestine motions. It came suddenly upon them with torrents of water, there was a rush, a crash, and it was gone."

Prof. Loomis, in an article on the Mayfield tornado, which occurred Feb. 24th, 1842, says:

"Mr. Halsey Gates, standing near his mills in a shed open to the north, saw the tornado pass and observed it very attentively. The entire heavens, he says, were covered with dense, black clouds, moving with great rapidity. No cloud seemed to descend to the earth, yet the progress of the tornado was marked by a huge column of a *dull yellow* or *smoky tinge*."

And in the same article he thus alludes to two which have occurred in the night.

"The tornado at Morgan, in 1823, is thus described by Deacon Beach. About 8 o'clock the sky became overcast with a dark cloud, attended with plentiful rain and some lightning. The rain suddenly ceased, but the cloud remained, covering the whole heavens, and producing intense darkness. The air was perfectly still, after the rain, for about an hour, and the heat unusually great. At half past 9, he heard a roaring, as of very heavy thunder, which called him to the door. Upon opening it he immediately discovered a *bright* cloud, having precisely the color of a *glowing oven*, apparently of the size of half an acre of ground, *lower* than the dark canopy which remained unbroken above, apparently within two or three miles, and moving rapidly in the direction of his house. The brightness of the cloud made the face of things light above the brightness of the full moon. Having turned into the house he was engaged in securing it, when the tornado passed, taking the roof and chamber floor, and many articles from below.

It was a log house. There was neither hail nor rain during the passage of the tornado, neither flashes of lightning nor distinguishable peals of thunder, but an *intense brightness* of the cloud and a continual and tremendous roar. The passage of the tornado seemed instantaneous, but the *light of the cloud* continued for more than a quarter of an hour. Deacon B. was able to read in his Bible, which he found many rods from his house, at least *ten minu'es* after the storm had passed.

"Judge Griswold saw the same phenomenon. The cloud appeared to him funnel-shaped, apex downward, from which a *stream of fire* apparently issued.

"The appearance of the cloud, as here described, corresponds very well with the account of the Shelbyville tornado, as given in this Journal, (Vol. xxxi, page 258.) The cloud is said to have been *permanently luminous*, and of the *color of red hot iron*."

Of a list of fifty or more which have occurred in this country, I know of no other which occurred at night except that at Stowe, and that was not observed. It is an important fact that fifteen out of sixteen have occurred in the day time, and in the afternoon. And this was generally true of the great number described by M. Peltier, but it cannot be necessary to cite other instances from our own records or from him. Those given are amply sufficient to exhibit the character of the organization, in relation to form, color, and substance. If all occurred in the night all would doubtless be luminous.

IV. APPARENT MANNER OF ACTION. Enough has been cited to show its *apparent* manner of action before coming in contact with the earth or when the contact was broken by contraction, and the apex was drawn up above it. All observers seem to agree in the fact that when occupying that position above the earth, there was a whirling motion distinctly visible, and the apex or extremity shifted about in every direction from a perpendicular line.

V. We come now to its ACTION AND MODE OF OPERATION WHEN IN CONTACT WITH THE EARTH, and here we can learn but little from actual observation of its movements at the time.

First, because dust, dirt, water, mud, and everything movable, including limbs of trees, leaves, parts of buildings and other objects detached and torn to pieces by its force, were embraced in it, exhibiting to the eye intense internal agitation, and apparent indescribable confusion. Second, because the apex of the spout undoubtedly whisks about at times in various directions while in contact with the earth, thereby creating much of the apparent confusion observable in its effects. Third, because it varies in form and extent, and its lateral motions are more or less masked where the breadth is considerable and the progress rapid. Fourth, because the force is exerted in veins irregularly, sometimes with enormous energy, and at the next instant feebly. Thus, where two buildings have stood near each other, and equi-distant from the center of the path, one has been destroyed while the other remained uninjured. And so also, of light objects easily moved, some have been carried away and never found, while others situated near by have been left undisturbed. We must endeavor then to discover their mode of operation at the surface of the earth, and when in contact, by analyzing their effects in cases where the cone and path were narrow, and at points where there was the least apparent deviation of the apex from a straight line, or other irregularity as it passed along.

Several of that precise character have been reliably described. That which occurred near New Haven, is one of them. It was carefully examined and charted by Prof. Olmsted, and Mr Haile. Their chart embraced four different sections. I copied and inserted a part of one of the sections on page 79. I also copy and insert as much of the rest of his chart as the smaller size of my page will permit.

The following is Prof. O.'s description of the effects observed in connection with the chart.

" Let us now trace more particularly those facts which have a bearing upon the laws which govern this storm.

1. The first great fact that strikes us is, that all the trees and other objects that mark the direction of the wind which prostrated them are, with very few exceptions, turned inwards on both sides,

FIG. 42.

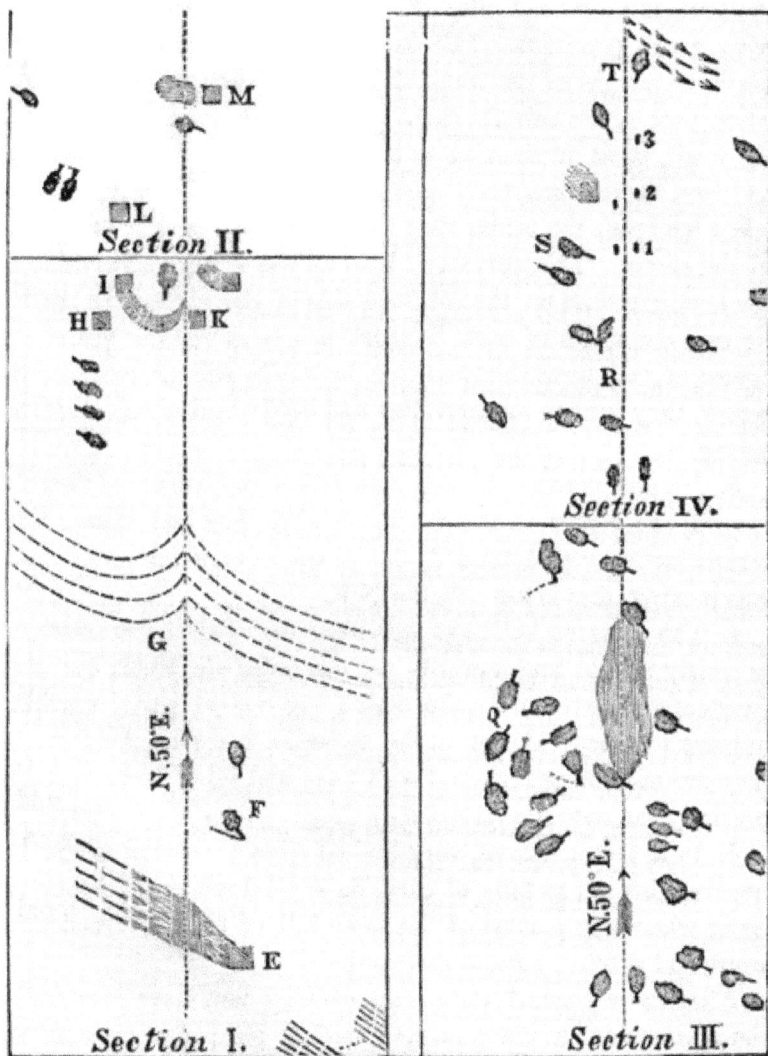

towards the center of the track, while near the center, the direction of the prostrate bodies is coincident with that of the storm.

2. On more minute inspection, we find prevailing a remarkable *law of curvature.* This is most favorably seen in cornfields, as the prostrate corn indicates the course of the wind at each spot, with great precision. The law is this. Commencing on the northern margin of the track, the stalks of corn are turned backward, that is, towards the S.E., proceeding towards the center of the track, their inclination to the south becomes constantly less and less, turning gradually towards the course of the storm, until when we reach the center they lie to the N.E. exactly in the line of the storm. This curvature is in all cases more observable on the northern than on the southern side of the track. In the latter case the stalks of corn lie more nearly at right angles to the course of the storm (but inclining forward,) still, on reaching the center, they turn to the northeast and become coincident with that course.

3. Numerous examples are seen where the bodies as they fell toward the center of the track, or after they had fallen, were turned farther round towards the direction in which the tornado was moving, that is towards the N.E.

4. The ruins of buildings that were demolished, are scattered in nearly a right line towards the center of the track, but they frequently are strewed quite across the central parts, reaching, in some instances almost to the opposite margin. In this case they are often found covered with trees and other bodies lying in precisely the opposite direction.

5. In a few instances, very limited spots are found where the prostrate bodies, as hills of corn, lie in all directions. Examples occur where one portion of the same hill of corn is turned westward, and another portion eastward.

After these general statements we may now have recourse to the accompanying diagram, and review particular cases of the foregoing laws or modes of action. For this representation of the phenomena of the tornado, I am indebted to Mr. A. B.

Haile, who took the bearings of the various prostrate objects with a compass. In most cases, I have been able to attest the accuracy of the representations by actual inspection, and in regard to the few instances where my attention has not been particularly attracted to the fact represented, I entertain no doubt of the entire accuracy of the delineation.

The diagram commences at A, (not copied,) at a mulberry grove, half a mile from the spot where the tornado first formed. The dotted line bears N. 50° E. It will be perceived that the trees which lie in the center of the track generally coincide with it, and that those which lie on either side, are turned inwards towards the center. Yet several examples are seen where trees lie pointing outward from the center, both in the middle and in the marginal portions of the track, as at N, Q, R, S. These exceptions, moreover, are all on the north side of the track. Examples of the remarkable *law of curvature* referred to, as seen at G and O, (see page 79,) where the figures represent the direction of stalks of corn, in two fields nearly a mile distant from each other. It will be observed from the diagram that from the margin, the direction of the stalks inclines more and more inwards, and finally in the center, coincides with the course of the storm. It is also obvious from the figure that this law is more fully developed on the left than on the right side of the track. The same tendency to this curve is exhibited in the scattered fragments of a roof at I.

The dotted lines connected with the figures of some of the bodies that were thrown down, as at B, D, and F, show the position in which the bodies first fell, and from which they were moved round into the places they now occupy. In some cases a tree is seen to have commenced falling at right angles to the track, but during its fall to have been twisted round towards the course of the storm. Similar examples are found of limbs bent around the trunks of the trees from which they were partially severed.

At E are represented the ruins of a building which was completely demolished, and its fragments carried in a right line far beyond the center of the track. According to Mr. Haile, the frag-

ments in the central parts of the track are arranged in parallel
lines, coinciding with the course of the storm, while in places
farther from the center, they lie promiscuously. In the parts most
remote from the building, the fragments are covered by corn
thrown down in the opposite direction. A more striking example of the same fact is seen near the eastern limits of the tornado,
where the fragments of a roof are scattered towards the west,
while a tree a few paces from the building is turned directly
towards the building, covering a portion of the fragments.

At C is represented a limited spot in a cornfield where the
stalks lie in every direction. While in a few places, at distant
points, particular spots seem to have been subjected to a peculiar violence, other limited spots exhibit a remarkable exemption
from the effects of the tornado. In a garden near H are a few
rows of pole beans, apparently untouched by the storm, while
within a few feet on either hand, the most violent effects are exhibited. Near L a barn was demolished, and a dovecote scattered in fragments, while a hen-roost which stood feebly on blocks
was unharmed. Large trees in the immediate vicinity were torn
up by the roots. A house that stood between I and L, was completely torn in pieces, leaving nothing but the southern half of the
ground floor. In the room of this floor, a woman was washing,
and another was at work in a basement room immediately below,
while her child was asleep in a cradle in a room above, at the
northeast angle of the house. They saw the tornado approaching ; the woman in the basement ran up and caught her child in
her arms, and immediately afterwards found herself and child in
an open field a few paces north of the house, the child having been
carried only a few feet from the spot where they were, while the
mother was carried eighteen or twenty feet farther to the westward. The other woman meanwhile was swept off from the floor
where she was standing, and carried northward and deposited in
the cellar, the floor of the northern half of the house having been
borne away along with other parts of the building. None of the
persons were seriously injured."

I refer next to the description, from which I have already cop-

ied, of the Brandon tornado, by Prof. Stoddard. The following
are Sections II and III, taken from his chart, showing the manner in which the trees were prostrated where the tornado entered
successively two different standing forests. A more favorable opportunity for the exhibition of the precise manner in which the
force was exerted could not well exist, the tornado crossing cleared
fields and striking perpendicularly into a forest which stood upon
rising ground.

Fig. 43.

Section II.

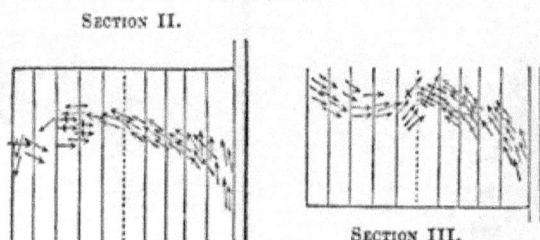

Section III.

After describing the effects of the tornado upon the village of
Brandon, he describes the place where Sections II and III were
surveyed, as follows :

" About one-fourth of a mile east of Brandon it struck a dense
forest. At this point a careful survey was made across the track
represented by Section II. For nearly three miles its course was
mainly through the forest, with intervals of cleared land, uprooting or breaking almost every tree, and crushing the buildings
which unfortunately stood in its way. Crossing the Newark and
Mt. Vernon railroad, it swept over cultivated fields, destroying
the few trees which had been left, and razing to the ground a
stable and brick house. Three-fourths of a mile beyond this, an
open grove of very large trees, mostly oak, stood on rising ground
and in the line of the storm's axis. They seemed like an advanced
guard, to the forest a little farther in advance. The tornado
struck them with appalling fury, and appeared well-nigh irresistible. Scarcely one was left standing, some were uprooted, others
broken and split into fragments. Near this place where it entered the forest another survey was made, (see Sec. III.)

I take next two sections of the chart of the Harrison tornado, also surveyed, charted, and described by Prof. Stodard.

FIG. 44.

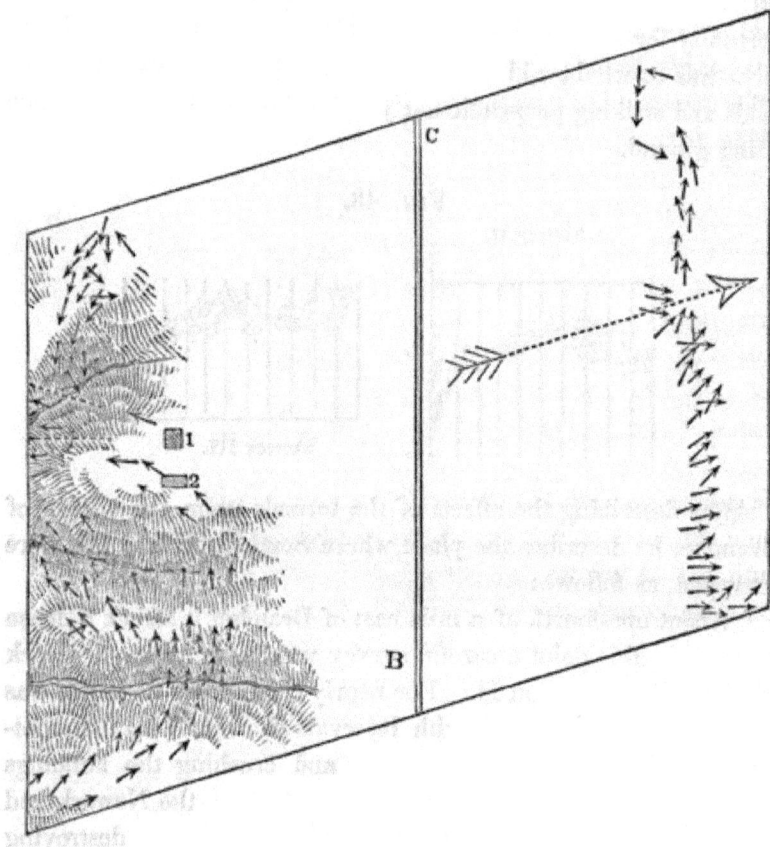

Section B is a chart of the Graham place, which the tornado demolished. The topography of the ground was peculiar, and the effects do not precisely correspond with those produced in other parts of the path. The curvature of the right hand lateral currents, was nearer the axis than in other portions of its path, and the house and barn, No. 1 and 2, were stricken by them on the east side and the fragments carried in the direction of the dotted lines. But the distinct and peculiar operation of the lat-

eral currents, and particularly those upon the north side, are
clearly observable.

In Section C, where the currents struck the edge of a forest, the
action of the lateral currents, and especially those which crossed
on to the north side, is distinctly shown.

We come next to a consideration of the most violent effects
produced, and the measure of the force required to produce them.
I take first, an extract from the description of the Schuyler
tornado which has been alluded to.

" But the climax of the storm is yet to be seen. About three
hundred feet directly southeast of the apple trees, upon an abrupt
eminence, say thirty feet high, there stood a first-class farmer's
barn, thirty five by fifty feet with a stone foundation, and with
posts sixteen feet high. Its length was east and west, and its
roof sloped towards the north and towards the south. The south
portion of the roof was carried north over a garden with fruit
trees and deposited in a field beyond, some portion 200, some 300,
and a large portion at least 500 or perhaps 600 feet from the barn,
while the north side was carried just about as far towards the
south. A threshing machine of iron and wood, said to weigh
400 pounds, was carried 230 feet south-southeast, and a sill or
plate of the barn, weighing probably twice as much, lies near it,
while huge timbers are beyond it, one 300 feet to the south, nine
inches square by twelve feet long. Thus for 500 feet or more all
around towards the north, east, and south of this barn, lie its fragments, every timber and board removed from the greater portion
of the foundation, *and even the two inch plank hemlock floor of the
basement of the west end of the barn, saturated with water, any
one of them a load for two men, were lifted up and thrown together in a heap.* This it must be understood is at the *bottom of
the basement* some *ten feet deep, protected by the stone foundation
wall,* as it should seem, from the direct action of the wind."

I copy next, the following from Mr. Stoddard's description of
the Brandon tornado :

" Another mode of estimating the force of the wind may be
adopted. Among the oaks previously named as standing on ris-

ing ground, was one, a giant among giants. Its trunk was three feet in diameter and straight, its top symmetrical, and the whole sound to the core. It was shivered to fragments, near the ground. Let us estimate the force necessary to break it. We will call the diameter $2\frac{1}{2}$ feet, the height 80 feet, the outline of the top a rectangle 30 by 40 feet, and let us suppose the whole surface exposed to the wind to equal one-fourth of that included in the outline, or 500 square feet. Under these conditions the resultant of all the forces would act on the tree at a distance of 55 feet from the point of fracture. Taking the strength of oak at the usual standard, a force of 73,636 lbs. would be required to break the tree. If the surface exposed to the action of the wind be estimated at 500 square feet, the pressure upon each square foot would be 147 lbs. This gives a velocity of 172.9 miles per hour, equal to 253.5 feet per second. This is about one-fourth the velocity of a cannon ball. Though the above estimate gives an enormous force to the wind, yet I cannot perceive that any of the assumptions are exaggerated. The most doubtful point is the amount of surface supposed to be presented to the wind by the limbs, &c. The estimate is believed to be a large one, as the trees were at the time destitute of foliage.

Other circumstances are not wanting to sustain this view of a high rate of velocity in the tornado. A mass of brick cemented together 4 feet by 3, and one foot thick, containing at least 12 cubic feet, and weighing more than 1000 lbs., was carried 15 feet from the wall of a house. *A board was driven 3 feet into a charred oak stump.* The writer pulled a *shingle* from an oak tree which had been driven into it *one inch*. An estimate of the force in the aggregate, of this tornado, if made clear to the mind by comparison with some well-known standard, would excite astonishment. If any one will take the trouble to make the calculation, he will probably assent to the following: A section of it one-half mile wide and 100 feet high, exerted a force equal to half the steam power of the globe. More than 50,000 trees were prostrated or broken by it in less than one-half hour."

And from the same writer respecting the Harrison tornado.

" This storm was remarkable for the occasional exhibitions of extreme violence. The Graham place afforded the most striking examples. The house was not merely thrown down, but shivered. A teaspoon was carried half a mile, and a piece of stove 30 rods. A heavy wagon was carried a few rods and dashed to fragments against a tree. A piece of scantling twelve feet long, and three by four inches, was taken thirty rods east, and driven into the ground three and a half feet. Four men tried in vain to pull it out, and the writer in connection with Mr. Graham, succeeded only after digging around it to the depth of 3 feet. The earth thus penetrated, was, with the exception of the first six inches, a stiff, yellow clay. What renders this incident more striking was the fact that the end of the scantling was not pointed. The cross section presented a resisting surface of eight square inches, and the whole amount of earth displaced was equal to 336 cubic inches. According to experiments made at the United States Navy Yard, a shot 68 inches in diameter would penetrate earth five feet, nine inches. A shot of 3 inches diameter, nearly the resisting surface of the scantling, would under similar circumstances penetrate three feet, nine inches. *This is but three inches deeper than the scantling was driven.* While the weight of the timber in question would have been greater than the weight of a three inch shot, on the other hand the form of the end was not so favorable for penetration. What portion of its velocity was due to falling cannot be told with accuracy. As it fell within thirty rods of the building from which it was taken, it could not have ascended to a great height. It entered the ground at about an angle of 45°, and if from this we estimate the velocity acquired from falling, to equal one-half of the whole, there is still left a velocity of 500 feet per second, due to the wind. The effects of such a wind would be fearful indeed; it would move at the rate of 340 miles per hour."

One other extract will suffice in relation to the power and effect of this class of conditions. It is from Prof. Loomis' description of the one which occurred at Mayfield, Feb. 1842.

"The velocity of the wind's motion, however, at points of the

most destructive violence, was far greater than this   A tolerable
idea of its velocity may be gained from the distance to which
light objects were driven into the ground.   Small pieces of clapboards with square ends were driven into turf land 18 inches, and
with sharp ends, two feet.   *What charge of powder is capable of
producing the same effect?*   According to the experiments of Dr.
Hutton a pound ball of cast iron fired from a gun of 2 inches
calibre into solid blocks of elm wood, in the direction of the fibres,
penetrated the following distances :

	VELOCITY.	PENETRATION.
With a charge of 2 oz. powder,	800 feet per second.	7 inches.
" " 4 " "	1200 " "	15 "
" " 8 " "	1500 " "	20 "

Dr. Hutton estimates the resistance of elm timber $7\frac{1}{3}$ times
more than that of firm earth.   A pound ball with a velocity of
800 feet, should then penetrate the earth 51 inches.   The depth
penetrated being assumed to be as the square of the velocity, a
pound ball fired with the velocity of 550 feet would penetrate 24
inches.   The space penetrated is said to be as the specific gravity
of the ball.   A wooden ball two inches in diameter, specific gravity 75, fired with a velocity of 550 feet, would then penetrate firm
earth 24 inches.   As this last result is a deduction from principles somewhat doubtful, I desire to verify it by experiments of
my own.   A six-pounder was accordingly charged with $1\frac{1}{4}$ lbs. of
powder.   Two or three short pieces of oak board, three inches
wide and one inch thick, were added, and the gun pointed toward
a steep hill, distant about a rod.   The boards penetrated the
ground a few inches, were badly shivered, and bounded some
distance up the hill.   A second experiment was tried with nearly
the same result.   The hill was of usually firm earth but not
strong.   The greatest penetration did not exceed six inches.
Velocity computed 1000 feet per second.   According to the
former data, the penetration should have been nearly 8 inches.   But
the ground at Mayfield was saturated with water.   I have no precise data for estimating the allowance required by this circum-

stance. I judge, however, that it would not increase the penetration more than threefold. We arrive then at the conclusion that the clapboards at Mayfield were driven into the earth with a velocity of 1000 feet per second, *or 682 miles per hour.*"

Perhaps I ought not to omit the fact alluded to in a general way, in a former chapter relative to the ploughing of the ground by the movement of an iron plough during a tornado at Stowe. The following is a description of it in Prof. Loomis' words:

"There is another fact which appears to my mind still more remarkable. A very heavy cast iron plough was lying between the two houses C and D; a massive iron chain was attached to it, and there was little wood work about it. This plough was dragged along about four rods and ploughed into the ground in several places. In one spot it appears to have been carried almost entirely around, removing all the turf from a space about four feet square, and throwing up the earth to the distance of six feet. The plough was broken so as to be worthless."

This fact is quite as astonishing as any of the other almost incredible facts contained in the foregoing extracts.

A few words in passing relative to the places where they occur and their frequency An examination of the many conditions of this description which have been recorded in the newspapers, as well as the forty or fifty which have been described in scientific books, or other publications, will show, that although occurring occasionally over all the eastern states, in the Atlantic system of conditions, nearly all have occurred at or south of the then location of the focal path. In the two or three cases mentioned where they have occurred north of the focal path, as at Brandon and Harrison, Ohio, they have occurred during very warm and intense southeast thaws. Thus of four which have occurred at, or in the immediate vicinity of Natchez, Mississippi, all have occurred in May. About the same number which have occurred in the valley of the Ohio, have occurred in the same month or in April. And they have visited the states north, northwest, and northeast of that valley later in the season, and in the summer after the focal path had passed up over or beyond them. Under that focal path, or southerly of it, they have occurred every year since the country was settled, and will continue to occur while the atmospheric system remains as now.

It is impossible to estimate satisfactorily the average number which have occurred in each year. The forests of the Mississippi valley are scarred with them. Five are known to have occurred in different states on the 15th of August, 1787, and several formed under the same belt of showers in which the New Brunswick tornado appeared on successive days, as the belt drifted to the eastward. Probably an average of ten a year would be a low estimate. Some of them have been very destructive to human life, and they constitute one of the dangers of the east, as earthquakes do to California, and perhaps the greater danger of the two. They do not occur in the Pacific system.

Their rate of progress over the earth varies with their size and intensity. That described by Mr. Chappelsmith, which was about one mile in width, and had a connection with the earth for 240 miles, moved with an estimated velocity of 60 miles an hour, prostrating trees at the rate of 7000 per minute. The progress of the greater number where the width has been less than 100 rods, has been at the rate of from 30 to 45 miles an hour.

Let us now review and examine the facts thus stated, and deduce from them inductively, the mode in which the organizing and sustaining force operates, constituting the law of the organization.

Turn now first to page 81, and examine Prof. Loomis' chart of the effects produced by the tornado at Stow, Ohio. We see at a glance by the arrangement of the prostrated trees that the force was exerted laterally on each side of the supposed central line, R, S. The general effect is obvious, and the exceptional cases are few and easily accounted for. It is not necessary to dwell upon it. Prof. Loomis sums it up in the following sentence:

"We have then, I think established that there were two powerful currents of wind blowing from the opposite sides of the track; that is, within a few rods of each other, and with such violence that the stoutest oaks fell before it."

Turn now to a representation of a part of the track of the New Haven tornado, as inserted on page 79. Here we have the same result, as shown by the manner in which the corn was prostrated in the cornfield, and by the direction in which the trees fell, and the fragments of a building were carried from the right hand side of the path. We have also some prostrated trees on the left hand side of the path, with their tops toward the center, and some trees also, and the fragments of other buildings carried exceptionally and obliquely outward to the northwest. It will be seen hereafter that these exceptions are perfectly consistent with the general effect we are considering.

Turn now to the remainder of Prof. Olmsted's chart, and we will commence with Section I. Here we have first, on the right of the track at the lower corner, the debris of buildings thrown inwards toward the center; then we have at E the representation of a building destroyed, and the fragments carried inwards toward the center and across to the opposite side of the path; then we have at F trees thrown inward and forward; at G corn prostrated toward the center, and further on, near K, another building destroyed and its fragments thrown toward the center of the path. On the left-hand side of the path, in this section, we see the corn prostrated toward the center, the trees also, farther on, and the contents of the building, I, in like manner. The same exhibitions of force are seen in Section II, on either side the track, with the exception of two trees thrown to the westward. In Section III we have the same exhibition with a similar exception at Q. The same may be said, in a single sentence, of Section IV.

Turn now to Sections II and III of Prof. Stoddard's chart of the Brandon tornado. On these we need not dwell; they show the same operation of the force.

Turn next to Prof. Stoddard's chart of the Harrison tornado. On Section B of that chart we see the same state of things on both sides of the center, with a curving of the force at the Graham place, where buildings 1 and 2 were demolished, similar to that shown at Q of Prof. Olmsted's chart, and substantially the same result is also observable in Section C.

Without dwelling upon this very plain matter, we may state as our first deduced proposition, that the organization and motive force of the tornado operates in two lateral lines of force, or currents, tending towards a central line.

Turn again to the diagram of Prof. Loomis on page 81. The house of Mr. Sanford, D, was utterly destroyed by a vein or concentration of the force which crossed the central line from the right hand side, and the fragments were strewed in the direction of the barn, N. 29° E. from the house. The members of the family were carried in the same direction, and found in the debris. "An ox cart, before the storm, was standing close by, and in the rear of Mr. Sanford's house, and was loaded with potatoes. The cart was lifted up by the wind; it soon turned a somerset, so as to empty the potatoes upon the ground, and nearly all in a heap. The cart itself was dropped a few rods *behind the barn*, and at a distance of 30 rods from the house. If the cart moved in a straight line, it must have passed directly over the barn. Indeed it is quite probable that such was the case, for the cart struck flat upon one wheel, which buried itself to a considerable depth in the earth."

We are not informed of the direction in which the fragments of the unroofed buildings on the north side of the track (and most of them were unroofed) were carried. But as Prof. Loomis says, "everywhere there is the same evidence of two currents in exactly opposite directions, having passed over precisely the same spot," it is probable that the southern current crossed the center of the track at other places.

Turn again to the fragment of Prof. Olmsted's chart on page 79, and observe how the fragments of the building on the right hand side of the track were carried across the center on to the left hand side, and also how the fragments of the building and some of the trees standing on the left side, are thrown outward and backward by the southerly currents which crossed the center of the path.

Turn next to the remaining portion of his chart on page 321, and observe how the fragments are carried by the southerly current across the center at E in Section I, and at M in Section II, and the manner in which the trees are thrown backward and outward on the left of the path near L in Section II, at Q, in Section III, and at S in Section IV.

Turn next to Section III of Prof. Stoddard's chart of the Brandon tornado and see the manner in which the southern currents cross the center of the path, and curve backward as they pass outward. The fact that they thus passed in advance of the lateral currents of the left side, is shown by the arrows indicative of the left hand current, which have a cross on them, as well as by the simple arrows intended to show the left hand current ; for the cross upon the arrows of the left hand current indicates that the trees which they represent, lay over trees previously prostrated in an opposite direction. The same overlying may be observed near the center on Section II where two such crossed arrows are placed.

Turn now to Prof. Stoddard's chart of the Harrison tornado, and you may see on Section B the precise manner in which the southerly currents cross the center and curve backward in advance of the left hand currents, on the north side of the path. On this chart, the action of the currents is unmistakable. And in this connection, I copy the following important suggestion from Prof. Stoddard's account :

"In conclusion, the writer would take the liberty to suggest to observers that he has found it important to carry his observations beyond the track of greatest violence. Though no trees nor houses may be thrown down, yet valuable evidence to show the mode of action can oftentimes be obtained.

Mr. Laird's house in the vicinity of the Graham place, was on the left of the axis, but too far from it to suffer any injury. The wind was violent, but left none of the ordinary marks which could determine its direction. Mr. Laird, however, stated to the writer that the wind first blew in the south door, and two men were unable to shut it. A moment afterwards, and the north door, wh'ch had been locked, was violently driven in. The direct and reverse stroke of the loop seem pointed out here. The action of a tornado along the axis only, affords but confused data to elucidate the laws which govern it."

In order that the reader may understand what Prof. Stoddard means by the " *direct and reverse stroke of the loop,*" I here insert a copy of a figure made by him to represent his view of the cycloidal form of the action, but without endorsing it. Our business now is with the fact that the right hand current does cross the center line in advance of the other as a rule.

FIG. 46.

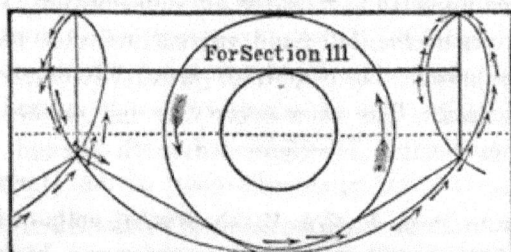

For Section 111

From this further review and examination, we deduce a second proposition, viz : that of the currents thus organized and moving inwards, the right hand or southerly current crosses the center in advance of the other, and curving backwards, *ascends over the left hand current,* constituting a part of the upward, ascending and visible whirl. The northerly or left hand current curves behind the other, having frequently a backward inclination, and being weaker and following the rapidly advancing spout it curves and rises over the right hand current, prostrating or turning trees so that they overlie those previously prostrated by the right hand current, and ascending over the right hand side of the path, constituting another portion of the whirl. These currents thus organized and operating when unaffected by the disturbing causes which produced the irregularity, constitute the law of the tornado.

I have made these deductions from a limited number of instances, but they are instances where the facts were carefully observed

by intelligent men, and I know of no instance that is in conflict
with them. The same facts do not appear so clearly in the New
Harmony tornado, and for obvious reasons. A mile square of
the track of that tornado was surveyed and plotted by Mr. Chappelsmith, exhibiting the direction in which every prostrated tree
fell, and is a valuable acquisition to the science. But, as we
shall see hereafter, there is a *tendency* in the wider organizations to
the *straight line gust*, rather than the limited whirl, and that tendency affected the New Harmony tornado. So it undoubtedly
was affected by the unusual rapidity of its progress, and it is natural that the law we have deduced should be more or less masked
by the causes named.

We are all familiar with the straight line thunder-gust, a few
miles in width only, which occurs beneath the summer belts of
showers. And we have seen in the Cincinnati hurricane, which was
from 30 to 40 miles in width, and moved with the speed of 70
miles per hour, that the lateral currents were from the S.W. on
one side, and the N.W. on the other, as they are in our summer
belts of showers, without observable evidence of whirl in either.
But notwithstanding the width and speed of the New Harmony
tornado, its lateral currents were clearly discernible amid the
confusion which other causes produced, although it is not probable that a distinct whirl, embracing the entire organization existed
at the surface or above it, at any instant of time.

AND WE MAY TRACE THIS LAW OF THE SMALL, AS THE LAW
OF THE GREAT, THROUGH ALL THE VARIED ATMOSPHERIC OR-
GANIZATIONS.

We have seen that those lateral currents are clearly developed
in the wide as well as the narrow belts of summer showers. In
the wider and less intense southeaster, and on the margins of the
still wider and still less intense northeasters, and in the great central condition where they constitute the trades.

Let us look at them, first, as we have seen them developed in
belts of showers. That organization, if regard is had to the conditions of both hemispheres, is undoubtedly the most frequent of
all; indeed, in the Southern Hemisphere nearly all of their organ-

izations are of that character.   There are upon record a few
southeasters, corresponding to our northeasters, in which the wind
blew continuously from the S.E., in opposition to the progress of
the storm, as it does in our northeasters; but they are occasional,
exceptional, and in high latitudes only.   We have seen from the
log of the China, that that vessel encountered and ran through
five distinct conditions—long, narrow, elliptical, and trough-like,
with N.E. or N.W. lateral winds on the northerly side, and S.W.
winds on the southerly side.   Through these conditions she held
her way, averaging 201 miles per day, and in no one of them did
she meet with a storm wind hauling round through the E. and
S.E. to the S.W. as she would have done had they been cyclones.
but in each and all of them she had the wind hauling from the
N. through W. to S.W. or shifting suddenly to the same quarter.

In Maury's Sailing Directions are given the logs of 130 ships.
or voyages around Cape Horn.   I have carefully examined those
logs.   I find four or five instances where, in the vicinity of the
Falkland Islands or in the Pacific west of lon. 78°, the wind went
around from N.E. through E. and S.E. to S.W.   But they were
all during variable weather without depression of the thermometer or serious gale, and not even such an instance occurred in any
of the 130 cases between lat. 56° and 59°, and east of lon. 75°.
Off the Horn and south of it, those 130 vessels all had their storm
winds and low barometer from N. to W., and their fair weather
winds with rising barometer from S.W., showing a succession
of passing elliptical belts with a wind setting in from some northerly point and ending at S.W.   Col. Reid assumes every one of
those belts, and all other passing organizations in the hemisphere
to be revolving gales, but it is a mistaken assumption.   Let us
examine.

The following is a copy of the cut given by Col. Reid, pur-
porting to represent the manner in which the wind changes in
what he assumes to be revolving gales at the Horn.

FIG. 46.

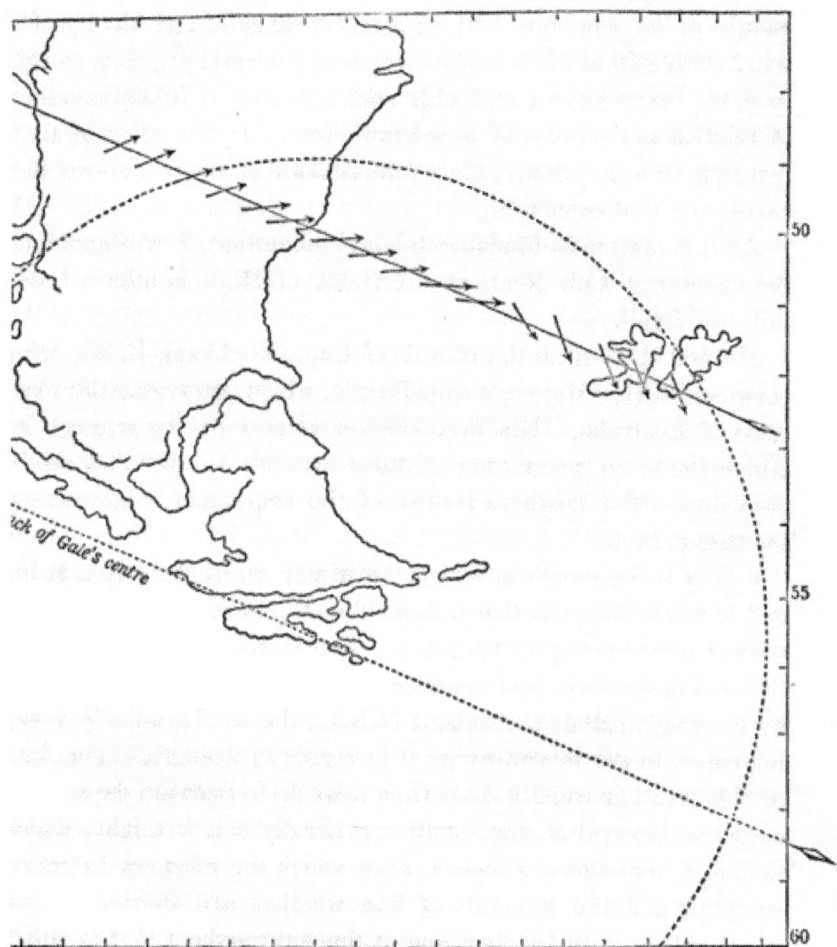

It was an easy matter for Col. Reid to sweep a circle upon the
map representing a revolving storm, so that in passing from the
S.E., a vessel in the vicinity of the Falkland Islands, having
a wind N.E. or N., should have the wind veering round from
W. to S.W. as the storm passed over the vessel. He has drawn
his line across the circle, to represent the manner in which the
storm passed over a vessel. Many similar circles, representing
the vessel as in the *northerly* half of the gale, are scattered
through his chapter on the gales of high southern latitudes. But

not in a single instance does he represent a vessel as having been
caught in the southern half of a revolving gale, and having the
wind setting in at N.E. and veering round through E., S.E. and S.
to S.W. Nor have I met with such an instance in my reading
in relation to the gales of that hemisphere. If the gales of that
hemisphere are cyclones, the southern half of them is above the
earth, and that cannot be.

As this matter is fundamental and important, I will examine
the chapter of Col. Reid, on the Gales of High Southern Latitudes in detail.

He first cites from the record of Capt. Wickham, R. N., who
commanded Her Majesty's ship Beagle, whilst surveying the west
coast of Australia. His first citation relates to the manner in
which the south polar zone of rains extends up over that continent during the northern transit of the zones, and is interesting
for that reason.

" May is the month in which the winter weather fairly sets in,
and it rarely happens that the middle of this month passes without the rains having commenced. This season seems to vary but
little as to the time and manner of setting in. It is ushered in
by blowing weather from about N.N.E., the wind gradually veering round to the westward, as it increases in strength. The first
of this weather usually lasts from a week to fourteen days, then
comes an interval of fine weather generally of a fortnight's duration, and sometimes a month, after which the rains set in more
constant, and the intervals of fine weather are shorter. This
weather lasts until October, and at times throughout that month."

We here see the manner in which the south polar zone of rains
extends up with the general transit, and that its peculiarities are the
same as exist in the northern zone. There is the same interval of
fine weather occurring between its first appearance and its steady
existence, " *the early and latter rain*," and sometimes the difference of a month in the time of its disappearance. When the
latter is traced out, it will be found to be connected with, and
conformed to the irregularity of the transits of the central belt.

Capt. Wickham sums up the character of the gales on the S.W. coast of New Holland, as follows:

" The N.W. gales that occasionally occur during the winter months, on the southern parts of the west coast of New Holland, are probably felt as far north as Sharks Bay. They blow with great violence, and are accompanied by dark, gloomy weather and rain. It is then unsafe to be near the land, as the gale that commences at N.N.E. *invariably veers to the westward,* making a lee shore of the whole line of coast, and between W.N.W., and W.S.W. blows the hardest."

Col. Reid copies from the record of Capt. Wickham, descriptions of a number of storms, but they all, with a single exception, commenced with wind from N.E. or N.W., and a falling barometer, and with a cloud bank coming from the west or northwest, ending with the wind southwest and a rising barometer. There was one exception which was evidently a straight line southeaster, corresponding to our northeaster. The record is as follows:

" On the 8th of the same month, the barometer was 30.05 ; at 8 P. M. with fine weather, wind S.E. by E. it then commenced to fall, and at 8 P. M. on the 9th, was 29.80, and blowing a heavy gale at S.E., which continued all night and until 8 P. M. on the 10th, at which time it became more moderate, and the barometer began to rise." That was clearly a straight line southeaster which did not veer, and not a revolving gale.

The last case cited was exactly like the northeasters of the northern hemisphere where they commence at N.E. in opposition to the progress of the storm, and blow in a straight line for 36 hours or more, and cease when the storm has passed by. Such, precisely, was the southeaster described by Capt. Wickham.

If the reader will remember that the storms of the southern hemisphere travel to the S.E instead of the N.E., that the lateral storm winds are on the northerly or tropical side, and the lateral fair weather winds on the south or polar side, directly the opposite of what occurs in the northern hemisphere, he will clearly understand that all the instances, so far, conform to the character of long, elliptical belts, rather than revolving gales.

Col. Reid next cites from the record of Sir J. Ross, of two
voyages to the south seas, with an account of one or more storms
which commenced with the wind from a northerly point, veering
through the W., and ending in S.W. One of these was in lat.
62° 42′, and another in 65° 48′. Not a storm is cited from either
voyage of Sir J. Ross which set in from the N.E. or E., and veered
through the S.E. to S.W. Col. Reid next cites from Capt. Fitzroy with relation to the storms on the western coast of South
America, and I copy a part of the citation.

"There is much less difference between the climate, the prevailing winds, and the order in which they follow; the tides and
the currents on the outer coasts of Chiloe, and at the west entrance
of Maghalaens Strait, including the intermediate coasts, than
persons would suppose, who judge only from their geographical
positions. Northwesterly winds prevail, bringing clouds and rain
in abundance. Southwesters succeed them, and partially clear the
sky with their fury; then the wind moderates and hauls into the
southeast quarter where after a short interval of fine weather, it
dies away. Light airs spring up from the N.E., freshening as they
veer around to N., and augment the store of moisture which they
always bring. From the N. they soon shift to the usual quarter,
N.W., and between that point and S.W. they shift and back,
sometimes for weeks before they take another turn round. When
the wind backs (from S.W. to W.N.W., &c.) bad weather and
strong winds are sure to follow. On that coast the wind never
backs suddenly, but it shifts with the sun (with respect to that
hemisphere) very suddenly, sometimes flying from N.W. to S.W.
or S. in a most violent squall. *Before a shift of this kind there
is almost always an opening or light appearance in the clouds*
TOWARDS THE S.W. *which the Spaniards call an eye (ojo), and for
that signal the seamen ought to watch carefully.* As the sudden
shifts are always with the sun, no man ought to be taken aback
unexpectedly; for so long as a northwester is blowing with any
strength, accompanied by rain, so long must he recollect that the
wind may fly round to the S.W. quarter at any minute. It never
blows hard from E.; rarely with any strength from N.E., but an

occasional severe gale from S.E. may be expected, especially about the middle of winter (June, July, August). In summer southerly winds last longer, and blow more frequently than they do in winter, and the reverse. The winds never go completely round the circle ; they die away as they approach E., and after an interval of calm, more or less in duration, spring up gradually between N.E. by E. and N."

In this extract we have the following facts: first, the northwesterly winds prevail, bringing clouds and rain in abundance, southwesters succeed them, and when they die away, fine weather ensues. Second, the wind blows steadily from the northward, sometimes hauling slowly from N.W. to S.W., and sometimes flying suddenly from one to the other in a violent squall. Third, it is in the interval between the passing by of the stratus cloud of the storm, and the obscuration of the sky by the scud of the S.W. wind that what is called the ojo or supposed eye of the storm is seen, and it is obvious that this is all there is of it. Fourth, an occasional severe southeaster, precisely like the exceptional one described by Capt. Wickham, corresponding to our northeasters, is experienced. Fifth, IT NEVER BLOWS HARD FROM E., *rarely with any strength from N.E.*, and it is perfectly evident that a cycloidal storm setting in between N.E. and E., and veering through S.E. to S.W., was never witnessed there by Capt. Fitzroy. All the phenomena point conclusively to the passage of successive long, narrow, elliptical belts, with northerly winds upon their tropical sides, and southerly winds on the polar sides, moving to the southeast and *drifting to the east*, precisely as they occur in this country, and that the "ojo," of which so much is made, is simply the lighting up, at the close of the storm, in the S.W. just as we frequently see it in the N.W.

Col. Reid next cites the case of a French vessel in lat. 38° S., long. 22° E., in which it was said the "*ojo*" was seen. The ship took the gale from the N.E., and it veered round through the N.W. to the W, and then again the "*ojo*" was nothing more than what we frequently see when the obscuring stratus clouds of the belt have drifted past the zenith, and the heavy westerly scud are

coming up in the westerly clearing off wind. I have seen hundreds of just such "*ojos*" in the course of fifty years, and an
average of at least ten may be seen any year, and in any part of
this country or off the coast, by any one who will take the pains
to look for them, when the storm ends by "lighting up" in the
northwest.

Col. Reid next cites from the record of Capt. Sullivan, R. N.,
who was employed making a survey of the Falkland Islands.
The substance of his observations is contained in the following
paragraph:

"We had, on an average, during the five months I was there,
as many days of gales as of moderate weather, the usual round
being this :—The gale commenced at N. or N.W., and after having blown for some hours from those quarters drew round to *west*,
then *southwest* and *south*. But there were some gales that blew
for several days from S.W. without having commenced from the
northward of west. Some also commenced at N.E. and blew
strong from that quarter for several hours before they drew round
to the westward." This, allowing for the different direction in
which the storms move in the two hemispheres, is just what takes
place in the interior of this country, and all over our hemisphere ;
and Capt. Sullivan *met with no gales* which veered through E.
and S.E. to S.W.

Captain Sullivan had a method of keeping his record of the
wind which was peculiar. It was by mapping arrows on an east

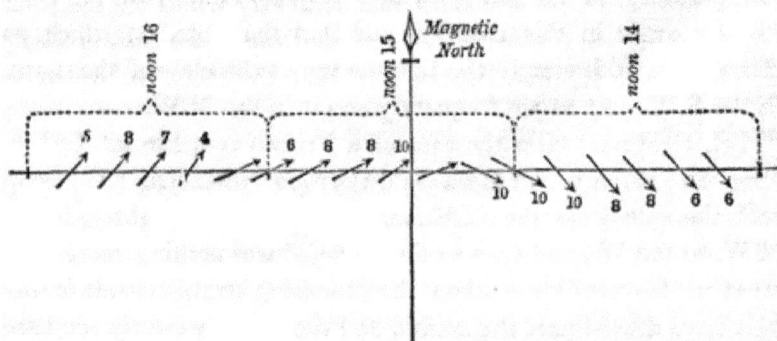

FIG. 47.

and west line, with dotted lines for the days and figures to show the strength of the wind, as shown by Fig. 47.

Col. Reid drew his imaginary circle around *such a record*, in mapping the diagram on page 339. But it should be observed, 1st, that he tilted up the line, so as to have it slide over the island from N.W. to S.E., and that changed the direction of the wind, and *falsified the record*. And 2d, that even then, the record showed steady winds from the two points, not gradually veering ones.

Col. Reid next cites from an account of the voyage of the Albatross from Van Diemans Land to England, round Cape Horn. When in the South Pacific, and in lat. 48° S., long. 159° W., the master, Mr. J. M. Gill, remarks:

" It is here where the mariner may study and depend on the barometer. As regularly as the gale veered from the northward to the southward of west, so the column rose, and from the southward to the northward, the indicator was depressed." And it is added, " that light winds, veering to the north and east forebode the day of change."

There is subsequently a description of a gale in which the *"ojo"* appears again, and I copy: " The N.W. gale continued to rage with fury until 6 A. M., when it suddenly changed to W.S.W., and blew with increased force. The heavy cross sea, produced *by a sudden shift, is described as all foam and spray.* The little vessel was then near the gale's center, and here we have an example of that clear space in the heart of the tempest, the storm's eye." Mr. Gill adds : " The sun having been concealed during the day, *now shone out for a few minutes nearly in the wind's eye, with fiery brightness,* some of us thought for the last time. *The small archway that opened to windward through dense masses of cloud,* served the purpose of a funnel, and forced the gale over us in gusts, every one of which appeared determined to tear away the few yards of canvass we had set ; strong, stout and new as they were." The idea of an archway in the cloud operating as a funnel for the surface wind is an absurd fancy ; but not more absurd than the idea that the setting in of a fair weather wind— the storm clouds having all passed by, and permitted the sun to

16

appear—constitutes the *center of the storm*. Those archways of clear sky are common when the wind shifts to the N.W., and it lights up in that quarter, the storm c'ouds having passing by.

Let me now turn your attention to another important and satisfactory piece of evidence. And here I cite from the "*Handy Book of Meteorology*," by Mr. Buchan, the Secretary of the Scottish Meteorological Society, published in 1868, page 273.

" STORMS OF THE INDIAN OCEAN, SOUTH OF THE EQUATOR. Through the activity and well-directed efforts of the Meteorological Society of Mauritius, the storms of the Indian Ocean have been submitted to a fuller examination than those of any other ocean on the globe. Since the formation of this Society in 1851, it has devoted a large share of its attention to the collecting of Meteorological statistics of the Indian Ocean, and the tabulating of them in chronological order. Upwards of 500 synchronous weather charts have been constructed under the direction of its able and energetic secretary, Charles Meldrum. I have through his courtesy, examined a considerable number of these charts, which, if the isobarometric lines were filled in, would leave nothing to be desired. At the meeting of the British Association, held in Dundee, 1867, Mr. Meldrum gave an account of these storms, and it is chiefly from this paper, a copy of which he has kindly sent me, that the following facts regarding them, have been taken. Of these storms there are two sorts, viz : tropical and extratropical storms."

"EXTRATROPICAL STORMS. These storms occur at all seasons, but are most violent during the winter months, from May to August inclusive, in this respect resembling the extratropical storms of the northern hemisphere. They are generally characterized by the presence of two currents of air, the one from the southward and the other from the northward. Sometimes the two currents exist side by side, the one from the N.E., the other from the S.W., each occupying a belt of 5° to 30° in longitude, and stretching from 30° S. lat., as far south as the observations extend, viz : to lat. 45°. In the space between the two winds, light airs, calms, and a high cross sea, with heavy rain, thunder and lightning, generally prevail, and there the barometer is lowest. The belt of southerly winds lies invariably to the west of the northerly winds, and the two travel laterally to the eastward, preserving their relative position often for several days. Instead of blowing in parallel belts, however, the winds are often inclined, and sometimes directly opposed to each other. The barometer stands higher and the thermometer lower, in the southern than in the northern gale, being in these respects quite analogous to European and other storms of the northern hemisphere. There cannot be a doubt that the form of by far the majority of these storms, is that of elongated ellipses, or trough-like, their length being very much greater than their breadth. On this account the shifts of the wind are generally sudden from N.E. to S.W. or from N.W. to S.W. The veering is from N.E. to N., N.W.W., and ending at S.W. or S.E. They last from one to seven days, and travel at the rate of from 4 to 20 miles an hour—their progressive motion being thus generally slower than that of European storms. Mr. Meldrum is of opinion that they are not revolving gales, like the storms which take place in the tropics—an opinion with which I am not prepared to concur for the following reasons."

These reasons are unimportant, and I omit them. They are simply doubts as to the sufficiency of the evidence, such as might be expected from one prepossessed in favor of a different theo·y.

And now, practical reader, let me beg you to read and re-read that concise description, so perfectly conforming *in every particular* to our long narrow belts and southeasters, and having the same lateral winds which are found in them.

It is pleasant to find a meteorologist like Mr. Meldrum, residing on an island in an ocean, and devoting his time as secretary of a meteorological society, to the investigation of storms which render the navigation of that ocean dangerous to an immense mercantile interest, thus arriving at correct results in relation to the extratropical storms of that hemisphere. It is to be hoped that his example may not be lost, and his facts unheeded, by the meteorologists of the northern hemisphere. His description of these extratropical belts conforms precisely to the character of

our summer belts of showers, and autumn southeasters as they
form and travel to the northeast and drift laterally to the eastward, in this country and in Europe, with the lateral wind on the
tropical side crossing in f.ont of the belt, and the opposite lateral
wind on the polar side, crossing in its rear. Such are, *in fact*, the
gales of the southern hemisphere.

From about the year 1834 till after the death of Mr. Redfield,
the rival theories of Messrs. Redfield and Espy were persistently
and strenuously urged by them and their friends, upon the attention of the public. The theory of Mr. Espy assumed, first that
there was an inward and upward current in all large cumulo-
strati, by which they were constituted, and the following diagram is a copy of that by which he illustrated it.

FIG. 48.

I have seen thousands of them and never saw any evidence of
such a current. He next claimed that the winds in all storms
blew in towards a central point or line, and he established the
fact in relation to many of them, and in respect to some which
were claimed by Mr. Redfield to be cyclones. Mr. Redfield on
the other hand claimed that all storms revolved in involute spirals
around a center, contrary to the movement of the hands of a

watch in the northern hemisphere, and with their movement in the southern hemisphere. Many storms were investigated, and charted by each. The following is a copy of the card of Mr. Redfield, representing the manner in which he first assumed that the winds revolved in the storms of the northern hemisphere.

FIG. 49.

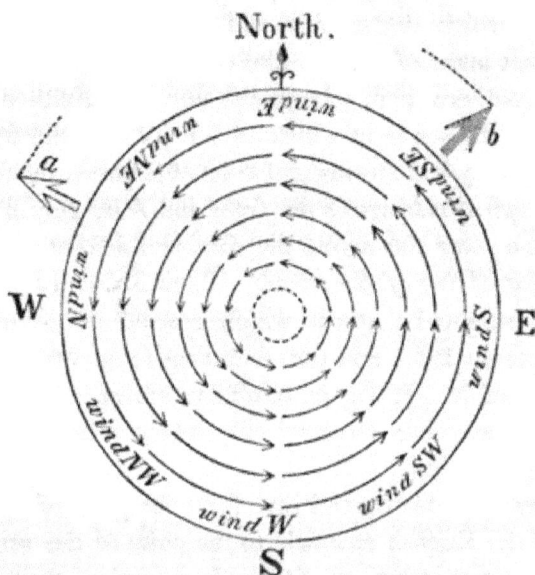

He subsequently modified his views and came to the conclusion that the wind revolved spirally inward. A great majority of the storms investigated and charted by Mr. Redfield were ocean storms, but he claimed that the tornadoes were of the same character. It cannot be said that Mr. Redfield established the fact that all storms and gales were vorticose, or that any such storm of an extensive character ever occurred upon the interior of this continent *or anywhere over the land.* So on the other hand it cannot be claimed that Mr. Espy proved that all the ocean storms charted by Mr. Redfield were not cyclones, and that the wind in them blew in toward a central point or line.

From my own observation of the storms of this country, and an examination of many claimed to be vorticose, I came to the conclusion that the theory of Mr. Redfield could not be sustained, and so expressed myself in the "Philosophy of the Weather." Since then, however, I have examined the subject with greater care, and in the light of new developments in relation to the character of the tornado and other storms, and think I can see clearly that the truth lies in the propositions before stated, and that many of the ocean storms investigated by Mr. Redfield were cyclonic, in the sense in which the narrow and distinct tornado is cyclonic. But the great mass of storms, like those we have been considering in the southern hemisphere, are not so. Neither are they centripetal in the sense in which Mr. Espy used the term, for in all our long, elliptical storms and belts of showers, the wind blows from the *tropical side* across the *front* and *front part* of the storm, and from the *polar side* across the *rear*, and *neither ascend at the center or central line of the storm.* When Mr. Espy was in England, he investigated a storm which crossed the Islands on the 6th of January, 1839, and the following (Fig. 50,) is a copy of his diagram of it. It was an extensive storm, and not a belt of showers, and seeming to prove his theory, it made a strong impression on my mind.

Mr. Espy drew his central line from the point of the arrow at No. 24, on the English channel, to the point of the arrow at No. 11, off the east coast of Scotland, assuming that the storm traveled in that direction, which would be from S.S.W. to N.N.E. Now I long since became satisfied that storms do not travel in that direction over the British Islands, at that if at any season of the year, and that his central line should have been drawn from the S.W. point of Ireland near 30, to the crossing of the arrows No. 10, from W.S.W. to N.N.E. Then if the direction of the wind had been given as it in all probability existed, over the North Sea, it would have been found that the wind crossed the front of the storm and curved backward on the North Sea to the east of Scotland, and the map shows but little more than half the storm. Such at least was true of a storm which occurred on the 2d of November, 1863, and which was carefully investigated and charted by Mr. Buchan, and may be found in his work on

Fig. 50.

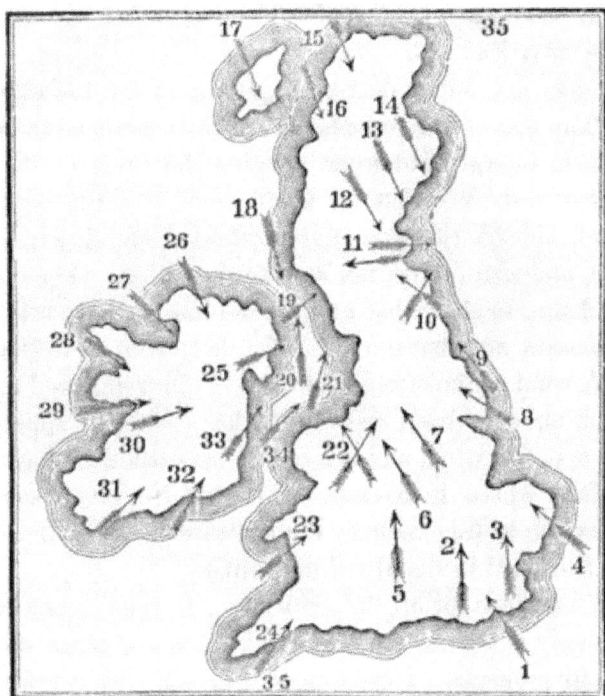

Meteorology, facing page 242. In that storm the wind on the tropical side crossed its front and curved to the westward on the north side of it.

The sharpest issue between Mr. Redfield and Mr. Espy was in relation to the storm of 1821, which was a violent hurricane, which, originating in the West Indies, came up the coast and crossed New England through the center of Connecticut, and the eastern part of Massachusetts. Both agreed that the wind was S.E. on the tropical or southern side of the storm, veering as the storm passed, through S. to S.W. Both agreed too, that the wind was N.W. on the northerly or polar side of the storm, west of the center; and the sharp point of the contest was whether the S.E. wind which blew across the front and front portion of the storm, curved to the west, becoming an east and northeast wind on the north front ending at N.W. I have not room for the data collect-

ed by either, but I am now satisfied that the lateral S.E. wind
did so curve, and it did so curve in many other storms investigated by Mr. Redfield.

But it does not follow that those storms were literally whirlwinds. The law of the tornado, as we have deduced it, requires
that the S.E. lateral wind after crossing the front of the storm,
and curving to the west, on its north front, constituting there a
N.E. wind, should rise over the northerly lateral wind of the
hurricane, and cease to be felt as a surface wind. There is very
much evidence to show that such is the law of that class of violent hurricanes, and that the lull which is spoken of, exists where
that N.W. wind of the storm when it is in the tropic, and the N.E.
wind when off our coast, rises from the surface to superimpose
upon the opposite wind, and before that opposite wind has arrived
at the place where it ascends. The lull, in other words, occurs where the loop is made by the contrary winds as represented
by Prof. Stoddard in the Harrison tornado.

I have not space for all the evidence. It requires an examination of many storms and of the observations of many observers
during their progress. I must content myself with a reference to
two or three facts.

The brig Chas. Heddle, sailed on the 21st of February, 1845,
from Mauritius in the Indian Ocean, bound north, and was met
in lat. 16° 42′ S., long. 57° 45′ E., by a violent hurricane, which
was moving slowly to the S.W. She was put before the wind on
her bare poles, and scud around the storm five times in as many days.
The following diagram (Fig. 51) shows the supposed track of the
storm, and path of the Chas. Heddle, as plotted by Mr. Piddington, from the log of the vessel. I have not room for that log.

Assuming the chart to be substantially correct, the reader will
see at a glance the contrast between the path of the vessel on the
right hand and left hand sides of the assumed track of the storm,
and the similarity between this chart and that made by Prof. Stoddard to represent the course of the wind in the Harrison tornado,
page 336. In both, a *loop* and not a circle is made upon the western side of the track. In my judgment the Chas. Heddle was,

FIG. 51.

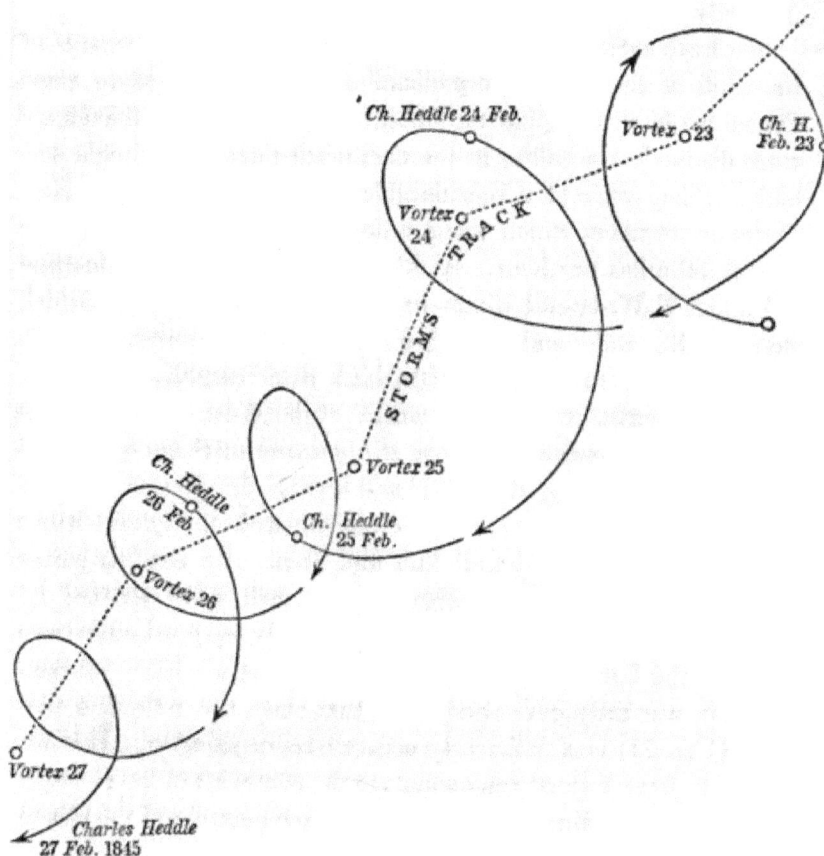

each time, driven across the front of the storm by the left hand lateral wind, which curved backward until it met the opposite lateral wind of the right hand side, when she was driven by that opposite wind back into the left hand lateral wind again. She was then driven by the left hand lateral wind onward and across the front and backward, and was again met by the right hand lateral wind, at a sharp angle, and again driven back to the left. This process was repeated until she was driven out of the storm or until what seems more probable, the storm moving slowly, gradually contracted and dissolved. I think we may clearly discern in this case, that the distinctive law of the small, was the law of the great. I could give other similar examples but not so distinctly marked.

We have seen that there is great variety in the movements of the wind of the various organizations, even of the same class. Hence we have seen that the lateral currents and their movement were distinctly traceable, in the narrow, distinct, and simple tornado. They were to a considerable extent masked in the New Harmony tornado, which was a mile in width, and moved at the rate of 60 miles per hour. We have also seen that they became S.W. and N.W. lateral winds in the Cincinnati hurricane which was 40 miles wide, and moved at the rate of 70 miles per hour. It is reasonable to assume, and I think demonstrable, that a similar variety exists in hurricane storms of the West Indies. There are undoubtedly some which are *straight line* hurricanes, like that of Cincinnati, moving with rapid and terrible destruction. There are doubtless others which are elliptical, with the lateral winds blowing in towards a central line, and frequently entirely across the front and rear of the storm. Some such were charted by Prof. Espy. There is another class, probably the most numerous, in which the law of the distinctive tornado as we have contemplated it, was fully developed. Of that class, the following diagram (Fig. 52) is a sufficiently accurate representation. It is assumed to have formed somewhere to the southeast of its position, within the dotted line where the summer temperature of the Ocean is 80°, and to be moving northwest.

The peculiarity of that class is that the action of the storm commences on the front, with a marked increase in the strength of the N.E. trades, and a gradual veering of the wind on the south front to the N.W., followed by a lull of longer or shorter continuance, according to the speed with which the storm progresses. That the N.W. wind is *suddenly* succeeded by the equally violent left hand lateral wind from the S.W. As the storm passes on to the N.W., that lateral wind is succeeded by the regular S.E. trade, and pleasant weather. In this class of storms we may see clearly exemplified the " law of the small," as in the distinct tornado.

FIG. 52.

Several of this class were investigated and charted by Mr.
Redfield. The hurricane of September, 1821, in relation to
which Mr. Redfield and Mr. Espy were at issue, I am now satisfied, was of this class. From the time it struck the coast of
South Carolina, after curving to the N.N.E., and until it passed
New York, and entered over Connecticut, the wind upon its left
hand front veered to the N.E. as it doubtless had done in lower latitudes. When, however, it had entered bodily over the land in
New England, it soon lost that distinctive peculiarity, and when
it crossed the valley of the Connecticut river, the right hand lateral S.E. wind blew up the valley the entire length of it, and
*across the entire width of the storm during all the time it was pass-
ing.* As I then lived in the valley, and noticed the storm particularly, that fact went far to induce the belief that it was not,
in any sense, a revolving storm. Nor did the lateral, left hand
northerly wind continue with any destructive strength, after it
entered upon the land in its N.N.E. course except for a limited
period in Litchfield County. We have, in the case of that storm,
satisfactory evidence that revolving, or to speak more correctly,
semi-revolving storms, sometimes lose their semi-revolving character, when they pass on to continents. The fact is demonstrably

true of all the hurricanes which enter bodily upon the Gulf States, and curve to the N.E. over them, or the Mississippi valley.

A hurricane substantially similar to that of 1821, passed over Rhode Island and Eastern Massachusetts, on the 8th September, 1869, pursuing a N.N.E. course. I was sitting upon the Court in Hartford on that day, and was interested to observe the approach of the storm. The wind was S.E. in the forenoon, the atmosphere very clear, and the sky very blue, but there were very large and peculiar scud floating in the fresh S.E. wind, and dropping dashes of rain, which I well understood to be the scud of a S.E. wind blowing across the front of a hurricane storm off the coast. I went two or three times in the course of the forenoon into the State Library, from the windows of which a fine view could be had to the east and southeast, for the purpose of observing whether the wind changed to E. and N.E., so as to indicate that the hurricane would pass over us. I called the attention of Mr. Hoadly, the State Librarian, to the character of the scud and their indication that there was a hurricane storm off the coast. The wind and scud did not change their course, or veer during the forenoon, to the E. or N.E., and I was satisfied that the storm would pass by to the S.E. of us, and gave it no more attention. The wind freshened somewhat in the early part of the afternoon; and during the course of it, and about the time when the gale was focal at Providence, which is about 80 miles to the E. of Hartford, the wind suddenly drew more to the east, and freshened with squalls of rain, and soon after shifted suddenly to about W. by N., and blew for a half hour or more with considerable violence, resembling in that respect the action of the Harrison tornado, at the *point of its loop*, at the house of Mr. Laird as described by Prof. Stoddard.

That storm of September 8th, 1869, was very violent in Rhode Island and Massachusetts, and I subsequently watched the marine reports of the New York newspapers, for the purpose of tracing its progress from the tropics. The gale was felt from the northward, in Cuba on the 6th, but was not severe. The next notice I found of it was from the report of the Fanny Lincoln at Boston, which took the gale in lat. 29°, long. 79° 30', on the 7th. The schooner Anna A. Holcomb from St. Martins, had the gale September 7th, lat. 31°, long. 73°, from the S.E., lasting twelve hours. The steamer Saragossa, had the gale on the 7th at Charleston Bar from the N.E. The schooner Emma D. Fenny had the gale September 8th, lat. 33°, long. 75°, from the N.E. The schooner Ben Borland had the gale September 8th, lat. 35° 8', long. 72°, from S.S.E. to S S.W. The schooner Julia D. had the hurricane on the 8th, lat. 36° 30', long. 73°. The bark Hannah H. had the hurricane on the 8th, lat. 37° 40', long. 73°. Ship Tamerlane had the hurricane on the 8th, lat. 39° 47', long. 70° 19'. Bark Joanna Kepler had the gale Sept. 8th, lat. 40° 25', long. 70° 30', from S.S.E. Schooner Hartstein had the gale the same day off Montauk point. And it crossed Rhode Island and Massachusetts in the afternoon and evening of the same day, and the next was felt severely on the coast of Maine and Nova Scotia.

These notices are sufficient to show that the hurricane came from the north of Cuba on the 6th, and moved rapidly up in a N.N.E. direction, crossing Rhode Island and eastern Massachusetts on the 8th, where it was very destructive, unroofing buildings, throwing down steeples and uprooting trees. Its path lay between long. 75° and 80°, at lat. 29° and between 70° and 75° at lat. 40°. At some points in its progress the wind was N.E., on its left front, but whether all the way while upon the ocean, the accounts are too meagre to enable us to determine. On its right the lateral wind was everywhere from E.S.E. around by the S. After it entered upon Rhode Island the lateral winds were S.E. or W.S.W. I passed from Boston through Providence, and over the Shore Line Railroad home some weeks afterwards, and looked carefully for its effects. Between Boston and Providence, most of the prostrated trees lay with their tops to the N.W. From a little this side of Providence, along the shore the prostrations were nearly all towards the N.E. or E.N.E. I have no doubt but that the hurricane was like that of September, 1821, semi-revolving, pursuing a path about 100 miles farther east, and in form like the next preceding diagram, and that it lost that semi-revolving character as soon as it entered over the land upon Rhode Island.

The European Meteorologists seem generally to adopt the idea that the European gales all revolve, conforming to the theories of Mr. Redfield, Mr. Piddington and others. Prof. Henry and our own meteorologists adopt the theory of Prof. Espy. There is a certain amount of truth in both theories. There is, as we have seen, a class of semi-revolving storms, but they are not frequent. Nearly all the storms of the Southern Hemisphere and most of those in the Northern, are more or less elliptical, with the right hand lateral wind blowing at right angles or obliquely across the front and front portion of the storm, and the left hand lateral wind blowing in like manner across its rearward portion and rear. The straight line northeasters of the northern hemisphere, and the corresponding straight line southeasters of the southern hemisphere, are comparatively infrequent except in a few localities, and under special circumstances. What they are has been sufficiently shown.

Of the great central organization, little need here be said. It is simple in its elements, and we have hereinbefore analyzed them with care and at sufficient length.

Having thus as carefully, and critically as our limits permit, examined the structure of the special and general organizations constituting the atmospheric system, and the mode of operation and phenomena peculiar to each, we come now to inquire what inference they authorize in relation to the organizing and motive force of the system. What agent is it that creates the general and special organizations and controls their action and mode of operation?

We see that each organization has its law, and that the whole is governed by a controlling agency; we see too that this agent has IMMENSE POWER; that it is capable of moving the atmosphere or exerting a force equivalent to moving it, at the rate of 682 miles per hour. What power is there in nature adequate to produce such results? Meteorologists say that it is produced by the ascensive force of heated air, but when we come to inquire into the strength of that ascensive force, we find it is but the infinitesimal power, even in air confined so as to prevent the operation of the laws of expansion and diffusion, and when heated 100° above the temperature of the surrounding air, of *but ⅓ of an oz. to the square foot.* We further see that some of the greatest exhibitions of that force have been made in limited tornadoes and hurricanes where the temperature of the air was below 70°, AND NO CONTRAST BETWEEN THAT AND THE SURROUNDING AIR EXISTED and the exertion of ANY ASCENSIONAL FORCE WHATEVER was UTTERLY IMPOSSIBLE. In the light of such facts and all the other facts we have noticed, the Halley theory becomes mere "trumpery and trash."

Prof. Henry and a few other meteorologists in this country, undertake to account for the awful exhibitions of organizing and motive force, by attributing it to an increase of ascensive power, derived from the condensation of the vapor of the atmosphere by the expansion of the air, as it attains a higher elevation, and an increase of momentum. The following is his doctrine as borrowed from Espy and scattered broadcast over the country in the Patent Office Report for 1858.

"As a general rule, previous to the commencement of an extended storm, during winter, the surface current is from the S.W. or some southerly direction, the temperature rises, and the pressure of the air diminishes as indicated by the fall of the barometer. This state may continue for several days, and we think it is produced by the southerly current, increasing in quantity, in velocity and depth, thereby rendering the stratum of air next to the surface of the earth abnormally warm and moist, and consequently lighter, while the upper current remaining the same, the atmosphere above the surface of the earth gradually assumes a state of tottering equilibrium. This condition, according to Mr. Espy, is not brought about by the gradual diminution of the density of the lower stratum, but by the increased density of the upper strata, due to the radiation into space of the latent caloric which has been evolved during a previous storm. This instability or tottering equilibrium will first take place at the far west, on the western plains east of the Rocky Mountains, since, as we have said before, the commotions on the western side can slowly be propagated across the high mountain system. A storm then consists of the ascent of the lower current into the upper, and the gradual transfer of the commotion of the air eastward. To take the simplest case, let us suppose the storm to be of circumscribed character, like that of a water-spout or thunder storm. In this case, after the unstable equilibrium has been produced, the slightest disturbance, such as the passage of the lower current over a slight elevation, or over ground more highly heated than the adjoining, will tend to establish an upward current. The light, warm and moist air below will be buoyed up with great rapidity, and as it ascends will come under less pressure, and will expand into a larger bulk. If it were perfectly dry it would again be in equilibrium, its bulk would be increased, its density would be diminished to that of the air to which it had ascended, and its temperature would be the same as that of the surrounding stratum. But since it contains moisture, and in expanding becomes colder, a portion of the vapor will be condensed, and in this condensation will give out its latent heat. Hence the air of the column will be warmer than that of the surrounding atmosphere; it will consequently rise to a greater height, again expand, again become colder; another portion of vapor will be condensed, and another amount of latent heat evolved, and so on; the air will rush up with accelerated velocity, and probably gather momentum sufficient to carry it to a height greater than that due to its buoyancy alone. The condensed vapor will fall in rain through the base of the cloud; on either side the air of the storm will be forced out from the uprising column into the surrounding air, and while the pressure at the base of the column will be diminished, that on each side will be increased; hence the barometer will be frequently found to rise slightly before the approach of a storm, and to sink rapidly as the center of the uprising column approaches the place of observation."   *   *   *   *   *

The following is the figure and description by which he illustrated the foregoing theory in relation to the tornado:

FIG. 53.

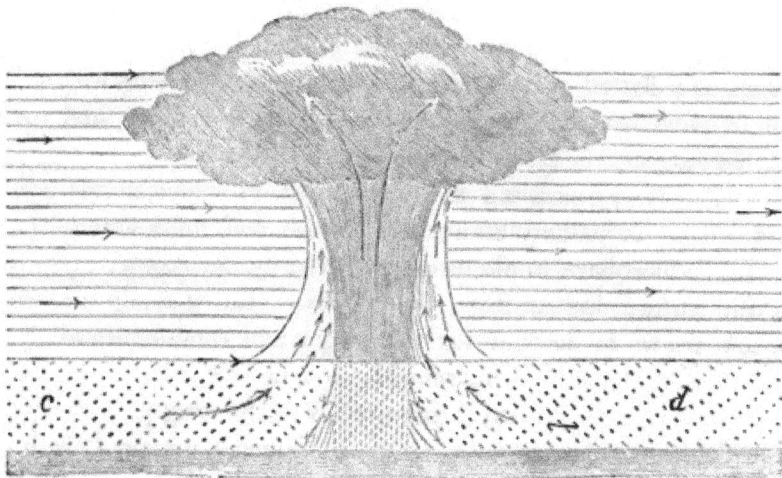

"The tremendous ascensional power which is exhibited in storms of this kind, although almost exceeding belief, is nevertheless in accordance with the established dynamical principle of the accumulation of momentum in cases of the continued action of a constant force. We are all familiar with the velocity given to an arrow, by a simple propulsion of the breath along the interior of a blow-gun. In this case, the air presses against the end of the arrow, at first with just sufficient force to move it; but the momentum it has thus acquired is retained, it receives another pressure from the air, retains the effect of this, and so on, until it leaves the other end of the tube with the accumulated momentum acquired during its whole passage through the interior of the gun. In the same way the air, as it approaches the uprising column below, commences its ascent with an amount of momentum which is constantly increased by continued pressure behind."

The foregoing is substantially the theory promulgated, ex-cathedra and semi-officially, by the Secretary of the Smithsonian Institution, in the Patent Office Report for 1858, and in a lecture before the American Association at Springfield, August, 1859, and endorsed by Prof. Loomis and an associated clique of meteorologists. It is doubtless in their estimation, to use the characteristic

language of Prof. Henry, " a scientific generalization deduced by
the profound reflections of men who *think*, in contradistinction to
those who *act*."* The italics are his, not mine. But in my judgment, and that of many other practical, intelligent men whom I
might name, it is inexcusable error. Let us examine it and see
if this language is either too strong or in bad taste.

I. The theory purports to be applicable to winter storms and
tornadoes. It assumes an abnormal warm and moist state of the
surface atmosphere in all cases, produced by the flowing in of a
warm surface current from the tropics. How then, it may be
asked, are we to account for the origin and continuance of the
cold northeast snow storm, commencing and continuing to the end
with a thermometer many degrees below the freezing point? The
question cannot be answered, or the contingency provided for, by
the theory, and there is a fundamental error at the start.

Again, it is untrue, that the warm southerly air which constitutes the southern lateral current of the winter storm, flows up
from the tropics, covering the country, before the storm originates.
It is an incidental, connected, and essential part and element of an
organization, as we have already seen and shall presently see
again, existing under a *portion only*, and traveling as part of a
whole. The southeast trades sometimes extend up over the West
Indies and Gulf of Mexico, and into the Mississippi Valley in
midsummer, as we have seen, and as surface winds, but the northeast trades prevail at the surface in the West Indies and Gulf
in winter, and the southeast trades which reach us, come not as
surface currents, but as the upper trade of the second story, as
any one who will take the trouble to *act* as well as *think*, may
readily *see*. That upper trade was thus seen by Mr. Fendler,
flowing over the surface trade 8,000 or more feet above the level
of the sea, coming towards us, as he stood on the mountains of
Venezuela. It has been visible in every winter storm which has
crossed the country since its settlement. That assumption is also
demonstrably untrue.

II. The next prominent idea of the paragraph is that the existence of this warm stratum of air next the surface, is accompanied

* Patent Office Report, 1857, p. 421.

at the same time by an abnormally cold and dense state of all the atmosphere above it, produced by "radiation into space of the latent caloric which had been evolved during a previous storm."

The idea of an abnormally cold state of the upper atmosphere is sheer assumption unsupported by any fact, and contrary to many known facts, as we shall see. The upper atmosphere or third story is always colder materially than that of the surface. Aeronauts cannot live where its peculiar cloud, the cirrus, forms. Moreover in winter storms the surface atmosphere is not abnormally warm. It is always below the freezing point in continuous snow storms, and sometimes at or near zero. It rarely rises above 60° *in the focus of winter thaws.* There is then no evidence of, nor a possibility of, such an abnormal contrast of temperature as is assumed. And suppose there was, what then? What and where is the *totter?*

If the upper atmosphere and all of it above is abnormally cold and dense, over an extended surface, will it not press to the same extent and *equally* on the inferior stratum? How then can a small portion of either *totter?* That idea is both absurd and untrue.

III. The next prominent idea is that this tottering equilibrium takes place, and the storm originates on the plains east of the Rocky Mountains. This too is demonstrably untrue. Storms in the winter season do not originate there. Those plains, and all the country east and northeast of them to the 95th meridian, and the northwestern states still further east are in their dry season in the winter, and do not receive an average of more than two inches of rain or snow for the three winter months, as we have seen on the diagram, page 89. Where the storms originate during that season, and how they travel, has been hereinbefore demonstrated. Prof. Henry might have learned both if he had consulted the charts of Prof. Espy, the Army Meteorological Register, the Climatology of Mr. Blodgett, or the Philosophy of the Weather.

IV. The next prominent idea is that a storm is constituted by the tottering ascent of the surface atmosphere into the atmosphere above in consequence of an unstable equilibrium in relation to temperature, between the surface stratum and the rest of the atmosphere above it, and that this is a mere mechanical effect—a mere "*commotion*" as he terms it, and that all clouds are produced by condensation consequent upon such ascent.

That this idea is untrue and absurd, and that such ascent never takes place in large volume or rapidly, is shown conclusively by all the evidence we have accumulated.

But there is a further view of the matter which should not be overlooked. The theory involves as a consequence, the formation of a single mass or stratum of cloud of uniform appearance, character, and function. The following diagram is also given by him to show the structure of a winter storm:

FIG. 54.

But such a storm never existed. No man living or dead ever saw one, or ever will. It is a *closet conception.* Now let us turn back and look again at a winter storm as produced by the Creator, and as represented by Fig. 41, on page 313, and as I have seen them a thousand times, and note the difference. Read also again the following description by Professor Stoddard of the storm at Brandon, in January, 1854.

The barometer fell gradually during the 19th, and rapidly on the 20th, and at 12.40 M., the time when the storm began at Oxford, it stood 28.21—lower than at any period during the last twelve months. The air was saturated with vapor, and the walls of brick buildings were dripping with moisture. Three strata of clouds were distinctly observed—the highest cirri, light and fleecy, moving toward the N.E.—the second, the proper storm cloud, in dark, heavy masses, moving rapidly in the same direction—the third and lowest, the scud of sailors, flitting violently past, a little east of north. Along the track of this wind, there were at different times during the day, violent rains, vivid lightning, and heavy thunder, though not in great quantity. In the N.E. part of the state the storm assumed the form of a tornado of great violence."

Such is a storm as organized by the Creator. Meteorologists utterly fail to comprehend it. Now note the difference. The storm as it exists in the *"profound reflection"* of such meteorologists, consists of an ascending column of air, spreading out like an umbrella overhead, with condensation consequent upon expansion and reduction of temperature, with a single cloud stratum or mass, of homogeneous character, appearance, and color. The storm of the Creator consists of three strata, each having its peculiar cloud-formation, and each description of cloud differing in *form, appearance,* and FUNCTION. What those strata and their appearance are, you have frequently learned from me, and you now learn from Prof. Stoddard. And it is open to the observation of these men of "profound reflection," if they would but appreciate their duty as public teachers, and become observers of the book of nature which is open before them, that the *general storm* begins, not at the *surface of the earth,* but in the *formation of cirrus cloud in the upper story,* and long before the surface story becomes warm and moist and the supposed unstable equilibrium occurs.

V. The next idea is that the *tornadoes* commence at the surface of the earth and in consequence of a limited portion of the surface atmosphere passing over some heated surface, or coming in contact with some "slight elevation"—for aught we are told, a horse-block or corn-crib—and being in unstable equilibrium, is thereby made to *totter* (which in common language I suppose means to trip and stagger) upwards "*with great rapidity,*" its vapor condensing, after it arrives in the upper cold stratum, and its progress increasing from the evolution of latent heat and the acquisition of momentum until it spreads itself out above as represented in the diagram, and thus the tornado is produced.

Now here we have another of those extraordinary conceptions of a " profound reflection," which is in opposition to all observed truth as heretofore developed, and contains several intrinsic and fatal errors.

The first is, that the tornado does not commence at tne earth, but forms at the inferior surface of the stratus clouds of the second story, and extends downward to the earth, sometimes reaching it, and sometimes not; sometimes after reaching it drawing itself back, breaking its connection with the earth, and subsequently elongating downward again, and renewing that connection, and ending by drawing itself up gradually, and ceasing at last, at its place of beginning, the undisturbed stratus cloud. Between a commencement at the cloud, one or two thousand feet above the surface of the earth, and an extension downward, and a commencement at the earth, by staggering against some slight elevation and sweating upwards, practical men cannot avoid seeing an irreconcilable and fatal antagonism. Perhaps Prof. Henry does not see it. He will live to see, and I hope correct, his error.

In the second place, the theory requires for the acquisition of the required tremendous force, a condensation in the upper strata, and the acquisition of momentum by progress there, its ascent through the surface atmosphere being due to the *totter or stagger alone*. But the greatest exhibition of force is in fact at the surface of the earth, when the distinctly marked apex of the spout comes in contact with it, and before *any ascensional force* can be organized by the *evolution of heat*, or any *additional force* by the *organization of momentum*, and when there is no force applied but the *tripping* caused by a " *slight elevation*," or a " *heated spot.*" Moreover, neither the " *slight elevation* " nor the *heated spot* are found at sea, where the water spout is seen desending from the cloud and acting with equal apparent violence.

These two facts are sufficient disproof of the 5th idea also, and

it must be apparent to the reader in view of these facts and all
the other facts developed in this volume bearing with equal force
on the point, that there is not a healthful or truthful spot in
any part of the theory, or of the cited paragraphs, and that I
have not characterized it too strongly by the language which I
used. The figures, too, are *caricatures*, and that fact, as well as
the utter untruthfulness of the descriptions, Prof. Henry might
easily have learned by a few weeks actual observation, or a just
regard for the testimony of those who had tested them, which he
had before him. In sending them to the country *untested* and as
unquestioned truths, he failed, in my judgment, to appreciate the
responsibilities of his position.

Let no man say I am unjust or err in this. Prof. Henry fills
a public office, and is amenable to public criticism. When he
sent the Espyan theories to the country with *his endorsement*, he
had before him conclusive evidence of their falsity in the cited
work of Prof. Coffin, and observations of Mr. Fendler,—in the
published assertions of Mr. Redfield, that he found the caloric
theory untrue; in a *demonstration* of the same fact in the "*Philosophy of the Weather*·" in the investigations of Hare, Peltier,
Faraday, and others; and in my published statement, that I had
watched for the claimed uprising currents for thirty years, and
knew they *never occurred*,—but he ignored it all. Such errors,
from such men, should be *stripped of influence as authority.*

There is a prevalent error upon this subject which should be
reformed. There are many intelligent scientists in the country
who know and admit that the Halley theory is untrue, and the
public deceived, but because its advocates are many and powerful in position, they shrink in weakness and cowardice from a contest with it. The late Mr. Redfield was one of those men. For
the sake of peace, he avoided all allusion to the cause of storms
in his publications, and confined his inquiry to the question, "what
are storms," evading the question, "how are storms produced."
Near the close of his life, in two notes to his articles on the Cuba
Hurricane, he gives us an inkling of his views, and his reasons

for his reticence.   The first occurs in the January number, 1846, page 13, of the American Journal of Science and Arts, in which, after giving an extended account of the winds and cloud currents within the tropics, he adds :

"These, with a vast extension of similar phenomena, merit the serious attention of those naturalists who rest on the caloric theory of the general and trade winds; and they seem fully to account for the northwesterly courses of storms in the West Indies."

Again in the September number of the same year, he says in another note:

"It may be proper again to state, that the results of the author's inquiries on the courses of winds and their relations to temperature, in different regions and at different elevations, have constrained him to relinquish the common theory that heat is the sole or main cause of wind or progressive motion in a planetary atmosphere.   He has been aware of the disadvantage in which this avowal may tend to place him, in the minds of many votaries of science whose approbation it would be his happiness to obtain. The proper elucidation of this question, he conceives, will belong to the future."

And he closed a later article by saying, "That the current theory or hypothesis for explaining the general winds of the globe is essentially erroneous, and defective in its application, and greatly obstructs the path of scientific enquiry."

Now in my judgment the interest of society demanded that a current theory which obstructed the path of scientific enquiry in relation to so important a matter, should be forthwith attacked, and by any and every fact, and by any and every person who possessed the necessary information wherewith to attack it ; and it was the duty of Mr. Redfield, because more than any other man possessed of such information, to grapple with it to the death. And it was a weakness, to hesitate through fear of losing the approbation of those who sustained it.

At any rate such was my conviction and sense of duty as I sat listening to the jumble of fallacies which Prof. Henry detailed in his lecture at Springfield, when there was convincing and conclusive evidence that such was their character, in sight from his position and m'ne, and within 2,000 feet of us, and I obeyed the impulse of that sense of duty when I denounced it as such to the audience, and made then and there a challenge which I afterwards renewed in the newspapers in the following form :

"I desire to renew it (the challenge) in the following precise and distinct form ; I propose to meet Prof. Henry or Prof. Loomis, or both of them, before a board of three persons, mutually chosen, sometime in June or July of next year, for the purpose of discussing the position taken by Prof. Henry in his lecture, and endorsed by Prof. Loomis. On such discussion I will assume the burden of disproving them, by proving the following propositions, viz:

First, the normal circulation of the atmosphere is not a series of systems as alleged in the lecture, and shown on the diagrams, but consists of one universal, unchanging system, with its base at the atmospheric equator, and its apex at the poles.

Second, that the assertion that the air, the earth, or the water, are hotter under the belt of condensation between the trades, and where the air is supposed to be rising and sucking in the adjoining air, thereby making the trades, is without foundation. The fact is demonstrably the other way.

Third, The Halley theory in all its parts and however applied, is shown by conclusive data to be an error.

Fourth, There is not an "immense amount of latent heat" developed during the formation of a cloud, which heats the air and causes it to rise.

Fifth, That storms and wind are not produced by currents of uprising air, and the Espyan theory advanced by Prof. Henry and endorsed by Prof. Loomis, is demonstrably untrue.

And I propose to deposit with the umpires, $500, of which $100 is to be paid over to the Professors or either of them, for each and every one of the foregoing propositions I shall fail to maintain.

I thus give the gentlemen every advantage ; I ask for myself only an opportunity to present the truth and develop the subject in such a manner and under such circumstances, that it cannot be

17

smothered in order to give one man an opportunity to monopolize discovery and 'live in science,' to the detriment of the country."

That challenge the gentlemen did not accept, and I renew it.

Doubtless that discussion and challenge placed me at a disadvantage, as Mr. Redfield expressed it, with the votaries of science, but that, which seems to have been an important matter with him, was unimportant to me. My happiness did not rest on their approbation, but then, as ever, in respect to my public conduct, in the consciousness of having done my duty to the community.

That fearless attack upon a pernicious error, has not been without its fruit. In the minds of a great many men, that error then received a death wound; and I have a stack of letters, many of them from some of the best minds in the country, thanking me for the discussion. Ten years have gone by, and neither in lecture nor essay, have those gentlemen or any others, appeared before the public to vindicate the theory. "Truth is mighty and will prevail."

If, as I think, it conclusively appears from the foregoing examination and all that has been hereinbefore adduced, that there is one general, organized, atmospheric system, consisting of one great central organization with minor, connected systems of special organizations, from which all the atmospheric phenomena result, and the major and minor organizations have a common law, and are organized and controlled by a common force, we come directly to one of the most important unsolved problems of the science. What is that force?

That it emanates from or is excited by the sun, we cannot doubt. The annual transits of the system, from north to south and from south to north, are conclusive of that. So are the diurnal changes which follow his path. Right here, let us look at those diurnal changes and learn the lesson which they teach. The following diagram exhibits the character of those changes during a summer day, when no disturbing causes are in operation at the point; no storm existing within influential distance, and there is no unusual intensity or irregular action of the operating force or forces. Let us, I repeat, look carefully at the diagram, and see what lesson it teaches.

FIG. 55.

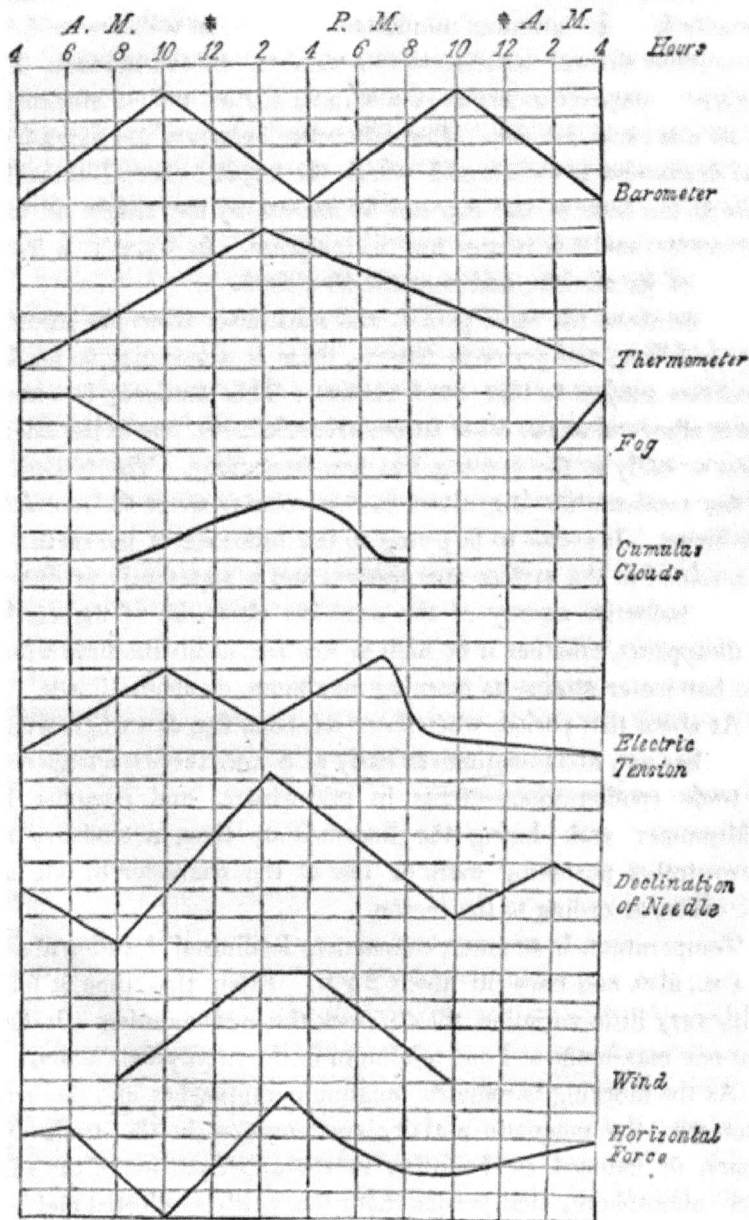

The day may be said to commence in midsummer at 4 A. M. The atmospheric day does at all seasons. At that hour the barometer is at its morning minimum. It has, as we have said, a perceptible diurnal variation of two maxima and two minima. Its periods of depression are at 4 A.M., and 4 P.M., and of elevation at 10 A.M, and 10 P.M. The difference between the elevation and depression is considerable within the tropics, where Humboldt tells us the hour of the day can be known by the height of the barometer, and it decreases toward the poles. At 4 A.M. it is then at one of its minima, and rises till 10 o'clock.

At, or about the same period, and sometimes when the barometer is falling, and previous thereto, there is a tendency to fog in localities subject to that condensation. This tendency is sometimes observed at the other barometric minimum, late in the afternoon or early in the evening, but less frequently. The tendency to fog condensation is greatest in this country about the morning minimum. It seems to be owing to the influence of the earth; it is confined to the surface atmosphere, and is apparently produced by the inductive agency of the negative electricity of the earth. It disappears, whether it be high or low fog, about the time when the barometer attains its morning maximum, or about 10 A.M.

At about that period, when there has been fog, or earlier, when there has not, and sometimes as early as 8 A.M., there is a tendency to trade condensation—cirrus in mid-winter, and cumulus in midsummer, and, during the intermediate time, a tendency to cirro-stratus, partaking more or less of the character of one or the other, according to the season.

Temperature, in summer, commences its diurnal elevation about 4 A.M., also, and rises till about 2 P.M. From that time it falls with very little variation till 4 o'clock the next morning. It has but one maximum and one minimum in the twenty-four hours.

As the morning barometric maximum approaches, and the heat increases the magnetic activity, condensation in the trade appears, or induced condensation in the upper portion of the surface atmosphere, that portion near the earth is affected and attracted—and the " wind rises," according to the locality, the season, and the activity of the condensation. The tendency to blow increases with the tendency to trade and cumulus condensation, and continues till toward night, when it gradually dies away, unless there be a storm approaching. As the heat increases, and stimulates magnetism into activity, the magnetic needle commences moving to the west, its regular diurnal variation, and continues to do so until about 2 P.M., when it commences returning to the east, and so continues to return until 10 P.M., when it moves west again until 2 A.M., and from thence to the east, till 8 A.M.

Similar variations also take place in the horizontal force, as evinced by the action of the magnetometer needle, and in the vertical force, as shown by the oscillations. So that it is evident that there are two maxima, and two minima of magnetic activity every day, shown by all the methods by which we measure magnetic action and force—more than double at the acme of northern summer transit over that of winter, and proceeding *pari passu*, with the other daily phenomena—evincing the same irregular action which the other phenomena evince. Still another phenomenon, which has a daily change, is electric tension, or the increase or decrease in the tension of the positive or true atmospheric electricity.

The following table shows the mean two hourly tensions for three years, at Kew, viz :

Hours	12 P.M	2 A.M	4 A.M.	6 A.M.	8 A.M.	10 A.M.
Number of observations	655	784	804	566	1,047	1,013
Tension	22 6	20 1	20.5	34.2	68 2	88 1

Hours	12 A.M.	2 P.M	4 P.M.	6 P.M.	8 P.M.	10 P.M.
Number of observations	848	858	878	874	878	1,007
Tension	75.4	71 5	69.1	84.8	102.4	104

From this it will be seen that the tension of electricity is at a minimum at 4 A.M., also, that it rises till 10, falls till 4 P.M., but not as rapidly, rises till 10, falls again till 4 A.M, or the close of the meteorological day—having two maxima and minima, as have most of the phenomena thus far considered.

In order to see what the connections between these ever-present, daily phenomena are, and their connection with other phenomena,

and that we may understand their normal conditions, I have traced
them approximately in the diagram.

It is obvious that the other phenomena do not all depend upon
temperature merely, if indeed any of them do.

Temperature has but one maximum and minimum, and that is
exceedingly regular, and does not correspond with any other.

The barometer has two; electric tension, two; magnetic activity, two; condensation, two—one the formation of cloud, and the
other the formation of fog and dew; wind, one—resembling temperature in that respect, but embracing a much less period.

Fog forms at one barometric minimum, and cloud at another.

Fog forms at one period of the magnetic variation, cloud at
another.

The formation of cloud corresponds with the greatest intensity
of magnetic action, and its associate electricities. But the oscillations of the barometer do not correspond with either. And thus,
then, we connect them:

CAUSE.	EFFECT.	EFFECT.
Increase of magnetic or mag-neto electric activity, as shown by declination and increase of horizontal and vertical force.	Decrease of pressure. Of positive electric tension Of surface condensation, i.e. fog and dew.	Increase of primary con-densation. Of wind. Of electrical disturbance and phenomena in the trade and its vicinity

This connection is equally obvious if the order is reversed—
thus:

CAUSE.	EFFECT.	EFFECT.
Decrease of magnetic or mag-neto-electric activity.	Increase of pressure. Of tension of atmospheric electricity. Of surface condensation, i e fog and dew.	Disappearance of primary condensation. Of wind, and Of electric disturbance in the trade and its vicinity.

The view which this diagram gives us of the diurnal changes,
shows again the absurdity of the Halley theory, and that there is
some other force operating upon the atmosphere, and affecting the
phenomena beside that of heat. We see an utter want of all
connection between the *degree of heat* and the *other changes.*
This is marked in relation to the barometer, and we have seen
before that the annual range of the barometer depended upon the

position of the focal path of the conditions, and not upon temperature.

But this group of diurnal changes, although they are instructive, do not solve the problem, and the question still remains. What is the organizing and motive force of the system?

That it is not heat, acting mechanically upon the atmosphere, we know. What then is it? Remember we are seeking a force which organizes and controls the great central condition which creates and continues the trade winds uninterruptedly through the night as through the day, with the storms and the hurricane in their bosoms or without them, to form the basis of other conditions, to pursue a stated and prescribed path, and to carry warmth, moisture, and fertility to their appropriate hemispheres.

And right here we may observe, that the pursuit by the upper trades and of the storms they embosom, of a stated and prescribed path, first to the N.W. in the northern hemisphere, and to the S.W. in the southern hemisphere, curving at the outer limits of the surface trade, and moving thereafter to the N.E. in one hemisphere, and the S.E. in the other, are important elements to be regarded in the problem. Is that curvature pursuant to a law of the force, or is it the result of something else? Meteorologists say that this curvature is produced by the rotation of the earth.

Thus Sir John Tyndall says:

" Were the earth motionless, these two currents would run directly north and south, but the earth rotates from west to east round its axis once in twenty-four hours. In virtue of this rotation, an individual at the equator is carried round with a velocity of 1000 miles an hour. You have observed what takes place when a person incautiously steps out of a carriage in motion. He is animated by the motion of the carriage, and when his feet touch the earth, he is thrown forward in the direction of the motion. This is what renders leaping from a railway carriage when the train is at full speed, almost always fatal. As we withdraw from the equator the velocity due to the earth's rotation diminishes, and becomes nothing at the poles. It is proportional to the radius of the parallel of latitude, and diminishes as these circles diminish in size. Imagine then, an individual suddenly transferred from the equator to a place where the velocity, due to rotation, is only 900 miles an hour; on touching the earth here he would be thrown forward in an easterly direction with a velocity of 100 miles an hour, this being the difference between the equatorial velocity with which he started and the velocity of the earth's surface in his new locality. Similar considerations apply to the transfer of air from the equatorial to the northern regions, and vice versa. At the equator the air possesses the velocity of the earth's surface there, and on quitting this position, it not only has its tendency northward to obey, but also a tendency to the east, and it must take a resultant direction. The farther it goes north, the more it is deflected from its original course; the more it turns toward the east, the more it becomes what we should call a westerly wind. The opposite holds good for the current proceeding from the north, this passes from places of slow motion to places of quick motion, it is met by the earth; hence the wind which started as a north wind becomes a northeast wind, and as it approaches the equator, it becomes more and more easterly."

Professor Loomis in his recent treatise, published in 1868, states it thus:

"UPPER CURRENT IN THE EQUATORIAL REGIONS. The mean temperature of the surface air at the equator is considerably higher than it is over the parallel of $32°$, while near the upper limit of the atmosphere, the temperature must be nearly the same in all latitudes. Now air is expanded by heat to the amount of $\frac{1}{491}$ part of its bulk for each degree of the thermometer. The atmosphere over the equator must therefore rise somewhat higher than it does over the parallel of $32°$, notwithstanding the difference in the height of the barometer. If the earth were at rest, the air thus expanded at the equator would flow over at the top, and descend as along an inclined plane, toward the middle latitudes. But while in the northern hemisphere, an upper current flows toward the poles, it crosses in succession parallels of latitude whose easterly motion is less than its own, and since it retains the easterly motion which it had at the equator, it has a relative motion from the west, which combined with the first northerly motion, carries it toward the northeast. Thus above the northeast trade winds we find an upper current moving from the southwest. For a similar reason in the southern hemisphere above the southeast trades, the upper current moves from the northwest."

A substantially similar opinion is expressed by Prof. Henry in his Patent Report compilations.

These gentlemen occupy distinguished positions as scientists, but we will remember that these views, thus expressed by them, are also traditionary assumptions of the Halley theory, and we can and should fearlessly examine them as such, regardless of the men or their positions. Let me ask you then to look at them and say whether they are not also false assumptions.

Look first at the looseness of these paragraphs. They all speak of an ascent and flowing over *at the equator*, meaning the *geographical equator*, yet elsewhere, they all, including Prof. Henry, speak of this ascent as between the trades, and causing the belt of rains. If there is any ascent at all, upon their principles, it must be there, but that belt of rains—the atmospheric equator as I call it—moves north of the geographical equator in summer, and south of it in winter. To say, therefore, that the air ascends at the equator, is to speak loosely and inaccurately, and shows their limited comprehension of the subject.

Again, as to the fact of the ascent, these paragraphs, and especially that of Prof. Loomis, involve a gross absurdity. He represents the atmosphere within the tropics as fifty miles high, as standing four miles higher there, than at latitude 32°—and the atmosphere as flowing over *at the top* along an inclined plane, and descending toward the middle latitudes. As that inclined plane has a descent of four miles, the elevation of the overflow at the middle latitudes, when it arrives there, must be 46 miles above the earth, or allowing the depth of the overflow to be four miles, the elevation is 42°. Of course it could not constitute the upper trade, which Lawson saw at Barbadoes, and Redfield satisfied himself by extended inquiry, existed all over the West Indies; which Fendler saw on the mountains of Venezuela, and which we know comes to us with its vapor and its storms, constituting our second story, and giving us our peculiar climatology and prosperity. It is high time that a man of Prof. Loomis' ability to *investigate* should cut loose from such false and traditionary theories, and look this subject *in the face*, unprompted and uncontrolled by men who *learn* and *teach* or *compile* by *rote*.

But the great question involved in the cited paragraphs, is whether the change of movement to the east is occasioned by the *rotation* of the earth. And here at the outset we are met by the question which these gentlemen upon their principles cannot answer, and which staggered and convinced Mr. Redfield, viz: *what occasions the initiatory movement through 20° of latitude to the northwestward before the curvature to the east commences?* If that could be satisfactorily answered we should come to the sharp point which is made by all of them, expressly or by implication, that the storms and currents curve to the east, by reason of the *greater motion* which they derived from the earth at the place of their origin at or near the equator, than they have when they curve. And the point narrows itself and hinges on the sharp and distinct question, whether at lat. 32° or 35°, wherever the storms or currents commence to curve, they "*retain*," to use the language of Prof. Loomis, the motion eastward derived from the rotation of the earth which they had at the equator, or their place of beginning, and continue to retain it while curving and after. Sir John Tyndal assumes that they do. Disregarding the fact that their first movement is to the northwest, and that they must thereby, or so far lose the velocity they had at the place of origin, unless they can be supposed to carry it, "*bottled up for use*," he assumes that they move directly north, and would continue to do so to the pole if they did not retain as they reach more northern parallels the rotary velocity which they had at their place of origin. The case he puts, and in which the other gentlemen concur, presents the question precisely as if the storm leaped at a bound, and over all obstacles which could affect its rotary veloci-

ty, from the equator to lat. 32°, and thus leaping, and thus retaining its equatorial velocity *and assuming* that it reaches the
earth again at 32°, the velocity there being less, it is thrown
forward precisely as a man is thrown forward who jumps from
a railway train. Now is not this the grossest absurdity of the
many contained in that conglomeration of gross absurdities known
as the Halley theory? Let us see.

Undoubtedly the rotary velocity of the earth is greater at the
equator than at 32°, and I do not question the accuracy of Prof.
Loomis' table of relative velocities, which is as follows:

"In lat. 0° the velocity eastward is 1036 miles per hour.

"	15°	"	"	1000 " "
"	30°	"	"	897 " "
"	45°	"	"	732 " "
"	60°	"	"	518 " "
"	75°	"	"	268 " "

But the question is, does the storm or the current which
changes its latitude, retain its primary velocity? Does the
steamer which steams from the equator to lat. 30° at the rate of
250 miles per day, retain the velocity which it had at the equator?
We know it does not, and that it loses that equatorial velocity instant by instant, and inch by inch, as it progresses to the north.
It has, wherever it may be, the velocity of the water in which it
is imbedded, and of the parallel at which it floats. Does the
migratory bird which takes its spring flight from the Gulf coast
of our country directly northward to the upper lakes, at a speed
of 100 miles an hour, "*retain*" at the lakes, the rotative velocity
which it had at the Gulf? and when at night it rests its bosom
upon the surface of the lake, and folds its wings, is it thrown forward by a retained excess of easterly velocity at the rate of 200
miles per hour, and as if thrown from a rail car. We know it is
not. The atmosphere through which it has projected itself, is an
aerial ocean which is retained in its position by the force of gravity as effectually as the ocean of water. The rotary velocity of
its mass at different parallels, conforms to that of the solid earth

to which it is bound by an attraction which rotation does not affect, in the same manner that the oceans and rivers conform. The bird and the current and the storm imbedded in it and moving north, are held and *supported by it* on *all sides*, as the ship is supported by the waters, or the rivers by their banks, and have at every instant the velocity of their supporting medium, at the parallel, precisely as if attached firmly to the earth.

The earth, ocean, and atmosphere, and all that they contain form but a part of an entire whole, which revolves *as a whole*, each particle of each held in its place by the superior force of gravitation, and no particle of either is disturbed in its relation to other particles by friction, or rotative force. If the doctrine had not been deemed an essential part of the Halley theory, it would have been abandoned long ago.

There are many other considerations which show that the doctrine cannot be true. The tracks of different storms originating over the same space, frequently differ materially. Some curve at a very low latitude; others, like those of 1821 and 1869, pursue a course almost directly north. Some have a speed of 4 to 6 miles an hour; others of 40, 60, and 70 miles per hour; and others move at intermediate velocities.

So too the storms and currents which curve on our continent over northwestern Mississippi, northern Louisiana and Arkansas in midwinter, when the central condition is far south, curve 10° or more, and with broader curve, to the northwest in midsummer when that central condition is to the north, for the force which curves them then operates so much further to the north and west.

No, reader, we cannot, without a disregard of facts and self stultification, accept the theories of these men. We must find a force capable not only of originating the currents and the storm, but of directing their pathway through the tropics, the temperate zones, and the arctic circle, unaffected by the rotation of the earth.

And we must find a force that is capable of producing the minor organized conditions; of directing their pathway, regulating their intensity and controlling their effects; a force that creates the cirrus in its appropriate stratum, the rain-bearing stratus in its appropriate stratum, the attendant lateral winds and their scud in their stratum, all working together harmoniously and continuously from day to day, scattering warmth, moisture, and fertility over an earth, which without such organizations would not be habitable for man. A force not only capable of organizing and controlling the system in all its parts, and in all their ordinary modes of operation, but of exerting those irregular and awful manifestations of energy and power which we have contemplated.

Now the agency of heat being found insufficient, the question comes up to us, what is this force that thus organizes and controls the system with an energy and power without a parallel in the physical world? Do we know of any such force? Do we know of more than one? The answer to these questions is, we know of one, and but one, such force, and that one is electricity. No logic founded on the relation of cause and effect of which an honest mind is capable—no honest reflection, however profound, can come to any other result, for no other adequate force is known. Deceive ourselves as we may—place confidence in the opinions and authority of other minds as we may—to that complexion we must come at last. That agency is sufficient. No other known agent is. Logically and irresistibly we come to the conclusion that this is the cause ; and here we may add, that every fact in nature rightly understood, confirms the view. I have not space remaining to examine *in extenso* the character of this agent, and the mode in which it operates as the organizing and motive force of the system. Nor is it of practical consequence that I should. If I have developed the system and the operation of its laws and their effects, as discoverable by our senses as matter of fact, and as they bear *practically* upon our interests, I have done all that I purposed, all that it is practically important to do. An inductive examination of the character of this agent and its mode of operation in every particular case, would require a collection and array of facts, and an investigation of its associated relations to heat, light, and magnetism, in order to a correct induction, which would extend this volume beyond any reasonable limit. For such an examination and investigation, I have accumulated much material, and it may be the work of future leisure hours. Yet I do not feel inclined to pass a subject of such engrossing interest unnoticed, and as if I had no conceptions of its mode of operation; and I will briefly allude, in a few general propositions, to the manner in which electricity may operate in controlling the system and its laws.

We are far from fully understanding the nature of this element, or of its associated relations to heat, light, and magnetism, yet we know enough of it to be able to trace in a general way, the agency we *know* it *must* exert.

There is electricity in everything, and it is associated with every law of the physical world, unless it be the attraction of gravitation, and Faraday never despaired of proving that it was identical, or associated with that. If it is not in itself the vital force of the animal and vegetable creation, it is certainly associated with it, or a component of it, and capable of controlling it. If it does not in itself constitute the law of chemical affinity, of crystalization, of evaporation and condensation, it is certainly associated with those laws, however they may operate. It is associated with light, if it be not light; for it will produce the most intense light known. It is associated with heat in all its manifestations and operations, if heat be not a phase or form of it, and its association with magnetism is such as to leave little doubt that magnetism is a manifestation of it by one of its modes of operation.

Manifesting itself in the ordinary operations of nature—primarily or by association, it is not regarded; but when it is disturbed and excited into abnormal activity, it becomes the visible and powerful agent we have assumed it to be. Thus, when excited into activity and drawn from the atmosphere by the electric machine it becomes a controllable but immensely powerful and destructive agent. So, when obtained by the agency of chemical affinity and decomposition, it becomes the same controllable and powerful agent, but with the slight modification of character pertaining to galvanism. So again, when electricity is communicated in a certain way to soft iron, it confers upon it temporary magnetism, and if communicated to steel, it attaches itself permanently to it, constituting it a magnet with opposite polarities, and with permanent electric currents around it.

Now with this brief résumé of some of the salient points in the character of electricity, so far as we at present understand it, let me state in a few concise propositions the manner in which this agent may, and so far as we can now judge, does organize and control the atmospheric system and its phenomena.

I. The earth is a magnet, but not a natural magnet. All bodies may be magnetised temporarily by causing currents of electricity to flow around them. A few can be magnetised permanently, and are naturally magnetic. The natural electricity of the earth and the atmosphere, or of any other body, is excited into activity and made to flow in currents, not only by the disturbance of chemical affinity, the decomposition of composite bodies, the disturbance of the laminæ of crystallization, but also by *the unequal heating* of them, the currents of electricity originating or being excited in the heated part and flowing to the colder. Electricity so excited and made to flow in currents is termed thermal electricity.

II. That portion of the earth upon which the sun shines at any given hour of the day, is warmer than the portion which lies so far west of it as to be still enveloped in night. According to the law of thermo-electricity, currents must be excited at the place where the earth is being heated, and flow to the west toward that portion of the earth which is coolest. The contrast between the heat of the earth and the atmosphere at 2 P.M., and that portion which is still enveloped in night at 4 A.M. is very great,—upon some surfaces not less than 60°. The electric currents which are constantly excited by the heat of the sun where it is day, flow around to the west where it is night, and thus form a permanent succession of currents flowing around the earth from east to west as it presents its surface to the action of the sun during its daily revolution. By a central belt of currents encircling the earth within the tropics, thus excited and operating, the earth is constituted a magnet. It is not, as I have said, my purpose to introduce here the proof of this. I will simply say that it is not new with me, but is a theory of Ampere, and any globular body may be magnetized temporarily, exhibiting all the elements of a magnet, such as dip, declination, and horizontal and vertical force, by passing currents of electricity around it. Illustrations of this are found in our text books.

How far the power of the solar rays assist in producing magnetism by direct action of the magnetic ray, is an unsolved question. Howard, whom Steinmetz styles the "Father of English meteorology," has the following note in his third volume:

Magnetism of the solar rays, as found in Milton.

> " The golden sun, in splendor likest heaven,
> Allured his eye ;   *   *   *
> *   '   *   *   where the great luminary,
> (Aloof the vulgar constellations thick,
> That from his lordly eye keep distance due,)
> Dispenses light from far.   They, as they move
> Their starry dance, in numbers that compute
> Days, months and years, towards his all-cheering lamp,
> Turn swift their various motions, *or are turned*
> *By his magnetic beam*, that gently warms
> The universe, and to each inward part,
> With gentle penetration, though unseen,
> *Shoots invisible virtue even to the deep.*"

What shall we say to this fine conception of our great poet, now that the philosophers have ascertained by direct experiment, that the violet ray of the solar spectrum is actually capable of rendering a needle magnetic which has never been touched by the loadstone, or by an artificial magnet? He seems to have had a thought (natural enough in the then state of science,) that the earth revolved from west to east in consequence of a peculiar attraction exercised on its substance by the sunbeams."

I am not satisfied that this conception of Milton's may not yet appear to be philosophically true.

III. Parallel currents of electricity have a tendency to converge toward each other. Doubtless, the primary central currents flowing to the west, exist under the entire central condition, converging toward the center where the currents are most intense, and where the great central belt of rains is found.

IV. As the sun in its transits is more vertical, and acts with greater heat'ng power on the *summer side* of the central belt, the currents gradually become more intense upon that side and less intense on the other, and thus the central condition with its belt of rains follows the sun in its transits, because the sun is continually creating a new focus of intense currents. And for the same reason the central condition continues to move north or

Fig. 56.

south after the sun has reversed his transit, and until it has heated up the waters and earth on the reverse side.

V. All successive currents of electricity induce secondary currents on each side of the primary one, and they flow in an opposite direction to, and parallel with, the primary current. Such currents are produced on either side of the central condition and in the temperate and polar zones; which gives to the atmosphere and all the conditions contained in it, a tendency or drift to the eastward.

VI. A second secondary current is sometimes induced by the first secondary, and that too moves parallel with and in an opposite direction to its inducing current. Prof. Coffin thinks he has shown that there is a tendency to an easterly movement of the atmosphere, in the arctic and antarctic latitudes. I do not consider the evidence sufficient to establish the fact, but if it be true it is doubtless the effect of a second induced system of secondary currents. Supposing it true, however, the systems of currents would be substantially as represented in the diagram, at the exreme northern transit. (See Fig. 56.)

VII. By this method of magnetization there is also produced in or over the earth, a class of lateral currents like those discoverable in all magnets by the aid of iron filings. The following diagram exhibits the currents as discoverable in a bar magnet, and we may assume it to be true that such lateral currents exist in and over the earth.

FIG. 57.

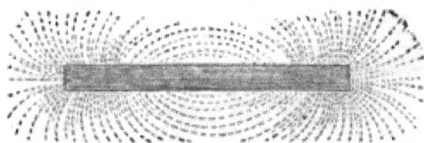

The following diagram exhibits their appearance as observed by Faraday in the globe magnet. It is taken from one of the plates appended to the third volume of his " Researches."

FIG. 58.

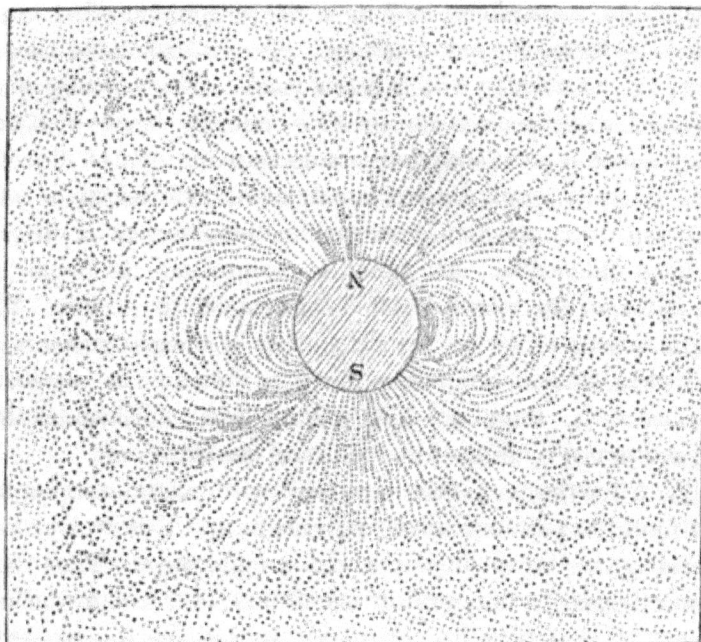

This is a section through the center of the magnet, and shows the manner in which the lines of force arrange the filings on each side. The reader will observe that round the center of the globe from east to west, shown in this *section* only at the sides, the currents are clearly exhibited, extending from one side or hemisphere to the other. It should be borne in mind that these are representations of the manner in which the currents of electricity flow around natural magnets.

VIII. The magnetic currents or currents of electricity which flow outwardly from the earth, are recognized by us in various ways. Faraday intercepted them by a revolving wire, which, by its revolutions cut them and obtained from the end of the wire currents of electricity, of low intensity but considerable volume, with which he experimented. The flow of these currents is variable in quantity, and when the quantity is excessive, the aurora is produced by them in the upper, attenuated hemisphere. Sometimes when the quantity is very excessive, constituting what

Humboldt calls magnetic storms, they produce an aurora in all parts of the atmosphere, and then the telegraphic wires can be worked by them, without a battery. Several such cases have occurred in this country within a few years. The working of the telegraph wires by magnetic currents indicates clearly that Prof. Henry erred in attributing to them no force but that of direction, in his meteorological articles, hereinbefore referred to, in which he assumed that magnetism exerted no force except a directive one upon the needle. He failed to comprehend the import and foundation of the views which he attempted to negate. Magnetism consists of, or has associated with it, electric currents, and all electric currents are lines of force, as Faraday has abundantly demonstrated. Faraday's last volume was published in 1855, and had probably escaped the notice of the Professor.

IX. All currents of electricity passing through the atmosphere tend to displace it, or to create currents in it. This is abundantly proved by the brush discharge, and in other ways. Thus we attribute thunder to the recoil of the air into the vacuum which the current of electricity has occasioned by carrying the air with it downwards in passing through the atmosphere. In substantially the same manner it carries the air upward in the tornado. To a certain extent it is undoubtedly true, that the recoil of the atmosphere into the vacuum makes the thunder, yet the concussion exerted by the displacing force, especially of its initiatory ramifications before they unite in a stream, is first heard, and constitutes a part of what we term thunder, and the recoil occasions the *heavy pound* which seems to descend upon us. The long-continued rolling of thunder dying away in the distance, is reverberation in the chamber between the upper and middle strata of cloud, when the discharge is between these two strata, from one to the other.

These magneto electric currents are constantly being discharged from trees, and mountains, and every object connected with the earth. They have much to do with animal and vegetable life. The traveler who moves rapidly in a railway car, may be invigorated by cutting them, while his nervous system will be excited. But if his nervous system is weak, he will become tired by cutting them, though the movement of the car is as easy as a drawing-room.

X. These lateral currents have much to do in constituting the great, permanent, general movements of the atmosphere; with the trade winds while surface trades, and when constituting the upper trade or equatorial current. But most of the special and local winds are the result of static electric induction and attraction.

XI. Evaporation is an electric process, aided by heat, but existing independently of it, for ice and snow evaporate in the arctic regions and everywhere and at all experienced temperatures. The vapor when evaporated is combined with electricity and oxygen, and exists by force of that combination in the atmosphere, and the disturbance of the electricity and the combination, by static induction or other action, occasions the condensation of the vapor, the formation of vesicles, and the constitution of a cloud, and the diffusion of the electricity thus set free over the surface of the vesicles. All these four are the effect of a disturbance of the electricity of the combination.

On this point I regret that my limits will not permit me to copy from Meissner's researches on "Ozone, Antozone, the electrization of oxygen, and the formation of cloud, etc." He has opened out the subject in a manner which should be followed up by those engaged in such pursuits. His discoveries and those of Faraday in relation to the magnetization of oxygen have laid the foundation for a clear understanding of the processes of evaporation, condensation, and the formation of cloud. When scientists get out of the Halley rut, the subject will soon be fully developed. (See Prof. Johnson's review of Meissner's work. American Journal of Science, vols. 37, 38, N. S.)

XII. Bearing these facts in mind, let us glance at the operation of the cause in producing various organizations.

The trade-winds are probably produced primarily by the lateral magneto-electric currents of the earth. Upon islands which lie near the outer limits of the northern trade in summer, the surface and upper trade constitute distinct strata of a different character.

When the surface trade is of sufficient volume to cover the elevation of the islands, they have unbroken drouth.   As the surface-
trade recedes in the fall, and the upper trade comes in contact
with the elevations, rain falls upon and to the leeward of them,
and the line of rain descends the slopes as the surface trade decreases in depth.   These facts indicate the initiation of the surface trade by the permanent magneto-electric currents.   As the
surface trades pass on beneath the stratum of cirrus condensation
which overlies them, they are affected statically, and storms and
showers and squalls are produced in them.   So the belt of rains
is constituted.   As they pass on beyond the belt of condensation
if they are in moderate volume, they become clear again, and pass
as upper or counter-trades into the opposite hemisphere, but partially deprived of their vapor.   Arriving in that hemisphere, they
are exposed to the static, electric induction of the positive atmosphere of the upper story, and the negative electric induction of
the earth.   Storms or showers are produced as one or the other
influence predominates, and operates with sufficient force.   Thus,
in most of our large, extensive storms, which originate upon the
level interior of our country, the incipient condensation is discoverable in the upper story in the form of cirrus, or in the upper
part of the trade-story in the form of cirro-stratus.   Subsequently
the stratus is formed in the trade-story or its inferior portion, and
after that the wind and the scud are by like induction, and by attraction, produced in the surface story.   In this class of cases,
the storm is originated by the positive inductive action of the
electricity of the upper atmosphere.

There is a class of cases where the inductive action of the
earth is clearly visible in producing condensation and rain, as in
the islands referred to, and as upon Table cloth mountain at the
cape of Good Hope, and in all other cases where the upper trade
is intercepted by mountain ranges.   There are undoubtedly instances, where the entire surface story is surcharged with vapor,
that showers and dashes of rain are produced by the inductive
action of the earth.   This is frequently true in England, but it
is not common here.   As a rule, our general storms are initiated

in the great central condition or in the polar zone, by the positive
electricity of the upper story, acting by induction upon the upper
trade of the second story, surcharged with the vapor of evaporation from the ocean surface where it originated. That the earth
may at the same time act inductively upon the inferior surface of
the upper trade where the country is level, may be true, and it
probably is true when the storm is passing over considerable elevations, and is particularly true in respect to our summer belts
of showers.

Induced electric excitement is felt, as we have seen, far in advance of all storms. It influences animal and vegetable life, as
we have seen. It checks and stops evaporation, and disturbs the
combination of electricity with the vapor of the atmosphere, producing partial condensation, and increasing humidity. It is a
mistake to assume that the increase of humidity, as indicated by
the hygrometer or by its deposition on surfaces, is owing to an
actual increase of the amount of water in the atmosphere. Evaporation decreases before an increase of humidity is apparent, and
humidity goes on increasing long after evaporation has ceased.
This increase of humidity is, as I have said, the effect of electric
induction, disturbing the electricity of the vapor, and causing the
apparent increase of humidity. There is, in fact, no sudden increase of the quantity of vapor in the air, it is simply *an increase
of that which is uncombined.*

The electric induction which thus causes humidity, and affects
the feelings of men and animals, may be from the more or less
perfectly formed cirrus in the upper story, or from the earth, affected by the still distant but approaching storm.

I do not think we have yet arrived at a satisfactory solution of
the cause of the barometric oscillations, and I think it quite as
likely as not, that it will ultimately appear that they are occasioned by electric induction also. There is, as we have seen, in
the finest weather, a group of diurnal changes, coincident with the
electric and magnetic changes, and apparently independent of
temperature. There is also an obvious tendency to the formation of the atmospheric conditions in the afternoon, and after the

maximum of temperature has been attained. This is so, when the trade current is in large volume, and the temperature is moderate or uniform, and the sun obscured by constant cloudiness. We have seen that the barometric changes are coincident with the electric and magnetic ones. I am aware that scientists now attribute the diurnal barometric changes to changes in the tension of the vapor of the atmosphere. I think it by no means certain that the cause assigned is the true one. And it is probable that the tension of vapor is an electric effect, but I have not now space for a discussion. It belongs properly to an examination of the mode in which electricity produces and controls other atmospheric phenomena, and must be left for future inquiry. Certain it is that the changes in the barometer and in the feelings of animals and plants, are in advance of thermometric or hygrometric changes which can account for them.

The advocates of the Halley theory have used every art and endeavor to make it appear that all the intense exhibitions of electric action which it is conceded attend all intense atmospheric conditions are incidental results, and not operative as *a cause* ; and yet they concede that when a cloud is formed, electricity is set free during the act of condensation, and diffuses itself over the globules or vesicles so that the cloud becomes a highly charged electric body. If electricity was not in combination with the vapor before its condensation, how could it be set free as a consequence of condensation? That it was so combined then, must be conceded. Is it any the less philosophical or logical to say that when the combination is disturbed by inductive action upon the electricity of the combination, the vapor must be condensed? And if, as I think, it is well established that the condensation consists in the formation of an infinite number of small vesicles which electricity can form and cold cannot, I think it very clear that the formation of cloud is an electric inductive process, as I have assumed, and that in all extensive storms the primary inductive action is by the positive electricity of the upper story, and the forming process may generally be observed upon the advance portion of the storm.

This forming process may also be observed on the easterly side of a belt of showers as it drifts to the eastward. The following diagram exhibits the location of a belt of showers near the focal path in summer.

Fig. 59.

The belt thus exhibited has a drift and a forming movement to the eastward, both of which have been heretofore illustrated. But let us re-examine for a moment.

Two hundred miles more or less to the eastward of the eastern edge of that belt the atmosphere is in a normal state, and beyond the influence of the approaching belt. A few hours later and when the belt has approached nearer, its influence begins to be felt. And how? First, evaporation ceases—the hygrometer begins to show an increase of humidity, and the barometer falls. There is a rise of temperature, and the air begins to feel sultry and close, and commences to move toward the belt. Men, animals, and plants begin to feel the influence thus exerted. Heads, which are accustomed to ache, when thunder showers are approaching, begin to ache. Now the Halley theory cannot account for these

18

changes. They are clearly the effect of electric induction and influence, extending from beneath the belt of cloud, and operating through the earth or the atmosphere. As the belt approaches, these states all increase in intensity—the barometer falls more rapidly—the air grows more sultry—its humidity increases—the deposition of moisture increases—headache increases—the wind also increases in strength, and patches of scud form and float along horizontally in it a few hundred feet above the surface of the earth, where the difference of temperature is trifling, and they could not be and obviously are not produced by the cold of ascension or expansion, for they are in distinct patches. The scud of the winds are also a refutation of the Halley theory.

All thus go on until the belt of showers appears above the northwestern horizon, and thereafter reaches the place. Then the belt of condensation may be observed, made up of masses of cumulo-strati, with an advance condensation agitated by lateral currents in which new cumulo-strati are continually forming, in part by induction upon the atmosphere in its front, and in part by the aid of the surface atmosphere which it attracts to itself. Sometimes that surface wind and its scud, when in small volume, may be seen losing themselves in the advance forming mass, at other times when in large volume passing in part under and across the entire belt, as they did at Springfield, in August, 1859.

By the forming process and the drift, the belt will make an easterly progress of 12 or 15 miles an hour, and sometimes more, according to its intensity and the season of the year, although the masses of condensation move E.N.E., and on the next day it may have the position represented by the most eastern of the two belts upon the diagram, Fig. 60, having passed in the manner described over the space between the two represented locations. I let the representation of its location on the first day remain, and add its location on the next, that they may be compared.

Upon the space over which it has passed there is a cool, clear, dry, northwest surface wind, a rising barometer, and a change in all the phenomena of the previous day.

FIG. 60.

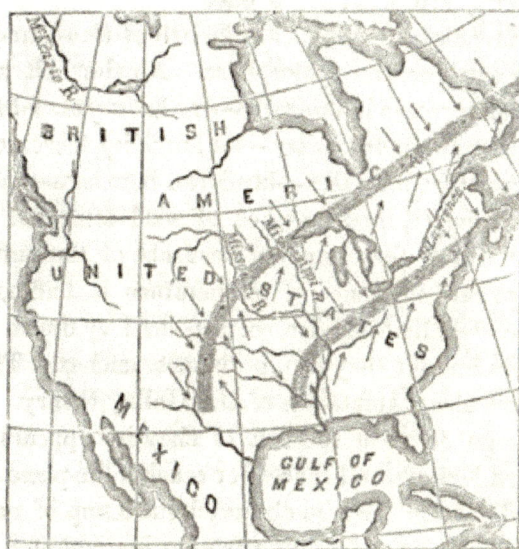

The woolen manufacturer had no occasion to take any precautions to prevent the action of electricity upon the wool, as it ran through his machines, on the day when the belt was approaching. and when the atmosphere was humid. But on the next day, when the belt had passed by and the atmosphere became dry and was left full of free electricity by the deposition in rain of the vapor with which it had been combined, he finds it necessary to neutralize that electricity by a supply of moisture in his card-room from sprinkling, or the letting in of steam. And that cool, dry northwest wind will evaporate the moisture from pools by the aid of its electricity, with a rapidity which it could not do if deprived of its electricity, and by the aid of heat alone

There is always more or less cirrus formation over the belt of showers, though sometimes existing in the form of misty cirrus only. Doubtless that misty formation aids in the induction which produces the belt, but I am inclined to think that in that class of conditions the principal induction is from the earth. Certainly the principal discharges of electricity are disruptive or explosive discharges—strikings as they are called—from the clouds to the earth, the earth being negative, and the inferior surface of the cloud stratum being positive. Such disruptive discharges to the earth indicate that the earth is then the inducing body, and draws to itself the positive electricity which its induction accumulates in the base of the cloud.

There is another kind of electric discharge termed the convective discharge, which is in a continuous current or stream without disruption or explosion. This is the most common method by which the negative electricity of the surface atmosphere and of the earth is discharged upward to the clouds. It is by such a discharge that the tornado, the hurricane, and the straight line gust are produced. The manner, as I think, is this: When the counter or upper trade of the second story is in unusual volume, and unusually charged with vapor, its electricity is disturbed by induction from the positive electricity of the upper story, and its stratum of cirrus clouds. The negative electricity of the trade is attracted into, and accumulated in the upper portion of the trade-story, and the positive electricity repelled and accumulated in the lower portion. That acts inductively upon the surface atmosphere and the earth, which become highly electrified negatively, humid, and sultry, and the surface atmosphere commences discharging its electricity convectively into the cloud. This discharge initiates the tornado. Gradually the discharge commences farther and farther from the cloud, the current is polarized, forming a whirl composed of double currents, and as the electricity of the atmosphere is thereby discharged, the currents originate lower down until their place of origin is from the surface of the earth, and the effects of its violence are experienced there as hereinbefore fully described.*

MEISNER has shown enough to establish the identity between the Antozonic or Atmizonic mist produced by electricity, when passed through moist air, and the strange, white, peculiar mist or cloud of the tornado. And the production of Antozone and its

---

*NOTE. While these pages were being printed, a terrible tornado occurred at Cave City, Kentucky, *in the night*, which is described by eye witnesses as composed of "*Electric Spirals.*" See N. Y. Herald, January 22, 1870.

peculiar odor by the lightning of thunder showers has long been known. It was formerly supposed to be a sulphurous odor. So the production of a *like mist* or vapor and even cloud, snow, and ice crystals by the currents of the earth, when sufficiently excited and intense to produce the Aurora in a marked degree, are well known. This connection and effect are evidence of identity that cannot be disregarded, and we may have it, not only in the tornado, but in the peculiar white scud of the thunder shower, flitting across in front of the dark mass of cloud beyond. So we may trace it in the careful description of intelligent observers in every form of atmospheric condition. On this point I could add fifty citations if I had space for them, but they must be left with a mass of other material for a future occasion.

In all cases of convective discharge from the atmosphere or earth to the clouds, produced by induction from the cloud, the currents assume polarity and constitute the opposite, lateral currents of the atmospheric conditions. The straight line hurricane and the straight line thunder gust are produced in the same way. The lateral currents produced by polarization may often be seen crossing each other in the *forming advance* of the thunder shower, and the straight line gust may be seen to be constituted by the attraction of the cloud drawing the atmosphere after it, and forward and upward to where that forming process is going on. Where the progress of the cloud and the forming process is slow the lateral currents are visible as they occur, and traceable by their effects. Where the progress of the cloud belt and the forming process are rapid, they are not so clearly visible or traceable.

In these intense conditions, *veins* of violence and of comparative inaction are traceable; but this is perfectly intelligible. The same thing is true in all storms and violent winds. The gust which threatens with destruction at one instant, and would, if it continued for any length of time, destroy, soon ceases almost entirely, succeeded very soon by others. These alternations are common in all storm winds. The gusts are constituted by portions of the atmosphere, more highly electrified and more strongly attracted than the portion which follows. So we see that the scud formed in the surface atmosphere by the induction of the cloud, and drawn towards and under it by attraction, are formed in portions which contain larger quantities of vapor, and move more or less rapidly, according to their size and density. They doubtless often move much faster than the less intensely electrified atmosphere by which they are surrounded.

Here we have another fact utterly inconsistent with the Halley theory. That theory calls for a *steady*, and not a fitful or gusty ascent of the heated air—a *steady* and not a *fitful*, gusty or puffy influx of the surrounding air. But we all know that nothing like such steady influx is ever experienced, and that verygreat alternations of strength and moderation attend all violent winds. Irregular action by puffs and gusts is always present.

But it is time to draw this chapter to a close. I greatly regret that I have not space for other facts, and for citations from Howard, Hare, Ampère, Peltier, Faraday, Meissner, and other distinguished men who have investigated this branch of the subject. I must close by repeating that the facts, illustrations, and citations accumulated upon this branch of our inquiry, and from which I have deduced the foregoing propositions, would fill a small volume. At some future day, if the cause of truth should require it, I may separately and more fully develop it.

There is one other topic, intimately connected with the climatology of our country, which I ought not to pass unnoticed.

The magnetism of the earth is not uniformly diffused over its surface. It is concentrated in excess in two lines, or rather on two areas upon opposite sides of the globe, known as lines of no variation, or lines upon which the magnetic needle points to the north. Departing either way from this line of no variation, the needle declines towards it, and ceases to point directly north. This line of no variation commences at the magnetic pole on the northern part of this continent in lat. 72°, lon. 97°, extending south, bearing to the east, entering the Atlantic Ocean above Charleston, crossing the eastern portion of South America, and pursuing the same course to a south, antarctic, magnetic pole. This course is shown, as well as the position of the magnetic equator, by the diagram, Fig. 61.

Upon this diagram, the dip of the needle at different parts of the earth, is also shown by the arrows. In the northern hemis-

FIG. 61.

phere, this dip is shown by the inclination of the *point* of the arrow. In the southern hemisphere it is shown by the inclination of the *feathered end* of the arrow. The location of the magnetic equator, where neither end of the arrow dips, is shown by the arrow which maintains its horizontal position. The arrows represent the position which is taken by a freely suspended magnetic needle at the points upon the earth's surface where they are respectively placed.

There is another corresponding line or area of no variation, extending from another magnetic pole on the opposite side of the earth, in the northern part of Asia, southerly to another magnetic pole within the antarctic circle. All the facts or peculiarities which apply to one of these lines or areas, is equally true of the other.

These areas of no variation are also areas of greater magnetic intensity. Thus, the intensity which is represented by 2, in the vicinity of our magnetic pole, is but .743 at the magnetic equator on the line of no variation in the South Atlantic.

The first important fact to which I will allude, is that the magnetic poles are poles of cold, and that all the isothermal lines of the hemisphere are affected by this area of no variation, and greater intensity. The following diagram shows the situation of these poles of cold and their effect upon the temperature of the earth, as shown by the isothermal lines or lines of equal temperature. It is a view of the earth as seen from above the North Pole.

FIG. 62.

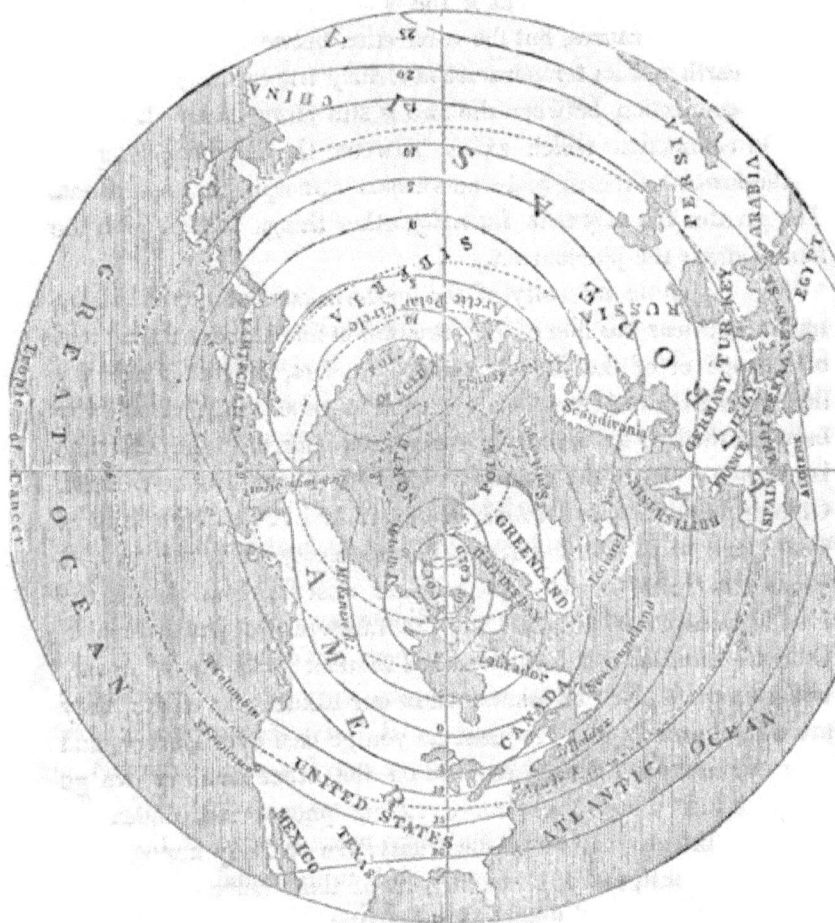

In the above cut the isothermal lines are Centigrade. The zero of the Centigrade thermometer is the freezing point of water, or 32° of Fahrenheit. The boiling point of water is 100° Centigrade, or 212° Fahrenheit. A degree of Centigrade is equal to one degree and four-fifths, Fahrenheit. The 0° line of the cut, therefore, is 32° of Fahrenheit—the line of 5° above is 41° Fahrenheit—the line of 5° below is 23° Fahrenheit, and so on. The dark lines are isothermal lines, and the dotted lines are lines of latitude.

The isothermal lines as shown upon the diagram, are undoubtedly affected by the warmth of the oceans, the elevations of the land and other causes, but the connection between the magnetism of the earth and its temperature is clearly traceable.

The connection between the two is still more clearly traceable in the connection which exists between the magnetic intensity of the lower latitudes, and the various atmospheric phenomena. The greater the *magnetic intensity*, other things being equal, the more *intense* the phenomena.

The magnetic intensity of the eastern part of our continent under and near the line of no variation is much greater than that of Europe or of the Pacific coast. In fact, magnetic intensity decreases from one line of no variation to some point near midway between the two lines, and then increases again until the other line is reached. Thus the intensity which is about 1.8 in Warren, Ohio, in lat. 41.16, lon. 72.57, decreases to 1.774 at New Haven, Conn., and to 1.348 at Paris. The intensity decreases in like manner as we follow the same latitude west from Ohio, on to the Pacific Ocean. Now while all the atmospheric phenomena in Ohio are intense, and tornadoes and thunder gusts are so proverbially common as to be mentioned in our elementary works, they are comparatively less frequent as you go east over Europe, and magnetic intensity decreases. So on the other hand as we go west, we find that all the more violent phenomena, dependent upon the intense action of electricity, are scarcely known upon the Pacific coast, and the eastern part of the Pacific Ocean, and understand the cause of that characteristic which gives the ocean its name. The influence of that excess of magnetism, and magneto-electricity, in intensifying our climate, our diseases, and the energies and activity of our people in the Atlantic States, is a most suggestive fact in relation to our climatology, and as evidence in relation to this branch of our subject. But I cannot dwell upon it.

In closing the volume, I desire to make a brief appeal to several classes of practical men. If the contents of it are true, and I think I *know* the first seven chapters to be substantially so, and I *believe* the last is so also, then an entire and radical revolution in the so-called science of meteorology is *inevitable*, and but a question of time. Whether that shall be early, so as to benefit the present generation or not, depends upon the extent to which the leading classes of practical men shall realize, and perform their duty in the premises. Nothing but opposition is to be expected from the men who now make meteorology a specialty, and little aid I fear, from the great body of scientists, who, although fair and worthy men and passive recipients and supporters merely of the Halley theory, are controlled to an extent they scarcely appreciate, by a minority who are actively and persistently engaged in sustaining it.

Let me then appeal in the first place to the members of Congress—to each of whom a copy of the book will be sent—to examine it with care, and give the country a telegraphic weather system, founded on the developments made in relation to the *paths of the storms*, at the various seasons of the year. Such a system, if employing a few intelligent observers, at *representative points* accessible to the telegraph lines, on cross sections of the paths of the storms, and at proper distances, and reporting to a central bureau twice each day the state of things on their respective cross sections, would not be a very cumbrous, nor a very expensive thing. The benefit of it, if properly organized, and placed under the control of men who understand, or who will inform themselves fully and practically in relation to the matter, can scarcely be over-estimated. But the European systems are not models for us, for the paths of our storms and our climatology differ essentially from theirs.

I appeal secondly to the Journalists of the country, who also are a power in the land, and accustomed to weigh and discuss great questions, to consider whether they do not also owe a like duty to the community with myself and others in relation to the matter, and ask of them an examination of the book, and a full expression of their views, that *discussion* may be had, the *truth elicited* and defended, and the smothering process be no longer possible.

And I appeal finally to the great body of professional and practical men in the country, to give this matter attention, and see whether an early and radical change in our colleges and schools and text books, is not imperatively called for by the interests of education and knowledge, and all the other material interests of the country, and if so, to institute such measures as may be necessary to effect that change. LET TRUTH PREVAIL.

FINIS.

APPENDIX.

NOTE I.

SINCE the foregoing pages were in type Congress has passed the following Resolution:

JOINT RESOLUTION to authorize the Secretary of War to provide for taking meteorological observations at the military stations and other points in the interior of the continent, and for giving notice on the northern lakes and seaboard of the approach and force of storms.

Be it resolved by the Senate and House of Representatives of the United States of America in Congress assembled, That the Secretary of War be, and he hereby is authorized and required to provide for taking meteorological observations at the military stations in the interior of the continent, and at other points in the States and Territories of the United States, and for giving notice on the northern lakes and on the seacoast, by magnetic telegraph and marine signals, of the approach and force of storms.

This is well as far it goes. But the Agricultural interest is quite as important as the Marine, and would be quite as much benefited by a proper weather telegraph system. Other special interests would be greatly subserved, and a great general interest gratified. We need, therefore, and sooner or later must have such a system. A few representative stations, on lines *cross-sectioning the focal paths of the conditions of the Atlantic System, as I have developed them,* reporting through a central Bureau twice a day, in season for the morning and evening newspapers, would suffice. No other benefit of equal importance is, or can be conferred by Government for less money.

NOTE II.

I was aware of the memoir and tables of Wolf when I wrote the pages respecting Sun Spots, but I had good reason to doubt the correctness of the tables, and of the principle on which they are constructed, and as the tables of Schwabe are sufficiently accurate and reliable, and for a considerable period, I made them alone the basis of my investigation.

(405)

www.ingramcontent.com/pod-product-compliance
Lightning Source LLC
Chambersburg PA
CBHW021348210326
41599CB00011B/790